# Advances in Pattern Recognition

**Advances in Pattern Recognition** is a series of books which brings together current developments in all areas of this multi-disciplinary topic. It covers both theoretical and applied aspects of pattern recognition, and provides texts for students and senior researchers.

Springer also publishes a related journal, **Pattern Analysis and Applications.** For more details see: http://link.springer.de

The book series and journal are both edited by Professor Sameer Singh of Exeter University, UK.

*Also in this series:*

Principles of Visual Information Retrieval
Michael S. Lew (Ed.)
1-85233-381-2

Statistical and Neural Classifiers: An Integrated Approach to Design
Šarūnas Raudys
1-85233-297-2

Advanced Algorithmic Approaches to Medical Image Segmentation
Jasjit Suri, Kamaledin Setarehdan and Sameer Singh (Eds)
1-85233-389-8

NETLAB: Algorithms for Pattern Recognition
Ian T. Nabney
1-85233-440-1

Object Recognition: Fundamentals and Case Studies
M. Bennamoun and G.J. Mamic
1-85233-398-7

Bir Bhanu and Ioannis Pavlidis (Eds)

# Computer Vision Beyond the Visible Spectrum

With 156 Figures

 Springer

Bir Bhanu, PhD, Fellow IEEE, AAAS, IAPR, Senior Honeywell Fellow (Ex.)
Center for Research in Intelligent Systems, University of California at Riverside,
CA, USA

Ioannis Pavlidis, PhD
Department of Computer Science, University of Houston, TX, USA

*Series editor*
Professor Sameer Singh, PhD
Department of Computer Science, University of Exeter, Exeter, EX4 4PT, UK

British Library Cataloguing in Publication Data
Computer vision beyond the visible spectrum. — (Advances in
   pattern recognition)
   1. Computer vision  2. Pattern recognition systems
   I. Bhanu, Bir  II. Pavlidis, Ioannis
   006.3'7
ISBN 1852336048

Library of Congress Cataloging-in-Publication Data
Bhanu, Bir.
   Computer vision beyond the visible spectrum / Bir Bhanu, Ioannis Pavlidis.
      p. cm. — (Advances in pattern recognition)
   Includes bibliographical references and index.
   ISBN 1-85233-604-8 (alk. paper)
   1. Computer vision.  2. Infrared detectors.  3. Synthetic aperture radar.  I. Pavlidis, Ioannis,
1961-  II. Title.  III.  Series.
TA1634.B46 2004
006.3'7—dc22                                                                                   2004048104

Apart from any fair dealing for the purposes of research or private study, or criticism or review, as
permitted under the Copyright, Designs and Patents Act 1988, this publication may only be repro-
duced, stored or transmitted, in any form or by any means, with the prior permission in writing of
the publishers, or in the case of reprographic reproduction in accordance with the terms of licences
issued by the Copyright Licensing Agency. Enquiries concerning reproduction outside those terms
should be sent to the publishers.

Advances in Pattern Recognition ISSN 1617-7916
ISBN 1-85233-604-8
Springer is a part of Springer Science+Business Media
springeronline.com

© Springer-Verlag London Limited 2005

The use of registered names, trademarks, etc. in this publication does not imply, even in the absence
of a specific statement, that such names are exempt from the relevant laws and regulations and
therefore free for general use.

The publisher makes no representation, express or implied, with regard to the accuracy of the information contained in this book and cannot accept any legal responsibility or liability for any errors
or omissions that may be made.

Printed and bound in the United States of America
34/3830-543210   Printed on acid-free paper   SPIN 10868565

# Preface

Traditionally, computer vision has focused on the visible band for a variety of reasons. The visible band sensors are cheap and easily available. They are also sensitive in the same electromagnetic band as the human eye, which makes the produced data more interesting from the psychophysiology point of view. In fact, computer vision was pre-occupied for a long time with the problem of understanding and imitating the human visual system. Recently, this obsession subsided and computer vision research focused more on solving particular application problems with or without the help of the human visual paradigm. A case in point is the significant progress achieved in object tracking.

It so happens that many imaging applications cannot be addressed in the visible band. For example, visible sensors cannot see in the dark; thus, they are not very useful in military applications. Visible radiation cannot penetrate the human body and, therefore, cannot be a viable medical imaging modality. Other electromagnetic bands and sensor modalities have been identified and developed over the years that can solve all these problems, which are beyond the reach of the visible spectrum. Initially, it was primarily phenomenological and sensory work that was taking place. Later came algorithmic work, and with that computer vision beyond the visible spectrum was born.

In this book, we explore the state-of-the-art in *Computer Vision Beyond the Visible Spectrum (CVBVS)* research. The book is composed of nine chapters which are organized around three application axes:

1. Military applications with an emphasis on object detection, tracking, and recognition.
2. Biometric applications with an emphasis on face recognition.
3. Medical applications with an emphasis on image analysis and visualization.

Although the chapters describe research, they are not written as typical research papers. They have a tutorial flavor appropriate for a book.

The book opens with the military applications since they represent the birthplace of CVBVS. All the major modalities used in military applications are represented in the first five chapters. These include SAR (Synthetic Aperture Radar), laser radar, hyperspectral, and infrared. The first five chapters also address fundamental issues with regard to object detection, tracking, and recognition, sometimes in more than one modality. This allows comparative evaluation of these important computational imaging questions across the electromagnetic spectrum.

In Chapter 1, Boshra and Bhanu et al. describe a theoretical framework for predicting the performance of object (target) recognition methods. The issue of identifying military targets in imagery is of great importance in military affairs. For years, target recognition was based purely on heuristics, and as a result performance was brittle. Boshra and Bhanu's work is representative of a more rigorous methodological approach, which promises to transform target recognition from art to science.

In Chapter 2, Bhanu and Jones unveil specific methods for improving the performance of an SAR target recognition system. SAR is probably the most successful imaging modality for military applications, because of its all-weather capability. Bhanu and Jones' methods conform to the model-based framework and involve incorporation of additional features, exploitation of a priori knowledge, and integration of multiple recognizers.

In Chapter 3, Arnold et al. present target recognition methods in a different modality, namely, three-dimensional laser radar. Three-dimensional laser radars measure the geometric shape of targets. The main approach described in this chapter is quite appealing because it bypasses detection and segmentation processes.

In Chapter 4, Kwon et al. deal with target recognition in the context of hyperspectral imagery. The basic premise of hyperspectral target recognition is that the spectral signatures of target materials are measurably different than background materials. Therefore, it is assumed that each relevant material, characterized by its own distinctive spectral reflectance or emission, can be identified among a group of materials based on spectral analysis of the hyperspectral data. Kwon et al. use independent component analysis (ICA) to generate a target spectral template. ICA is a method well-suited to the modular character of hyperspectral imagery.

In Chapter 5, Vaswani et al. close the sequence of military application papers by presenting a method for object detection and compression in infrared imagery. The proposed solution is guided by the limitations of the target platform, which is an infrared camera with on-board chip. The object detection method is computationally efficient, to deal with the limited processing power of the on-board chip. It is also paired with a compression scheme to facilitate data transmission.

Chapter 6 deals with biometrics and signals a transition from the military to civilian security applications. Wolff et al. present a face recognition approach based on infrared imaging. Infrared has advantages over visible imaging for face recognition, especially in the presence of variable lighting conditions.

Wolff et al. provide quantitative support for this argument by unveiling a system that performs comparative evaluation.

Chapter 7 opens the medical applications part of the book. It refers to cardiovascular image analysis of magnetic resonance imagery (MRI). While SAR is probably the most successful modality for military applications, one could make the case that MRI is the most successful modality for medical applications. Initially, MRI was treated much like x-rays. A radiologist, without any machine assistance, was interpreting the raw imagery. Increasingly, however, computer vision methods aid in this interpretation. In this chapter, Sonka et al. present techniques for 3D segmentation and quantitative assessment of left and right cardiac ventricles, arterial and venous trees, and arterial plaques.

In Chapter 8, Fenster et al. present segmentation and visualization techniques in another very important medical imaging modality, that is, ultrasound. Specifically, the authors describe methods to reconstruct ultrasound information into 3D images to facilitate interactive viewing. They also describe automated and semi-automated segmentation methods to quantify organ and pathology volume for monitoring disease.

In Chapter 9, Berry et al. introduce some very interesting image analysis work on a novel medical imaging modality, namely, terahertz pulsed imaging. Vis-a-vis the more established MRI and ultrasound modalities, terahertz pulsed imaging is the "new kid on the block". Berry et al. propose Fourier transforms and wavelets to analyze spectroscopic information of materials. They actually demonstrate that these methods perform as well as traditional analysis methods for material properties and predict a number of biomedical applications that stand to benefit form this technology.

The book can be used for instruction in graduate seminars or as a reference for the independent researcher. Although CVBVS is a broad and fast moving field, the balanced selection of key theoretical and practical issues represented in the chapters of the book will maintain their relevance for some time. It is our sincere hope that the book will serve as a springboard for the individual researcher who is interested in CVBVS research.

A number of people have contributed in our effort and we are deeply grateful to all of them. These certainly include the authors of the individual chapters and the reviewers who patiently went through three review cycles. We are especially grateful to Pradeep Buddharaju who handled most of the last minute editing and thanks to whom the book assumed its finished form.

Houston, Texas *Ioannis Pavlidis*
Riverside, California *Bir Bhanu*
USA
January 2004

# Contents

**Chapter 1 A Theoretical Framework for Predicting Performance of Object Recognition**
*Michael Boshra, Bir Bhanu* ........................................... 1

**Chapter 2 Methods for Improving the Performance of an SAR Recognition System**
*Bir Bhanu, Grinnell Jones III* ........................................ 39

**Chapter 3 Three-Dimensional Laser Radar Recognition Approaches**
*Gregory Arnold, Timothy J. Klausutis, Kirk Sturtz* .................. 71

**Chapter 4 Target Classification Using Adaptive Feature Extraction and Subspace Projection for Hyperspectral Imagery**
*Heesung Kwon, Sandor Z. Der, Nasser M. Nasrabadi* ................ 115

**Chapter 5 Moving Object Detection and Compression in IR Sequences**
*Namrata Vaswani, Amit K Agrawal, Qinfen Zheng, Rama Chellappa* ... 141

**Chapter 6 Face Recognition in the Thermal Infrared**
*Lawrence B. Wolff, Diego A. Socolinsky, Christopher K. Eveland* ....... 167

**Chapter 7 Cardiovascular MR Image Analysis**
*Milan Sonka, Daniel R. Thedens, Boudewijn P. F. Lelieveldt, Steven C. Mitchell, Rob J. van der Geest, Johan H. C. Reiber* ........ 193

**Chapter 8 Visualization and Segmentation Techniques in 3D Ultrasound Images**
*Aaron Fenster, Mingyue Ding, Ning Hu, Hanif M. Ladak, Guokuan Li, Neale Cardinal, Dónal B. Downey* .................................. 241

**Chapter 9 Time-Frequency Analysis in Terahertz-Pulsed Imaging**
*Elizabeth Berry, Roger D Boyle, Anthony J Fitzgerald, James W Handley* .................................................... 271

**Index** ............................................................. 313

# Chapter 1

# A Theoretical Framework for Predicting Performance of Object Recognition

Michael Boshra[1] and Bir Bhanu[2]

[1] Center for Research in Intelligent Systems, University of California, Riverside, California 92521, michael@cris.ucr.edu
[2] Center for Research in Intelligent Systems, University of California, Riverside, California 92521, bhanu@cris.ucr.edu

**Summary.** The ability to predict the fundamental performance of model-based object recognition is essential for transforming the object recognition field from an art to a science, and to speed up the design process for recognition systems. In this chapter, we address the performance–prediction problem in the context of a common recognition task, where both model objects and scene data are represented by locations of 2D point features. The criterion used for estimating matching quality is based on the number of consistent data/model feature pairs, which we refer to as "votes." We present a theoretical framework for prediction of lower and upper bounds on the probability of correctly recognizing model objects from scene data. The proposed framework considers data distortion factors such as uncertainty (noise in feature locations), occlusion (missing features), and clutter (spurious features). In addition, it considers structural similarity between model objects. The framework consists of two stages. In the first stage, we calculate a measure of the structural similarity between every pair of objects in the model set. This measure is a function of the relative transformation between the model objects. In the second stage, the model similarity information is used along with statistical models of the data distortion factors to determine bounds on the probability of correct recognition. The proposed framework is compared with relevant research efforts. Its validity is demonstrated using real synthetic aperture radar (SAR) data from the MSTAR public domain, which are obtained under a variety of depression angles and object configurations.

## 1.1 Introduction

Model-based object recognition has been an active area of research for over two decades (e.g., see surveys [1, 2, 3, 4]). It is concerned with finding instances of known model objects in scene data. This process involves extracting features from the scene data, and comparing them with those of the model objects using some matching criterion. Performance of the recognition process depends on the amount of distortion in the data features. Data distortion can be classified into three types: (1) *uncertainty:* noise in feature locations and other feature attributes; (2) *occlusion:* missing features of data object of interest; and (3) *clutter:* spurious data features which do not belong to the data object of interest. In addition to data distortion, recognition performance depends on the degree of structural similarity between model objects. The often-overlooked similarity factor can have a profound impact on performance. Intuitively, the difficulty of recognizing a specific object is proportional to the degree of its similarity with the rest of the objects in the model set.

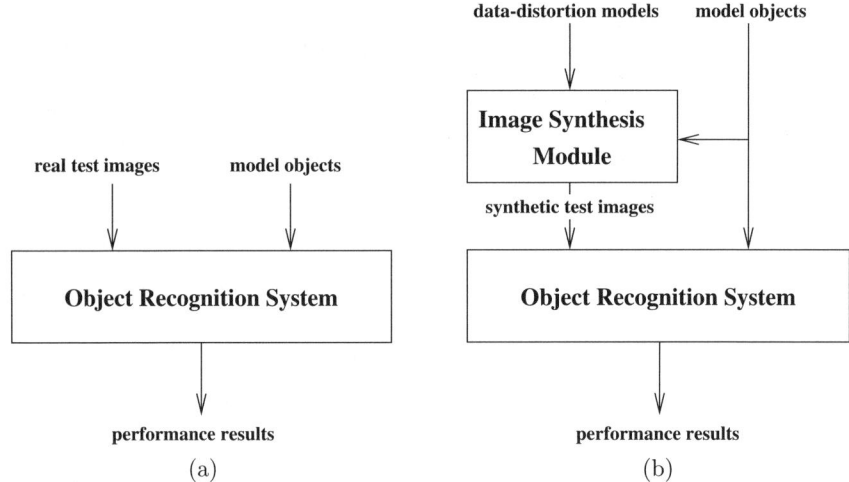

**Figure 1.1.** Empirical approaches for estimation of object recognition performance: (a) using real data, (b) using synthetic data.

Performance of object recognition is typically estimated empirically. This is done by passing a set of scene images containing known model objects to a recognition system, and then analyzing the output of the system. The set of scene images can be either real [5, 6, 7], or synthetic with artificial distortion introduced to them [8, 9, 10]. Both scenarios are illustrated in Figures 1.1(a) and 1.1(b), respectively. Empirical performance evaluation has a number of limitations:

1. It does not provide an understanding of the relationship between object recognition performance and the various data and model factors that affect it. In other words, the empirical approach can provide an answer to the question of *what* performance to expect, for a given set of model objects and specific data distortion rates. However, it does not explain *why* this is the expected performance. Such an understanding is critical for designing better object recognition systems, as it can provide fundamental answers to questions such as: (a) When does performance break down as a function of the amount of data distortion? (b) What are the

performance limits when using a specific sensor? (c) Is a given feature-selection scheme sufficient for achieving desired levels of performance? (d) What is the largest size of a model set that can be accommodated without significantly degrading performance? Fundamental understanding of the relationship between performance and the factors affecting it is essential for the advancement of the field of object recognition from an art to a science.
2. The performance estimated empirically is dependent upon the actual implementation of the object recognition system. This implementation can be based on recognition approaches such as alignment [11, 12], hypothesis accumulation [13, 14], or tree search [15, 16]. Note that the performance obtained using these approaches can be different, even if they use similar matching criteria. For example, systems that use a vote accumulator (Hough space) will generate different performance estimates depending on the resolution of the accumulator. Another example, alignment-based systems, achieve polynomial-time complexity by using a "looser" notion of data/model feature consistency.
3. Empirical evaluation requires the presence of an actual object recognition system. Obviously, this can considerably slow down the design process.

In this chapter, we address the performance–prediction problem in the context of a typical object recognition task. It can be described as follows. (1) Both model objects and scene data are represented by discretized locations of 2D point features. (2) A data object is assumed to be obtained by applying a 2D transformation to the corresponding model object. Notice that the space of possible 2D transformations is naturally discretized, since we are dealing with discrete 2D point features. (3) The data/model matching quality is estimated using a vote-based criterion. In particular, the quality of a given match hypothesis is estimated by counting the number of consistent data/model feature pairs, which we refer to as "votes."

We present a statistical method for formally predicting lower and upper bounds on the *probability of correct recognition* (PCR) for the task outlined above. The proposed method considers data distortion factors such as uncertainty, occlusion, and clutter, in addition to model similarity. Integrating these data and model factors in a single approach has been a challenging problem. The performance predicted is fundamental in the sense that it is obtained by analyzing the information provided by both the data and model features, *independent* of the particular vote-based matching algorithm. A schematic diagram of the prediction method is shown in Figure 1.2. It can be contrasted with the diagrams of the empirical approaches shown in Figure 1.1. The validity of the proposed method is demonstrated using real synthetic aperture radar (SAR) data from the MSTAR public domain. This data set is obtained under a variety of depression angles and object configurations.

The remainder of this chapter is organized as follows. The next section reviews related research efforts, and highlights our contributions. Section 1.3

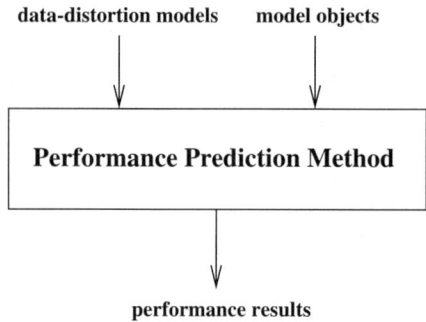

**Figure 1.2.** Formal estimation of object recognition performance.

presents an overview of the proposed method. Sections 1.4 and 1.5 describe the statistical modeling of the data distortion factors, and the object similarity, respectively. Derivation of lower and upper bounds on PCR is presented in Section 1.6. The validity of those bounds is demonstrated in Section 1.7, by comparing actual PCR plots, as a function of data distortion, with predicted lower and upper bounds. Finally, conclusions and directions for future research are presented in Section 1.8.

## 1.2 Relevant Research

Several research efforts have addressed the problem of analyzing performance of feature-based object recognition. Most of these efforts focus on the problem of discriminating objects from clutter. We present here a representative sample of these efforts. Grimson and Huttenlocher [17] presented a statistical method for estimating the probability distribution of the fraction of consistent data/model feature pairs for an erroneous hypothesis. They derived such a distribution using a statistical occupancy model (Bose–Einstein model), assuming bounded feature uncertainty and uniform clutter models. This distribution was used to determine the minimum fraction of consistent feature pairs required to achieve a desired probability of false alarm. Sarachik [18] studied the problem of predicting the receiver operating characteristic (ROC) curve for a specific recognition algorithm. The ROC curve described the relationship between the probability of correct recognition and that of a false alarm. The chosen algorithm used a weighted voting criterion based on Gaussian feature uncertainty. A statistical analysis was presented to determine the probability distributions of the weighted votes for both valid and invalid hypotheses, assuming uniform occlusion and clutter models. These distributions were used along with the likelihood-ratio test to predict the ROC curve. Alter and Grimson [8] used statistical knowledge about sources of data distortion to design a recognition criterion, based on the likelihood-ratio test. The likelihoods of observed data-feature set, conditioned on hypothesis validity and invalidity, were calculated by assuming both bounded and Gaussian feature uncertainty models, in addition to uniform occlusion and clutter models. Lindenbaum [19] extended the modeling of clutter to include background objects of known shape, in addition to uniformly distributed random features. This hybrid model was incorporated into a statistical analysis to predict the number of features needed to guarantee recognition of a data object at a given confidence level. The analysis considered bounded feature uncertainty, as well as structural similarity between a given data object and background ones. Irving et al. [20] derived a theoretical bound on the ROC curve of an object detection task. The generalized likelihood-ratio test was used to discriminate between model objects at various poses and random clutter. The likelihoods of both clutter and model objects were modeled using 2D Poisson processes. Modeling object likelihood as a Poisson process was based on the assumption of independence of object views at discretized poses. This work considered bounded feature uncertainty, and uniform occlusion and clutter models.

The problem of discriminating objects from other model objects has received considerably less attention than object/clutter discrimination. This problem is obviously central to integrated performance prediction of object recognition. It requires consideration of not only data distortion but also object similarity. In addition, it requires consideration of the *interaction* between object similarity and data distortion. Lindenbaum [21] presented a probabilistic analysis for predicting lower and upper bounds on the number of data features required to achieve a certain confidence level in object localization or recognition. It explicitly considered the similarity between different model objects, as well as the self-similarity between a model object and an instance of itself at a different relative pose. The data distortion factors considered were bounded uncertainty and occlusion. The analysis considered extreme cases in modeling the interaction between occlusion and similarity, thus resulting in the generation of relatively loose bounds. We note that the analysis presented in [19], outlined above, can be used in the context of object/object discrimination considering uncertainty and clutter, as well as object similarity. Grenander et al. [22] addressed the problem of predicting fundamental error in object pose estimation. In their work, objects were represented by templates at the pixel level. A minimum mean-square-error estimator, the Hilbert–Schmidt estimator, was used to estimate object pose in the presence of pixel uncertainty. Performance of object/object discrimination was determined partially empirically through the synthesis of distorted templates of one object, and then using the likelihood-ratio test, based on the Hilbert–Schmidt estimator, as a recognition criterion (refer to Figure 1.1(b)).

The methods outlined above are summarized and compared with our method in Table 1.1. This table highlights the main contribution of our work, namely the integration of uncertainty, occlusion, clutter, and similarity factors in a single approach for performance prediction. As shown in the table, previous methods considered only a subset of these factors. It can also be seen that our method is unique among other object/object discrimination methods

**Table 1.1.** Comparison between performance-analysis methods ($U$, $O$, $C$, and $S$ denote uncertainty, occlusion, clutter, and similarity, respectively).

| Work | Discrimination | Data/Model Features | Transform. | Factors $U$ $O$ $C$ $S$ |
|---|---|---|---|---|
| Grimson et al. [17] | object/clutter | 2D/2D lines | rigid | X  X |
| Sarachik [18] | object/clutter | 2D/2D points | affine | X X X |
| Alter and Grimson [8] | object/clutter | 2D/3D points & lines | weak persp. | X X X |
| Lindenbaum [19] | object/clutter | 2D/2D boundary pts. | affine | X  X X |
| Irving et al. [20] | object/clutter | 2D/2D points | 2D transl. | X X X |
| Lindenbaum [21] | object/object | 2D/2D boundary pts. | rigid | X X  X |
| Grenander et al. [22] | object/object | pixel-level templates | rotation | X |
| This work | object/object | 2D/2D discretized pts. | 2D transl. | X X X X |

in that it considers point features. Another unique aspect of this work is not just the new theory but also the validation using real data. We note that parts of this work have appeared in [23, 24, 25].

## 1.3 Overview

In this section, we present an overview of the proposed performance–prediction method. Our problem can be formally defined as follows. We are given the following:

1. A set of model objects, $\mathcal{MD} = \{\mathcal{M}_i\}$, where each object $\mathcal{M}_i$ is represented by discretized locations of 2D point features, $\mathcal{M}_i = \{F_{ik}\}$.
2. Statistical data distortion models.
3. A class of data/model transformations, $\mathcal{T}$.

Our objective is to predict lower and upper bounds on PCR as a function of data distortion. We consider recognition to be successful only if the selected hypothesized object is the actual one, and the difference between the hypothesized pose and the actual one is small. The pose error can be represented by the relative pose of the hypothesized object with respect to the actual one. It is considered acceptable if it lies within a subspace, $\mathcal{T}_{acc} \subset \mathcal{T}$. We assume in this work that $\mathcal{T}_{acc} = \{\mathbf{0}\}$, i.e., only exact object location is acceptable.

A block diagram of the proposed method is shown in Figure 1.3. The main elements in this diagram can be described as follows:

• **Data-Distortion Models:** The data distortion factors are statistically modeled using uniform probability distribution functions (PDFs).

1. *Uncertainty:* The location of the data feature corresponding to a model feature is described by a uniform distribution. Notice that the uncertainty PDFs are discrete, since the feature locations considered in this work are discretized. We further assume that the PDFs associated with different

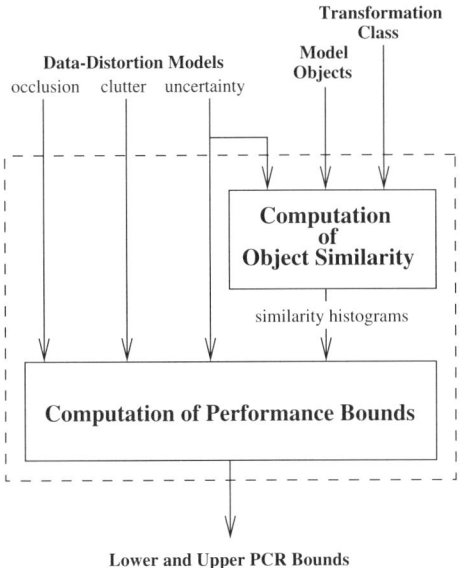

**Figure 1.3.** Block diagram of performance–prediction method.

model features are independent. We argue that such independence assumption is reasonable in most practical applications.
2. *Occlusion:* We assume that every subset of model features is equally likely to be occluded as any other subset of the same size. This assumption is more appropriate for modeling features that are missing due to inherent instability or imperfections of feature extraction. It is less suitable for modeling features that are missing due to being occluded by other objects, since it does not consider the spatial-correlation aspect among occluded/unoccluded features. Spatial correlation can be captured by using Markov random fields [9, 26, 27], at the expense of significantly increasing the complexity of the analysis. In Section 1.7, we outline a simpler approach that can implicitly consider the spatial-correlation factor, without increasing the analysis complexity.
3. *Clutter:* We assume that clutter features are uniformly distributed within a region surrounding the object. This distribution is useful for modeling random clutter, which does not have specific spatial structure. Modeling "structural" clutter requires analyzing its similarity with model objects. We note that the similarity-modeling concepts presented in this work can be used in modeling of structural clutter. This topic is a subject for future research.

• **Computation of Object Similarity:** The purpose of this stage is to compute the structural similarity information among all pairs of model objects. Our definition of object similarity depends on the amount of feature uncertainty. In particular, the similarity between two model objects is directly proportional to feature uncertainty. This agrees with the intuitive observation that as different objects become more "blurred," it becomes more difficult to differentiate between them, which is, in a sense, equivalent to saying that they become more "similar." The similarity between a model object, $\mathcal{M}_i$, and another one, $\mathcal{M}_j$, is defined as the number of votes for $\mathcal{M}_j$ given an *uncertain* instance of $\mathcal{M}_i$, i.e., an instance of $\mathcal{M}_i$ that is obtained by randomly perturbing its features. Accordingly, the number of votes for $\mathcal{M}_j$ is a random variable. The chosen definition of similarity depends on the relative transformation between $\mathcal{M}_i$ and $\mathcal{M}_j$, defined by transformation class $\mathcal{T}$. Accordingly, the similarity between $\mathcal{M}_i$ and $\mathcal{M}_j$ can be viewed as a probabilistic function, which we call the similarity function. The similarity information is encoded in two histograms, which we call all-similarity and peak-similarity histograms, to be described in Section 1.5.2. These histograms are used for predicting lower and upper PCR bounds, respectively.

- **Computation of Performance Bounds:** The objective of this stage is to compute PCR bounds. The computation is based on estimating the PDF of the votes for a specific erroneous object/pose hypothesis, given a "distorted" instance of a given model object. The estimation process takes into account the structural similarity between the model object and the erroneous hypothesis. The vote PDF is used to determine the probability of a recognition failure, which occurs if the erroneous hypothesis gets same or more votes than the distorted object. This information is integrated for potential erroneous hypotheses to determine the PCR bounds.

## 1.4 Data-Distortion Models

We formally model the effects of the three distortion factors considered in this work on a "perfect" model object. This modeling is used to determine the vote PDF in Sect. 1.6.2.

- **Uncertainty:** The effect of the uncertainty factor is to perturb locations of model features according to some PDF. Since this PDF is assumed to be uniform, it can be represented by a region. Let $F_{ik} \in \mathcal{M}_i$ be a model feature, and $\widehat{F}_{ik}$ be a distorted instance of it. Define $R_u(F_{ik})$ to be the *consistency region* associated with $F_{ik}$. Such region bounds the possible locations of $\widehat{F}_{ik}$ given $F_{ik}$, i.e., $\widehat{F}_{ik} \in R_u(F_{ik})$. Likewise, we can say that $F_{ik} \in \bar{R}_u(\widehat{F}_{ik})$, where $\bar{R}_u(\cdot)$ is the reflection of $R_u(\cdot)$ about the origin. Practically, $R_u(\cdot)$ is the same as $\bar{R}_u(\cdot)$, because $R_u(\cdot)$ is symmetric about the origin (e.g., circle, square). Accordingly, we assume in this work for simplicity that $\bar{R}_u(\cdot) = R_u(\cdot)$. An uncertain instance of $\mathcal{M}_i$ can be obtained by uniformly perturbing each of its features within corresponding consistency region. This can be formally represented as:

$$\mathcal{D}_u(\mathcal{M}_i, R_u(\cdot)) = \{P_u(R_u(F_{ik})) : F_{ik} \in \mathcal{M}_i\},$$

where $P_u(R)$ is a function that returns a feature selected randomly within region $R$.
- **Occlusion:** The effect of occlusion is the elimination of some model features. An occluded instance of $\mathcal{M}_i$ can be formally defined as

$$\mathcal{D}_o(\mathcal{M}_i, O) = \mathcal{M}_i - \mathcal{P}_o(\mathcal{M}_i, O),$$

where $\mathcal{P}_o(\mathcal{M}_i, O)$ is a function that returns a subset of $O$ features selected randomly from $\mathcal{M}_i$. For a fixed $O$, all subsets generated by $\mathcal{D}_o(\mathcal{M}_i, O)$ are equally likely to occur, since we are assuming uniform occlusion.
- **Clutter:** The effect of clutter on a model object is the addition of spurious features to it. They are assumed to be uniformly distributed within a *clutter region*, $R_c$, surrounding the model object. This region can have an arbitrary shape (e.g., bounding box of model features, convex hull, etc). A cluttered instance of $\mathcal{M}_i$ can be defined as

$$\mathcal{D}_c(\mathcal{M}_i, C, R_c, R_x) = \mathcal{M}_i \cup \mathcal{P}_c(C, R_c - R_x),$$

where $\mathcal{P}_c(C, R)$ is a function that returns $C$ features selected randomly within region $R$, and $R_x$ is a region that clutter features are excluded from falling into. The reason for including $R_x$ is explained below.
- **Combined Distortion:** Consideration of the combined effects of uncertainty, occlusion, and clutter on a model object raises an ambiguous situation. It takes place when a model feature gets occluded and then a spurious feature falls within its consistency region. The ambiguity arises from the fact that this situation can not be distinguished from the no-occlusion/no-clutter case. In order to simplify the analysis, we assume the latter case. This can be modeled by restricting the clutter features to lie outside region $R_x$, defined as the union of the consistency regions of occluded features. We refer to $R_c - R_x$, or simply $R'_c$, as the effective clutter region. A distorted instance of $\mathcal{M}_i$, $\widehat{\mathcal{M}}_i(R_u(\cdot), O, C, R_c)$, can be obtained by first occluding $O$ features of $\mathcal{M}_i$, perturbing unoccluded ones within their consistency regions $R_u(\cdot)$, and then randomly adding $C$ clutter features within the effective clutter region $R'_c$. This can be represented formally as:

$$\widehat{\mathcal{M}}_i(R_u(\cdot), O, C, R_c) = \mathcal{D}_c(\mathcal{D}_u(\mathcal{D}_o(\mathcal{M}_i, O), R_u(\cdot)), C, R_c, R_x),$$

where $R_x = \cup_k R_u(F_{ik})$, $\forall F_{ik} \in (\mathcal{M}_i - \mathcal{D}_o(\mathcal{M}_i, O))$. Figure 1.4 shows an example of the distortion process.

## 1.5 Computation of Object Similarity

In this section, we formally define a measure of the structural similarity between model objects, and outline the method used to construct the similarity histograms.

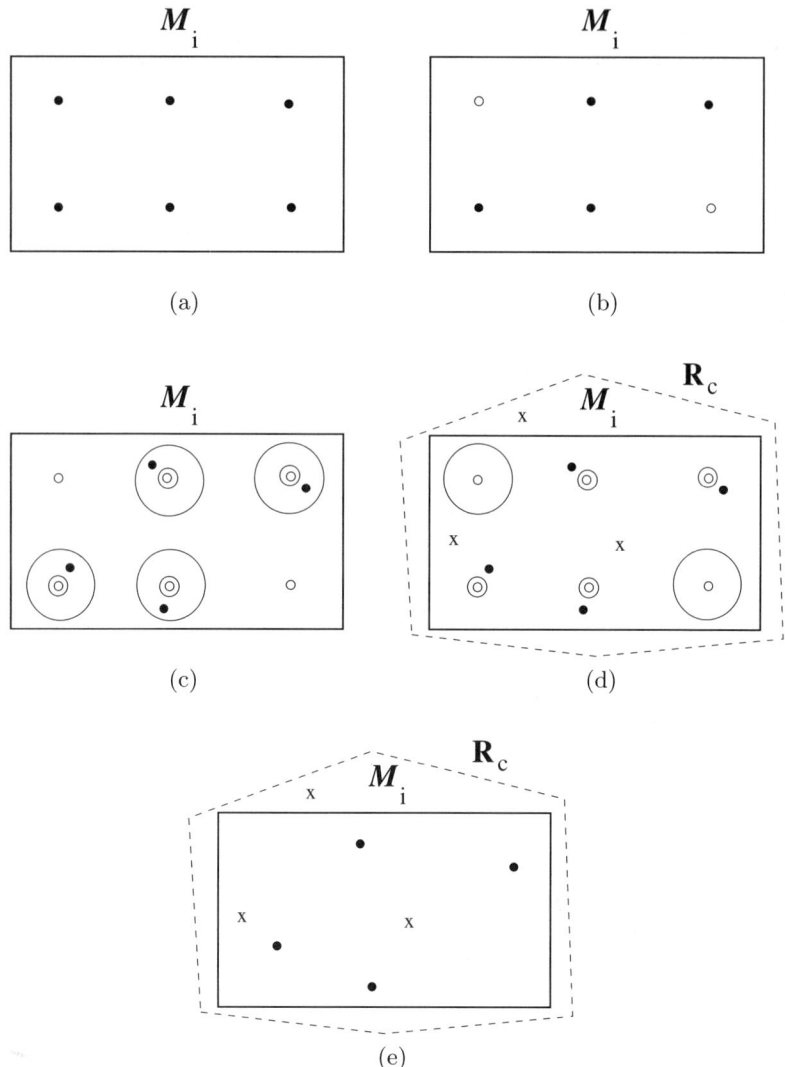

**Figure 1.4.** An illustration of the different stages of the distortion process: (a) original object consisting of six features (dark circles), (b) after occlusion ($O = 2$; small circles represent occluded features), (c) after perturbation ($R_u(\cdot) =$ a circle; small double circles represent locations of features before perturbation), (d) after clutter ($C = 3$; small crosses represent clutter features; notice absence of clutter features inside consistency regions of the two occluded features), (e) distorted object.

### 1.5.1 Definition of Object Similarity

We introduce a number of definitions that lead to a definition of the similarity between a pair of model objects.

- **Vote-Based Criterion:** Let $\mathcal{M}_i^{\hat{\tau}} = \{F_{ik}^{\hat{\tau}}\}$ be object $\mathcal{M}_i$ at pose $\hat{\tau} \in \mathcal{T}$ with respect to a data object, $\widehat{\mathcal{M}}$. We refer to $\mathcal{M}_i^{\hat{\tau}}$ as a *hypothesis* of object $\mathcal{M}_i$ at location $\hat{\tau}$. The votes for $\mathcal{M}_i^{\hat{\tau}}$, given $\widehat{\mathcal{M}}$, is the number of features in $\mathcal{M}_i^{\hat{\tau}}$ that are "consistent" with at least a data feature in $\widehat{\mathcal{M}}$. A model feature, $F_{ik}^{\hat{\tau}} \in \mathcal{M}_i^{\hat{\tau}}$, is said to be consistent with a data feature, $\hat{F}_l \in \widehat{\mathcal{M}}$, if $\hat{F}_l$ falls within the consistency region of $F_{ik}^{\hat{\tau}}$, i.e., $\hat{F}_l \in R_u(F_{ik}^{\hat{\tau}})$. Accordingly, we can formally define the votes for $\mathcal{M}_i^{\hat{\tau}}$ given $\widehat{\mathcal{M}}$ as follows:

$$\text{VOTES}(\mathcal{M}_i^{\hat{\tau}}; \widehat{\mathcal{M}}) = |\{F_{ik}^{\hat{\tau}} : F_{ik}^{\hat{\tau}} \in \mathcal{M}_i^{\hat{\tau}} \text{ and } \exists \hat{F}_l \in \widehat{\mathcal{M}} \text{ s.t. } \hat{F}_l \in R_u(F_{ik}^{\hat{\tau}})\}| \quad (1.1)$$

- **Feature/Feature Similarity:** Let us assume that we are given a pair of model features, $F_{ik} \in \mathcal{M}_i$ and $F_{jl}^{\tau_i} \in \mathcal{M}_j^{\tau_i}$, where $\mathcal{M}_j^{\tau_i}$ is a hypothesis of object $\mathcal{M}_j$ at location $\tau_i \in \mathcal{T}$ with respect to object $\mathcal{M}_i$. The similarity between $F_{ik}$ and $F_{jl}^{\tau_i}$, denoted by $S_{ff}(F_{ik}, F_{jl}^{\tau_i})$, is defined as the probability that an uncertain measurement of $F_{ik}$ is consistent with $F_{jl}^{\tau_i}$. Formally, we have

$$S_{ff}(F_{ik}, F_{jl}^{\tau_i}) = \frac{\text{AREA}(R(F_{ik}) \cap R(F_{jl}^{\tau_i}))}{\text{AREA}(R(F_{ik}))},$$

where $\text{AREA}(R)$ is area of region $R$. Obviously, $S_{ff}(F_{ik}, F_{jl}^{\tau_i})$ lies in the range $[0,1]$. It is proportional to the extent of overlap between the consistency regions of $F_{ik}$ and $F_{jl}^{\tau_i}$ ($R(F_{ik})$ and $R(F_{jl}^{\tau_i})$). Figure 1.5 illustrates $S_{ff}(F_{ik}, F_{jl}^{\tau_i})$ as a function of $\tau_i$, for a sample of three consistency regions. In some cases, we refer to feature pairs with overlapping/nonoverlapping consistency regions as *similar/dissimilar* feature pairs, respectively.

- **Object/Feature Similarity:** We define the similarity between an object, $\mathcal{M}_i$, and a feature, $F_{jl}^{\tau_i} \in \mathcal{M}_j^{\tau_i}$, as the probability that $F_{jl}^{\tau_i}$ is consistent with an uncertain measurement of *any* feature in $\mathcal{M}_i$. We can formally define object/feature similarity, denoted by $S_{of}(\mathcal{M}_i, F_{jl}^{\tau_i})$, as

$$S_{of}(\mathcal{M}_i, F_{jl}^{\tau_i}) = 1 - \prod_k (1 - S_{ff}(F_{ik}, F_{jl}^{\tau_i})).$$

- **Object/Hypothesis Similarity:** Let us denote the similarity between $\mathcal{M}_i$ and $\mathcal{M}_j^{\tau_i}$ as $S_{oh}(\mathcal{M}_i, \mathcal{M}_j^{\tau_i})$ or simply $S_j^{\tau_i}$. We define $S_j^{\tau_i}$ as the number of votes for hypothesis $\mathcal{M}_j^{\tau_i}$, given an uncertain instance of $\mathcal{M}_i$, which is $\mathcal{D}_u(\mathcal{M}_i, R_u(\cdot))$ (refer to Section 1.4). Formally,

$$S_j^{\tau_i} = \text{VOTES}(\mathcal{M}_j^{\tau_i}; \mathcal{D}_u(\mathcal{M}_i, R_u(\cdot))).$$

It is obvious that $S_j^{\tau_i}$ is a random variable. The minimum value of $S_j^{\tau_i}$ is the number of coincident feature pairs of $\mathcal{M}_i$ and $\mathcal{M}_j^{\tau_i}$. It can be expressed as

$$\min(S_j^{\tau_i}) = |\{F_{jk}^{\tau_i} : S_{of}(\mathcal{M}_i, F_{jk}^{\tau_i}) = 1\}|.$$

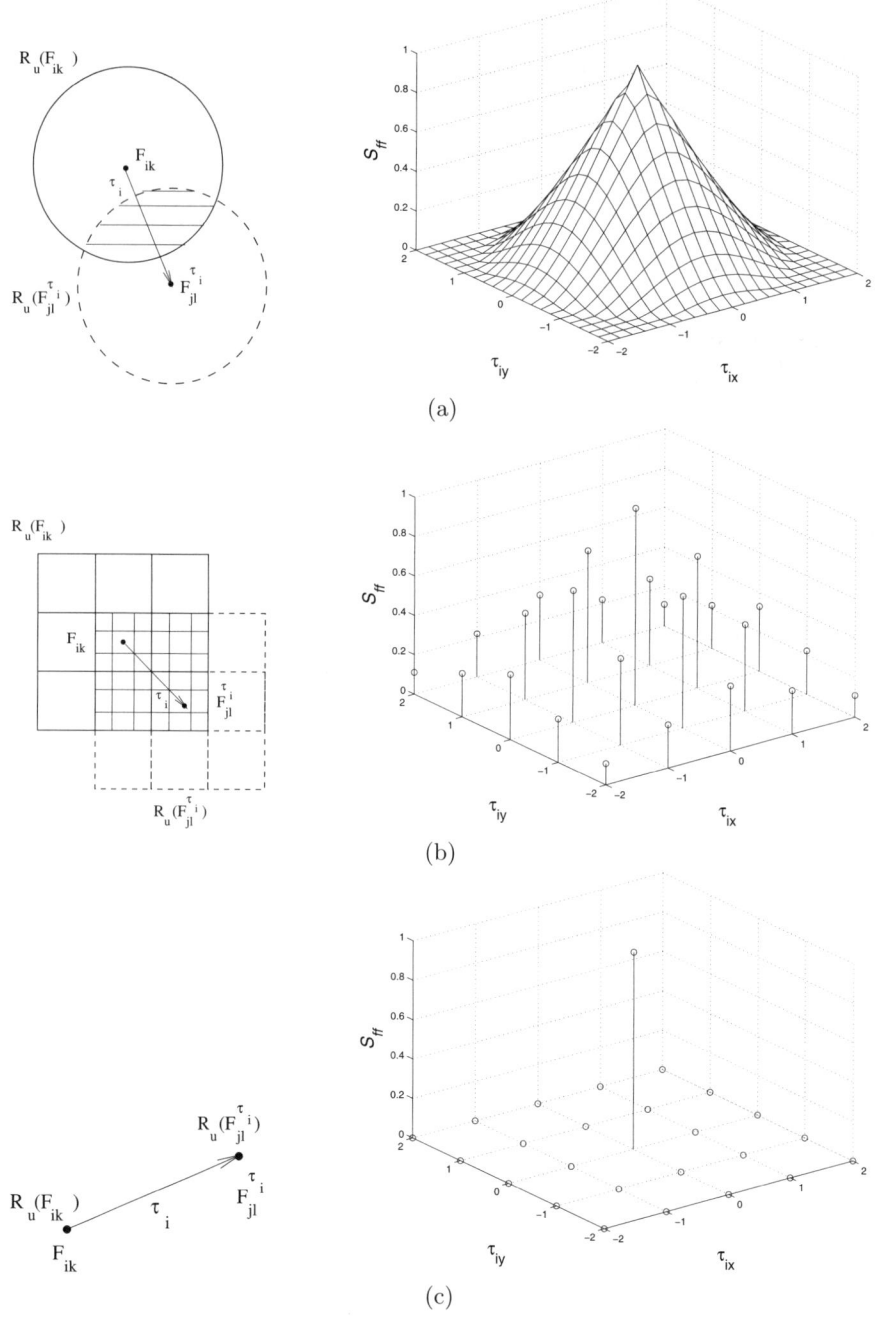

**Figure 1.5.** : An illustration of feature/feature similarity for a variety of consistency regions assuming $\mathcal{T}$ is the space of 2D translations: (a) a circle of unit radius, (b) a discrete eight-neighbor region, (c) a point region (implies absence of positional uncertainty). Note that the components of $\tau_i$ along the $x$- and $y$-axes are represented by $\tau_{ix}$, and $\tau_{iy}$, respectively. We assume here for simplicity that $F_{ik} = F_{jl}^{\tau_i}$ when $\tau_i = \mathbf{0}$.

On the other hand, the maximum value of $S_j^{\tau_i}$ is the number of features of $\mathcal{M}_j^{\tau_i}$ that are similar to features in $\mathcal{M}_i$ (i.e., whose consistency regions overlap with consistency regions of features in $\mathcal{M}_i$). Thus,

$$\max(S_j^{\tau_i}) = |\{F_{jk}^{\tau_i} : S_{of}(\mathcal{M}_i, F_{jk}^{\tau_i}) > 0\}|.$$

The expected value of $S_j^{\tau_i}$ can be approximated as

$$E(S_j^{\tau_i}) \approx \sum_k S_{of}(\mathcal{M}_i, F_{jk}^{\tau_i}),$$

where $F_{jk}^{\tau_i} \in \mathcal{M}_j^{\tau_i}$. Figure 1.6 shows an example of object/hypothesis similarity.

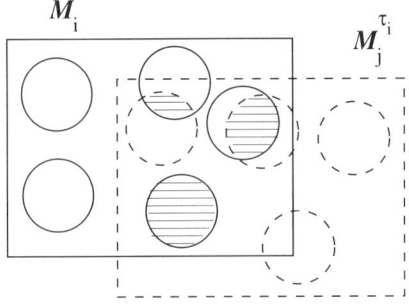

**Figure 1.6.** An illustration of object/hypothesis similarity. Notice that there are three similar feature pairs with feature/feature similarity values of approximately $\frac{1}{3}$, $\frac{2}{3}$ and 1. Accordingly, we have $S_j^{\tau_i} \in [1,3]$, and $E(S_j^{\tau_i}) \approx 2$.

- **Uniform Model of Object/Hypothesis Similarity:** In order to make the prediction of PCR bounds mathematically tractable, we make the following reasonable assumptions about $\mathcal{M}_i$, $\mathcal{M}_j^{\tau_i}$ and the structure of their similar feature pairs:

1. The consistency regions of the features that belong to each of $\mathcal{M}_i$ and $\mathcal{M}_j^{\tau_i}$ are not overlapping.
2. The correspondence between similar features in $\mathcal{M}_i$ and $\mathcal{M}_j^{\tau_i}$ is bijective (one-to-one).
3. The feature/feature similarity between every pair of similar features is a constant value, $P_j^{\tau_i}$. It is the average object/feature similarity of the features in $\mathcal{M}_j^{\tau_i}$ that are similar to features in $\mathcal{M}_i$.

The above assumptions result in a "uniform" view of the structural similarity between object $\mathcal{M}_i$ and hypothesis $\mathcal{M}_j^{\tau_i}$. As an illustration, Figure 1.7 shows the uniform model corresponding to the object/hypothesis pair shown in Figure 1.6. The uniform similarity model leads to the approximation of the PDF of $S_j^{\tau_i}$ by the following binomial distribution:

14   Michael Boshra and Bir Bhanu

$$P_{S_j^{\tau_i}}(s_j^{\tau_i}) = B_{S_j^{\tau_i}}(s_j^{\tau_i}; N_j^{\tau_i}, P_j^{\tau_i}),$$

where $P_X(x) = \Pr[X = x]$, $B_X(x; n, p) = K(n, x) p^x (1-p)^{n-x}$, $K(a, b) = \frac{a!}{(a-b)!\,b!}$,

$$N_j^{\tau_i} = \max(S_j^{\tau_i}), \text{ and}$$
$$P_j^{\tau_i} = \frac{E(S_j^{\tau_i})}{N_j^{\tau_i}}.$$

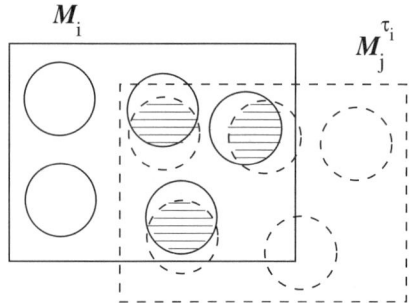

**Figure 1.7.** Uniform similarity model for object/hypothesis pair shown in Figure 1.6. Notice that similar feature pairs have constant feature/feature similarity, $P_j^{\tau_i} \approx \frac{2}{3}$ and $N_j^{\tau_i} = 3$.

- **Object/Object Similarity:** The similarity between a pair of objects, $\mathcal{M}_i$ and $\mathcal{M}_j$, is defined as the object/hypothesis similarity $S_j^{\tau_i}$, for all $\tau_i \in \mathcal{T}$. Thus, object/object similarity can be viewed as a probabilistic function. As an illustration, Figure 1.8(a) shows a pair of simple model objects. The corresponding expected-similarity function, $E(S_j^{\tau_i})$, is shown in Figure 1.8(b). Note that peaks in the expected-similarity function correspond to object hypotheses that have a higher degree of similarity with $\mathcal{M}_i$ than neighboring ones. A sample of these hypotheses, referred to as *peak hypotheses*, is shown in Figure 1.9. In our work, peak hypotheses are used for predicting an upper bound on PCR.

### 1.5.2 Construction of Similarity Histograms

As discussed in the previous section, we describe the object/hypothesis similarity between $\mathcal{M}_i$ and $\mathcal{M}_j^{\tau_i}$ by two parameters, $(N_j^{\tau_i}, P_j^{\tau_i})$. For our purpose of performance prediction, we add two more parameters:
1. The size of $\mathcal{M}_i$, $|\mathcal{M}_i|$.
2. The *effective size* of $\mathcal{M}_j^{\tau_i}$, $|\mathcal{M}_j^{\tau_i} \cap R_c|$, which is the number of features of $\mathcal{M}_j^{\tau_i}$ that lie inside the clutter region $R_c$.

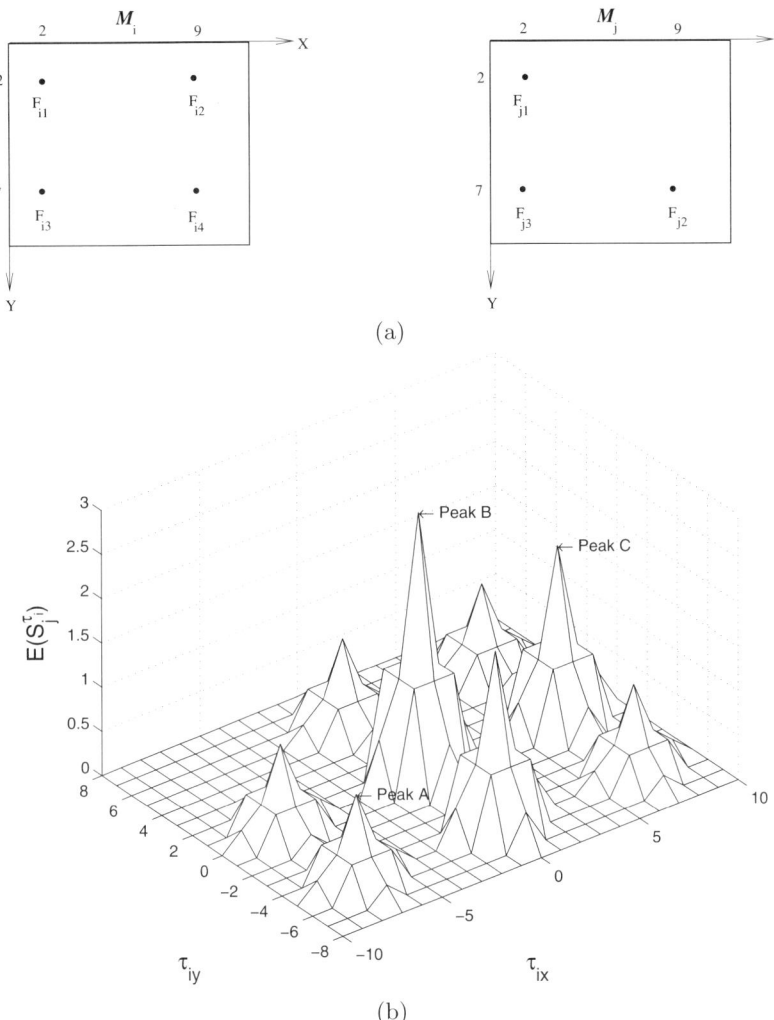

**Figure 1.8.** An illustration of object/object similarity: (a) model objects $\mathcal{M}_i$ and $\mathcal{M}_j$, (b) corresponding expected-similarity function, $E(S_j^{\tau_i})$, assuming four-neighbor consistency region, and 2D translation space.

Thus, we encode the information of object/hypothesis similarity using tuple $(|\mathcal{M}_i|, |\mathcal{M}_j^{\tau_i} \cap R_c|, N_j^{\tau_i}, N_j^{\tau_i} P_j^{\tau_i})$. Accordingly, the similarity information is accumulated in 4D histograms [3].

Two similarity histograms are needed in our work, one for storing similarity information corresponding to all erroneous hypotheses, and the other

---

[3] When calculating the effective size of $\mathcal{M}_j^{\tau_i}$, we have also included features of $\mathcal{M}_j^{\tau_i}$ that lie outside $R_c$ but are similar to features in $\mathcal{M}_i$.

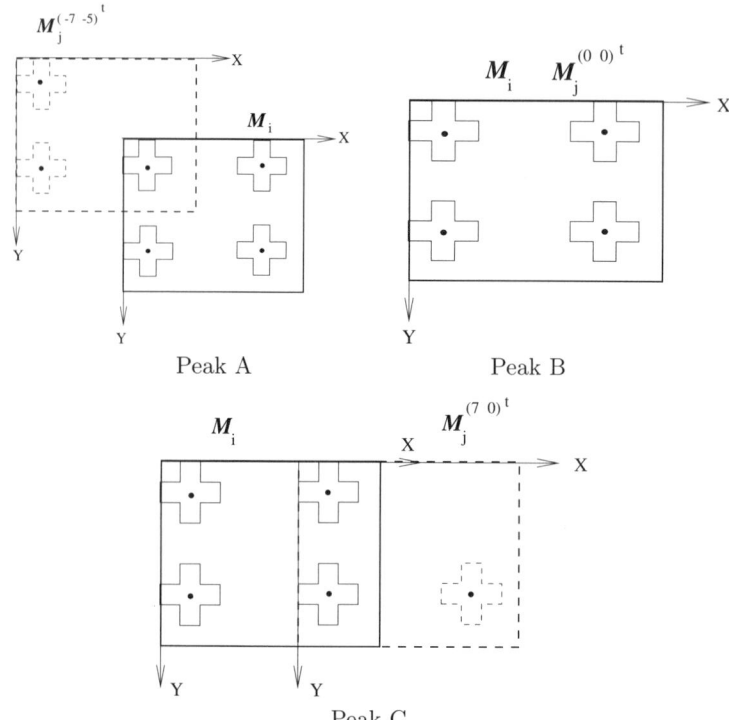

**Figure 1.9.** Three hypotheses corresponding to peaks $A$, $B$ and $C$ shown in Fig. 1.8(b), assuming four-neighbor consistency region.

for storing information corresponding to peak hypotheses only. They are referred to as all- and peak-similarity histograms, respectively. The algorithm used to construct these histograms is outlined in Figure 1.10. It calculates the similarity between every model object $\mathcal{M}_i$, and all the erroneous hypotheses competing with it. The erroneous hypotheses are selected to satisfy the following two criteria:

1. Each has at least one feature inside clutter region $R_c$.
2. For hypotheses that belong to $\mathcal{M}_i$, the relative pose, $\tau_i$, lies outside $\mathcal{T}_{acc}$, defined in Section 1.3.

The similarity information associated with $\mathcal{M}_i$ is accumulated in local all- and peak-similarity histograms, $ASH_i$ and $PSH_i$, respectively. These histograms, for all $\mathcal{M}_i \in \mathcal{MD}$, are then added to form global similarity histograms, $ASH$ and $PSH$, respectively.

```
Initialize global similarity histograms ASH and PSH
for each model object $\mathcal{M}_i \in \mathcal{MD}$ do
   Initialize local similarity histograms for $\mathcal{M}_i$, $ASH_i$ and $PSH_i$
   for each model object $\mathcal{M}_j \in \mathcal{MD}$ do
      for each $\tau_i \in \mathcal{T}$ such that $|\mathcal{M}_j^{\tau_i} \cap R_c| > 0$ do
         if $(i \neq j) \vee \neg(\tau_i \in \mathcal{T}_{acc})$ then
            Compute similarity parameters $(N_j^{\tau_i}, P_j^{\tau_i})$
            Increment $ASH_i(|\mathcal{M}_i|, |\mathcal{M}_j^{\tau_i} \cap R_c|, N_j^{\tau_i}, \lfloor N_j^{\tau_i} P_j^{\tau_i} + \frac{1}{2} \rfloor)$ by 1
            if $\mathcal{M}_j^{\tau_i}$ is a peak hypothesis then
               Increment $PSH_i(|\mathcal{M}_i|, |\mathcal{M}_j^{\tau_i} \cap R_c|, N_j^{\tau_i}, \lfloor N_j^{\tau_i} P_j^{\tau_i} + \frac{1}{2} \rfloor)$ by 1
            end if
         end if
      end for
   end for
   Add $ASH_i$ to $ASH$
   Add $PSH_i$ to $PSH$
end for
```

**Figure 1.10.** Similarity-computation algorithm.

## 1.6 Computation of Performance Bounds

In this section, we derive the PDF of votes for an erroneous hypothesis, and use this PDF for predicting lower and upper bounds on PCR.

### 1.6.1 Motivating Example

We start by presenting an example to illustrate the combined effects of data distortion and object similarity on the vote process. This example, illustrated in Figure 1.11, assumes the uniform model of similarity between $\mathcal{M}_i$ and $\mathcal{M}_j^{\tau_i}$, which is defined in Section 1.5.1. It can be described as follows:

- Prior to being distorted, $\mathcal{M}_i$ has five votes, since it consists of five features. On the other hand, $\mathcal{M}_j^{\tau_i}$ does not have any features of $\mathcal{M}_i$ within the consistency regions of its features. Accordingly, it gets no votes.
- The first distortion step involves occlusion of two features in $\mathcal{M}_i$. Obviously, this reduces the number of votes for $\mathcal{M}_i$ from five to three. At this point, $\mathcal{M}_j^{\tau_i}$ still does not get any votes. Notice that the number of similar feature pairs between $\mathcal{M}_i$ and $\mathcal{M}_j^{\tau_i}$ decreases from three (which is $N_j^{\tau_i}$; refer to Section 1.5.1) to two.
- The second step involves randomly perturbing the three unoccluded features in $\mathcal{M}_i$ within their consistency regions. This keeps the number of votes for $\mathcal{M}_i$ at three. On the other hand, observe that both of the two unoccluded similar features of $\mathcal{M}_i$ move to the regions that overlap with the consistency regions of their corresponding similar features in $\mathcal{M}_j^{\tau_i}$. This contributes two votes to $\mathcal{M}_j^{\tau_i}$.

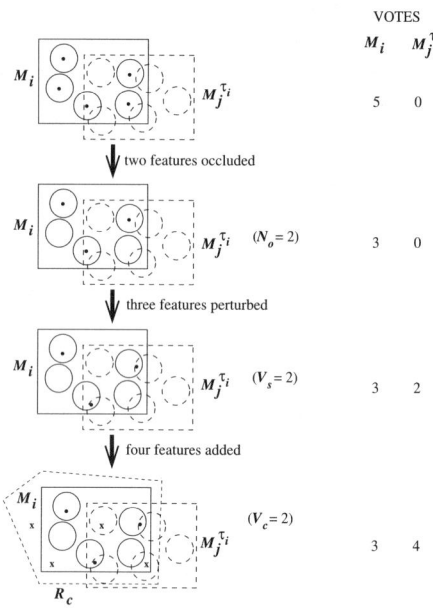

**Figure 1.11.** An example showing the vote process for object $\mathcal{M}_i$ and erroneous hypothesis $\mathcal{M}_j^{\tau_i}$, as $\mathcal{M}_i$ gets distorted.

- In the final distortion step, four clutter features are randomly added within clutter region $R_c$. Two of these features happen to fall within the consistency regions of two new features of $\mathcal{M}_j^{\tau_i}$. This contributes two extra votes for $\mathcal{M}_j^{\tau_i}$, thus bringing its total number of votes to four. The number of votes for $\mathcal{M}_i$ stays the same (recall from Section 1.4 that clutter features are excluded from falling into consistency regions of occluded ones).

The above example shows how data distortion and model similarity can result in a recognition failure by reducing the number of votes for the correct hypothesis, and increasing them for an erroneous one. It also provides us with the following valuable insight into the distribution of votes for both correct and incorrect hypotheses, as a function of data distortion:

- The number of votes for $\mathcal{M}_i$, denoted by $V_i$, is simply the number of unoccluded features of $\mathcal{M}_i$. That is, for a distorted instance of $\mathcal{M}_i$, $\widehat{\mathcal{M}}_i(R_u(\cdot), O, C, R_c)$ or simply $\widehat{\mathcal{M}}_i$, we have

$$V_i = |\mathcal{M}_i| - O. \qquad (1.2)$$

- The number of votes for $\mathcal{M}_j^{T_i}$, denoted by $V_j^{T_i}$, comes from two different sources: (1) object $\mathcal{M}_i$, due to structural similarity (second distortion step in Figure 1.11), and (2) clutter features, due to random coincidence (third distortion step in Figure 1.11). Thus, $V_j^{T_i}$ is a random variable that can be expressed as follows:

$$V_j^{T_i} = V_s + V_c, \qquad (1.3)$$

where $V_s$ and $V_c$ are random variables that represent *similarity* and *clutter* votes for $\mathcal{M}_j^{T_i}$, respectively.
- The number of similarity votes, $V_s$, is bounded by the number of similar feature pairs that remain unoccluded, which we denote by $N_o$ (obviously, $N_o \leq N_j^{T_i}$).

In the above example, it can be seen that $V_s = 2$, $V_c = 2$, and $N_o = 2$. In the next section, we use these three random variables to determine the PDF of $V_j^{T_i}$.

### 1.6.2 Probability Distribution of Hypothesis Votes

In order to determine the PDF of $V_j^{T_i}$, we need to determine the PDFs of $V_s$ (number of similarity votes), $V_c$ (number of clutter votes), and $N_o$ (number of unoccluded similar features). In the previous section, we have seen that $V_s$ depends on $N_o$. Accordingly, we can express the PDF of $V_s$ as

$$P_{V_s}(v_s) = \sum_{n_o} P_{V_s}(v_s; n_o) P_{N_o}(n_o), \qquad (1.4)$$

where $P_{V_s}(v_s; n_o) = \Pr[V_s = v_s; N_o = n_o]$. From (1.3) and (1.4), we can represent the PDF of $V_j^{T_i}$ as

$$P_{V_j^{T_i}}(v_j^{T_i}) = \sum_{n_o} P_{V_j^{T_i}}(v_j^{T_i}; n_o) P_{N_o}(n_o), \qquad (1.5)$$

where

$$P_{V_j^{T_i}}(v_j^{T_i}; n_o) = \sum_{v_s} P_{V_s}(v_s; n_o) P_{V_c}(v_j^{T_i} - v_s; n_o, v_s)$$

and $P_{V_c}(v_c; n_o, v_s) = \Pr[V_c = v_c; N_o = n_o, V_s = v_s]$. We estimate the PDF of $N_o$ and the conditional PDFs of $V_s$ and $V_c$ based on the uniform models of data distortion and structural similarity, presented in Sections 1.4 and 1.5.1, respectively.
- **PDF of $N_o$:** The process of occluding $O$ features in $\mathcal{M}_i$ can be viewed as picking $O$ balls from an urn, which contains $N_j^{T_i}$ white balls and $(|\mathcal{M}_i| - N_j^{T_i})$ black balls, with no replacement. In our case, the white (black) balls represent features in $\mathcal{M}_i$ that are similar (dissimilar) to features in $\mathcal{M}_j^{T_i}$. Based on the

uniform occlusion and similarity models, the PDF of $N_o$ can be described by the following hypergeometric distribution,

$$P_{N_o}(n_o) = H_{N_o}(N_j^{T_i} - n_o; O, N_j^{T_i}, |\mathcal{M}_i| - N_j^{T_i}), \qquad (1.6)$$

where $H_X(x; n, a, b) = \frac{K(a,x)K(b,n-x)}{K(a+b,n)}$. Note that

$$n_o \in [\max(0, N_j^{T_i} - O), \min(N_j^{T_i}, |\mathcal{M}_i| - O)].$$

- **Conditional PDF of $V_s$:** It can be easily shown that the conditional PDF of $V_s$ is represented by the following distribution!binomial binomial distribution:

$$P_{V_s}(v_s; n_o) = B_{V_s}(v_s; n_o, P_j^{T_i}).$$

This distribution is obtained based on the assumptions of uniform uncertainty and similarity models. Notice that $P_j^{T_i} < 1$ implies $v_s \in [0, n_o]$, while $P_j^{T_i} = 1$ implies $v_s = n_o$.

- **Conditional PDF of $V_c$:** The estimation of the PDF of $V_c$ is considerably more involved than those of $N_o$ and $V_s$. It can be outlined as follows. Let $R'_{V_c} \subset R'_c$ be the largest region such that a clutter feature falling within it will contribute a vote for $\mathcal{M}_j^{T_i}$. Region $R'_{V_c}$ is the union of the consistency regions of features in $\mathcal{M}_j^{T_i} \cap R_c$ that do not have any features of $\mathcal{M}_i$ within their consistency regions. They are basically all the features of $\mathcal{M}_j^{T_i} \cap R_c$ minus those that have similar features of $\mathcal{M}_i$ within their consistency regions. A slight complexity arises from our clutter modeling explained in Section 1.4: features in $\mathcal{M}_j^{T_i} \cap R_c$ that are similar to *occluded* features in $\mathcal{M}_i$ are effectively associated with "truncated" consistency regions. Figure 1.12 shows an example of $R'_{V_c}$. Based on the assumption of uniform similarity, we can show the following:

1. The area of a truncated consistency region is $\text{AREA}(R_u(\cdot))(1 - P_j^{T_i})$.
2. The numbers of potential vote-contributing features with truncated and full consistency regions are $n_t = N_j^{T_i} - n_o$, and $n_f = |\mathcal{M}_j^{T_i} \cap R_c| - v_s - n_t$, respectively.

Splitting the effective clutter region $R'_c$ into two subregions, $R'_{V_c}$ and $R'_c - R'_{V_c}$, we can approximate the conditional PDF of $V_c$ by the following binomial distribution,

$$P_{V_c}(v_c; n_o, v_s) \approx B_{V_c}\left(v_c; C, \frac{\text{AREA}(R'_{V_c})}{\text{AREA}(R'_c)}\right), \qquad (1.7)$$

where

$$\text{AREA}(R'_{V_c}) = \text{AREA}(R_u(\cdot))(n_f + (1 - P_j^{T_i})n_t), \text{ and}$$
$$\text{AREA}(R'_c) = \text{AREA}(R_c) - O \times \text{AREA}(R_u(\cdot)).$$

The lower bound of $v_c$ is 0, while the upper bound is either $\min(n_f + n_t, C)$ if $P_j^{T_i} < 1$, or $\min(n_f, C)$ if $P_j^{T_i} = 1$[4].

---

[4] The area of $R'_c$ is calculated by assuming, for simplicity, that clutter region $R_c$ totally covers the consistency regions of the features of $\mathcal{M}_i$.

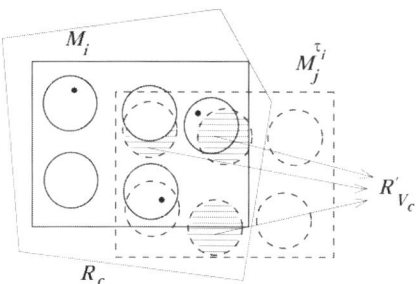

**Figure 1.12.** An illustration of the clutter vote region $R'_{V_c}$, assuming the uniform model of similarity between $\mathcal{M}_i$ and $\mathcal{M}_j^{\tau_i}$.

### 1.6.3 Lower Bound on PCR

Let $\mathcal{H}_i$ be the set of erroneous object/pose hypotheses corresponding to $\mathcal{M}_i$. It can be defined as

$$\mathcal{H}_i = \{\mathcal{M}_j^{\tau_i} : \mathcal{M}_j \in \mathcal{MD} \text{ and } \tau_i \in \mathcal{T} \text{ s.t. } |\mathcal{M}_j^{\tau_i} \cap R_c| > 0\} - \{\mathcal{M}_i^{\tau_i} : \tau_i \in \mathcal{T}_{acc}\}.$$

We can express the probability of misinterpreting a distorted instance of $\mathcal{M}_i$, $\widehat{\mathcal{M}}_i$, as any hypothesis in $\mathcal{H}_i$, as

$$\Pr[\mathcal{H}_i; \widehat{\mathcal{M}}_i] = \Pr[\exists\, \mathcal{M}_j^{\tau_i} \in \mathcal{H}_i \text{ s.t. } V_j^{\tau_i} \geq V_i]. \quad (1.8)$$

The probability that $\mathcal{M}_j^{\tau_i}$ "beats" $\mathcal{M}_i$ (i.e., $\mathcal{M}_j^{\tau_i}$ reaches or exceeds votes for $\mathcal{M}_i$) can be obtained from (1.2) and (1.5):

$$\Pr[\mathcal{M}_j^{\tau_i}; \widehat{\mathcal{M}}_i] = \sum_{v_j^{\tau_i} \geq |\mathcal{M}_i| - O} P_{V_j^{\tau_i}}(v_j^{\tau_i}). \quad (1.9)$$

From (1.8) and (1.9), we obtain the following upper bound on the probability of recognition failure:

$$\Pr[\mathcal{H}_i; \widehat{\mathcal{M}}_i] < \sum_{\mathcal{M}_j^{\tau_i} \in \mathcal{H}_i} \Pr[\mathcal{M}_j^{\tau_i}; \widehat{\mathcal{M}}_i].$$

The above inequality directly leads to the following lower bound on PCR:

$$\Pr[\mathcal{M}_i; \widehat{\mathcal{M}}_i] > 1 - \sum_{\mathcal{M}_j^{\tau_i} \in \mathcal{H}_i} \Pr[\mathcal{M}_j^{\tau_i}; \widehat{\mathcal{M}}_i]. \quad (1.10)$$

From the derivation of the vote PDF discussed in the previous section, we can observe that $V_j^{\tau_i}$ and, in turn, $\Pr[\mathcal{M}_j^{\tau_i}; \widehat{\mathcal{M}}_i]$ depend on only four object-dependent parameters: size of $\mathcal{M}_i$, effective size of $\mathcal{M}_j^{\tau_i}$, and the two similarity parameters $(N_j^{\tau_i}, P_j^{\tau_i})$. Define

$$W(a,b,c,d) = \Pr[\mathcal{M}_j^{\tau_i}; \widehat{\mathcal{M}}_i],$$

such that $a = |\mathcal{M}_i|$, $b = |\mathcal{M}_j^{\tau_i} \cap R_c|$, $c = N_j^{\tau_i}$ and $d = \lfloor N_j^{\tau_i} P_j^{\tau_i} + \frac{1}{2} \rfloor$. Equation (1.10) can be rewritten as

$$\Pr[\mathcal{M}_i; \widehat{\mathcal{M}}_i] > 1 - \sum_a \sum_b \sum_c \sum_d ASH_i(a,b,c,d) W(a,b,c,d). \quad (1.11)$$

Taking the average of (1.11) over all model objects in the model set $\mathcal{MD}$, we obtain the following lower bound on average PCR for $\mathcal{MD}$:

$$\text{PCR}(\mathcal{MD}) > 1 - \frac{1}{|\mathcal{MD}|} \sum_a \sum_b \sum_c \sum_d ASH(a,b,c,d) W(a,b,c,d), \quad (1.12)$$

where

$$ASH(a,b,c,d) = \sum_i ASH_i(a,b,c,d).$$

### 1.6.4 Upper Bound on PCR

In this section, we present three possible upper bounds on PCR. These bounds differ from each other in the degree of their tightness, and their reliance on assumptions.

The first upper bound can be obtained by observing that recognition fails if *any* hypothesis in $\mathcal{H}_i$ beats $\mathcal{M}_i$. The probability that this event takes place for a given hypothesis, $\mathcal{M}_j^{\tau_i}$, is $\Pr[\mathcal{M}_j^{\tau_i}; \widehat{\mathcal{M}}_i]$, which is defined in (1.9). The maximum of these probabilities among all hypotheses in $\mathcal{H}_i$ forms a lower bound on the probability of recognition failure. This directly leads us to the following upper bound on PCR:

$$\Pr[\mathcal{M}_i; \widehat{\mathcal{M}}_i] < 1 - \max_{\tau_i, j} \Pr[\mathcal{M}_j^{\tau_i}; \widehat{\mathcal{M}}_i], \quad (1.13)$$

where $\mathcal{M}_j^{\tau_i} \in \mathcal{H}_i$. Obviously, we do not expect this bound to be tight, since it involves only a *single* erroneous hypothesis.

One possible approach for obtaining a bound that is tighter than the one in (1.13) is to consider a subset of the hypotheses in $\mathcal{H}_i$, and make the assumption that the vote PDF's for these hypotheses are independent. Obviously, the vote-independence assumption is not reasonable among adjacent hypotheses due to the overlap of their consistency regions. We propose to consider hypotheses that correspond to *peaks* in the expected similarity function (refer to Section 1.5.1). The rationale behind such choice can be outlined as follows:

- Peak hypotheses tend to occur at random locations, which makes the vote-independence assumption among them more reasonable than when applied to adjacent ones.

- Peak hypotheses have a higher degree of similarity with given model object $\mathcal{M}_i$, than neighboring ones. Accordingly, we can say that a distorted instance of $\mathcal{M}_i$ is more likely to be misinterpreted as a peak hypotheses than an off-peak neighbor.

Let $\mathcal{H}_{pi}$ be the set of peak hypotheses associated with model object $\mathcal{M}_i$. That is,

$$\mathcal{H}_{pi} = \{\mathcal{M}_j^{\tau_i} : \mathcal{M}_j^{\tau_i} \in \mathcal{H}_i, \text{ and } \tau_i \text{ is a peak in } E(S_j^{\tau_i})\}.$$

Based on the vote-independence assumption, we can obtain the following upper bound on PCR:

$$\Pr[\mathcal{M}_i; \widehat{\mathcal{M}}_i] < \min_j \prod_{\tau_i}(1 - \Pr[\mathcal{M}_j^{\tau_i}; \widehat{\mathcal{M}}_i]), \tag{1.14}$$

where $\mathcal{M}_j^{\tau_i} \in \mathcal{H}_{pi}$. This upper bound is tighter than the one defined in (1.13), since it considers a representative subset of model hypotheses that belong to a single object, instead of just a single hypothesis as in (1.13).

In order to obtain a bound that is tighter than the one in (1.14), we need to consider hypotheses in $\mathcal{H}_i$ that belong to *all* model objects, not just a single object as in (1.14). One possible way of achieving this goal is to make the assumption that the vote PDFs for peak hypotheses that belong to *different* model objects are independent. This assumption leads us to the following upper bound,

$$\Pr[\mathcal{M}_i; \widehat{\mathcal{M}}_i] < \prod_{j,\tau_i}(1 - \Pr[\mathcal{M}_j^{\tau_i}; \widehat{\mathcal{M}}_i]), \tag{1.15}$$

where $\mathcal{M}_j^{\tau_i} \in \mathcal{H}_{pi}$. We note that the above assumption can be invalid in some extreme cases, such as when two of the model objects are identical or very similar. However, we argue that it is reasonable in many practical scenarios. For example, let us consider the domain of object recognition using SAR data, which is the main motivation behind our efforts (see the experimental results in Section 1.7). It is well known that the features extracted from a SAR image are extremely sensitive to the geometry of the object shape. Accordingly, even in the presence of similar objects, we can expect the corresponding feature sets to be considerably different [10].

In our experiments, we use the tightest bound defined in (1.15). This bound can be rewritten as

$$\Pr[\mathcal{M}_i; \widehat{\mathcal{M}}_i] < \prod_a \prod_b \prod_c \prod_d (1 - W(a,b,c,d))^{PSH_i(a,b,c,d)}. \tag{1.16}$$

Taking the geometric mean of (1.16) for all model objects in $\mathcal{MD}$, we obtain an estimate of the upper bound of PCR corresponding to $\mathcal{MD}$,

$$\text{PCR}(\mathcal{MD}) < \prod_a \prod_b \prod_c \prod_d (1 - W(a,b,c,d))^{\frac{PSH(a,b,c,d)}{|\mathcal{MD}|}}, \tag{1.17}$$

where

$$PSH(a,b,c,d) = \sum_i PSH_i(a,b,c,d).$$

## 1.7 Experimental Validation

In this section, we validate the proposed prediction method by comparing predicted bounds on performance with actual ones determined empirically.

### 1.7.1 Recognition Task

The recognition task that we consider in this work involves recognition of objects using SAR data [10]. We use real data from the MSTAR public domain [28]. The model set consists of a number of objects, which are military targets. Each object is represented by a number of SAR views that sample its signature at several azimuth angles and a specific depression angle. We consider each view as an independent 2D "object." Each model or data view is represented by locations of scattering centers, which are image peaks. These scattering centers correspond to eight-neighbor peaks. The features of each object are the 30 peaks that are strongest in magnitude of radar returns. Recognition involves comparing peaks extracted from data and model views assuming that the space of applicable transformations, $\mathcal{T}$, is discrete 2D translations in the image plane [5, 10].

### 1.7.2 Model and Test Sets

Our model set consists of three objects, which are T72, BMP2, and BTR70. The numbers of views for these objects are 231, 233, and 233, respectively, and so the total number of views is 697. All of these views are obtained at depression angle 17°. Figure 1.13 shows a representative sample of a single view for each object, along with associated scattering centers.

The test data consists of seven sets, $\mathcal{TD}_1$ through $\mathcal{TD}_7$. They are classified into two groups depending on whether the data distortion in the set is synthetic or real.

1. **Synthetic-Distortion Group:** This group consists of sets $\mathcal{TD}_1$ through $\mathcal{TD}_5$. It is obtained by synthetically distorting the model set, following the distortion process outlined in Section 1.4. We have added the constraint that no features are eight-neighbors, in order to simulate the process of peak extraction. The distortion in each set is characterized by specific consistency and clutter regions, and a number of occlusion/clutter $(O/C)$ values. Notice that the numbers of occluded and clutter features are always the same in our case, since we are considering a fixed number of features

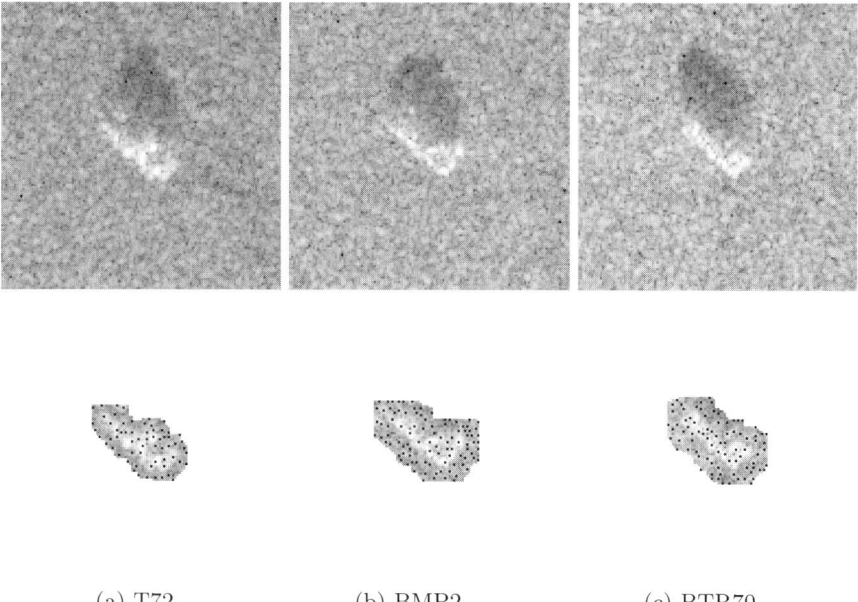

(a) T72  (b) BMP2  (c) BTR70

**Figure 1.13.** Examples of SAR images and associated scattering centers at depression angle 17°, and azimuth angle 132°.

for both model and data views. The test subset corresponding to a specific $O/C$ value consists of four randomly generated distorted instances for each model view. Accordingly, the size of each test subset is $4\,|\,\mathcal{MD}\,|$, and the total size of the test set is $4n_{oc}\,|\,\mathcal{MD}\,|$, where $n_{oc}$ is the number of $O/C$ values considered in the set. The upper section of Table 1.2 describes the five synthetically distorted test sets. Note that the fourth column in this table consists of two subcolumns: the first one describes the basic shape of the clutter region, while the second describes the scale of that shape. In particular, the clutter region is obtained by scaling the basic clutter region by some factor. Furthermore, notice that the consistency region $R_u(\cdot)$ for test set $\mathcal{TD}_1$ is the zero-neighbor region, which implies the absence of any positional uncertainty.

2. **Real-Distortion Group:** This group consists of two sets, $\mathcal{TD}_6$ and $\mathcal{TD}_7$. These sets correspond to variants of the model set involving real distortion. The first set, $\mathcal{TD}_6$, is obtained by changing the configurations of the model objects [5]. Examples of changing the object configuration are using different flash lights, changing numbers of barrels, etc. The second set, $\mathcal{TD}_7$, is obtained by changing the depression angle from 17° to 15°. Note that due to the nature of the SAR imaging process, such a small angle change can result in a significant variation in the object signature [28]. For both sets, the distortion parameters are determined as follows. The

Table 1.2. Description of the test sets used in the experiments.

| Set | Distortion | Consistency Region | Clutter Region Shape | Factor | Occluded/ Clutter Features | Size |
|---|---|---|---|---|---|---|
| $\mathcal{TD}_1$ | synthetic | 0-neighbor | convex hull | 1 | $18, 19, \cdots, 27$ | $4 \times 697 \times 10$ |
| $\mathcal{TD}_2$ | synthetic | 4-neighbor | convex hull | 1 | $9, 10, \cdots, 20$ | $4 \times 697 \times 12$ |
| $\mathcal{TD}_3$ | synthetic | 8-neighbor | convex hull | 1 | $0, 1, \cdots, 15$ | $4 \times 697 \times 16$ |
| $\mathcal{TD}_4$ | synthetic | 4-neighbor | convex hull | 2 | $9, 10, \cdots, 24$ | $4 \times 697 \times 16$ |
| $\mathcal{TD}_5$ | synthetic | 4-neighbor | convex hull | 3 | $9, 10, \cdots, 24$ | $4 \times 697 \times 16$ |
| $\mathcal{TD}_6$ | $\Delta$ config. | 4-neighbor | convex hull | 1 | estimated | 464 |
| $\mathcal{TD}_7$ | $\Delta$ dep. angle | 4-neighbor | convex hull | 1 | estimated | 581 |

consistency and clutter regions, $R_u(\cdot)$ and $R_c$, are empirically chosen to be four-neighbor region and convex hull of view features, respectively. The $O/C$ value is estimated for each test view through finding the best matching model view within a difference of $\pm 3°$ azimuth angles, and counting the number of unmatched features. In the case of the absence of a model view within $\pm 3°$ azimuth angles, the test view is matched with the available model one that is closest in azimuth. The lower section of Table 1.2 summarizes the two real sets.

Note that our models are defined from real data that may in fact contain some distortions. Thus, our notion of occlusion and clutter is relative, not absolute. That is, a spurious feature in the model view that does not match any feature in the test view is considered as a true model feature that is occluded. Furthermore, a true feature in the test view that is missing in the model view is considered as a clutter feature. Obviously, recognition performance would be improved through learning the true model features, or by giving a variety of weights to the model features depending on the probability that they correspond to true features. This topic is beyond the scope of this work.

### 1.7.3 Results

Our selected performance measure is PCR as a function of data distortion, in particular, the occlusion/clutter rate $(O/C)$. As mentioned, the consistency and clutter regions are assumed to be fixed for each test set. The empirical performance is determined using an object recognition system, which uses the vote-based criterion defined in (1.1). The recognition system examines *all* the relevant translations between a given test view, and each model one. The translations examined are defined by the bounding box on the translations that lead to at least a single match between a test feature and a model one. Accordingly, the performance determined by the recognition system is *optimal*, assuming the given vote-based criterion. The predicted bounds are obtained as described in Section 1.6 with two minor modifications. The first one involves the clutter region. Notice that in our experiments, the exact shape of the clutter region is not fixed, but differs from one view to another. Accordingly, AREA($R_c$) is replaced by the average of clutter-region areas for all model views. Recall that AREA($R_c$) is used when calculating the vote PDF of an erroneous hypothesis (refer to Section 1.6.2). The second modification involves the conditional PDF of clutter votes, defined in (1.7). Notice that in our recognition task, image peaks cannot be eight-neighbors. This fact needs to be considered in the estimation of the clutter vote PDF, in order to obtain more accurate performance bounds. An approximate method for estimating such a PDF is presented in the Appendix.

We first analyze the results involving the synthetic-distortion test group. Figures 1.14 and 1.15 show actual and predicted performance plots for the test sets $\mathcal{TD}_1$ through $\mathcal{TD}_5$. From these figures, we observe the following:

- In all cases, our method consistently succeeds in predicting reasonably tight bounds on actual performance.
- The predicted lower bounds consistently predict the actual breakpoint in performance with high accuracy. In all cases, the breakpoint in the lower-bound plot either coincides with the actual one, or occurs very slightly before it.
- The prediction method confirms the intuitive observation that performance degrades when the size of the consistency region increases. This can be observed by comparing Figures 1.14(a)–(c). The method also confirms the intuitive observation that performance improves when the density of clutter features decreases, which takes place when the size of the clutter region increases. Again, this can be observed by comparing Figures 1.14(b), 1.15(a), and 1.15(b).
- The predicted lower bound is consistently very close to the actual PCR plot along the knee section, defined as the section between the first two breakpoints in the plot. In all cases, the difference between the lower bound and corresponding PCR in the knee section is less than 3%. Beyond the knee section, the lower bound diverges. This observation can be explained as follows. Let $\mathcal{H}_i(D) \subset \mathcal{H}_i$ be set of erroneous hypotheses that simultaneously beat true hypothesis at distortion level $D$, where $D = O = C$. At relatively low levels of distortion, the probability that more than one erroneous hypotheses will *simultaneously* beat the true one is negligible compared to that of only a single erroneous hypothesis [5]. Accordingly, in case of recognition failure at low distortion levels, $\mathcal{H}_i(D)$ has only a single element in most cases. This makes the lower bound defined in (1.10) an estimate of the actual performance. However, as distortion increases, the above assertion ceases to be valid, which makes the lower bound in (1.10) a strict one.
- The predicted upper bound becomes less tight beyond the breakpoint of performance, but then becomes very tight towards the end of the plot. This can be explained as follows. It is easy to see that the size of $\mathcal{H}_i(D)$ is

---
[5] This is assuming we do not have identical or very similar model objects.

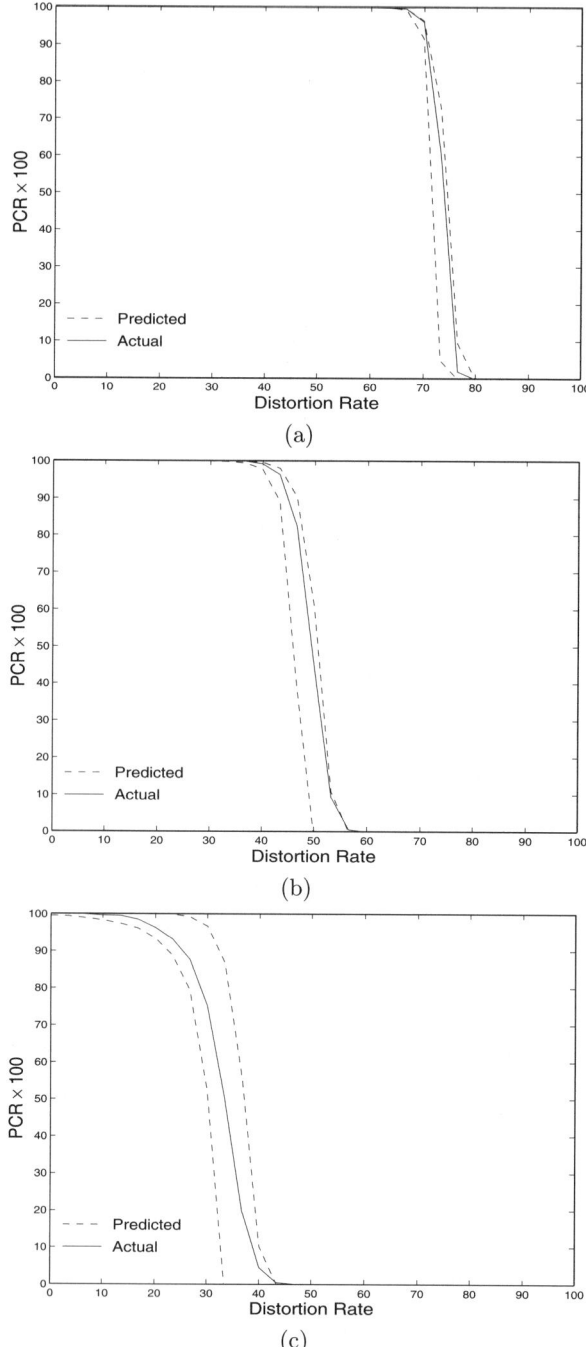

**Figure 1.14.** Actual and predicted PCR plots for synthetic-distortion sets involving different consistency regions: (a) $\mathcal{TD}_1$, (b) $\mathcal{TD}_2$, (c) $\mathcal{TD}_3$.

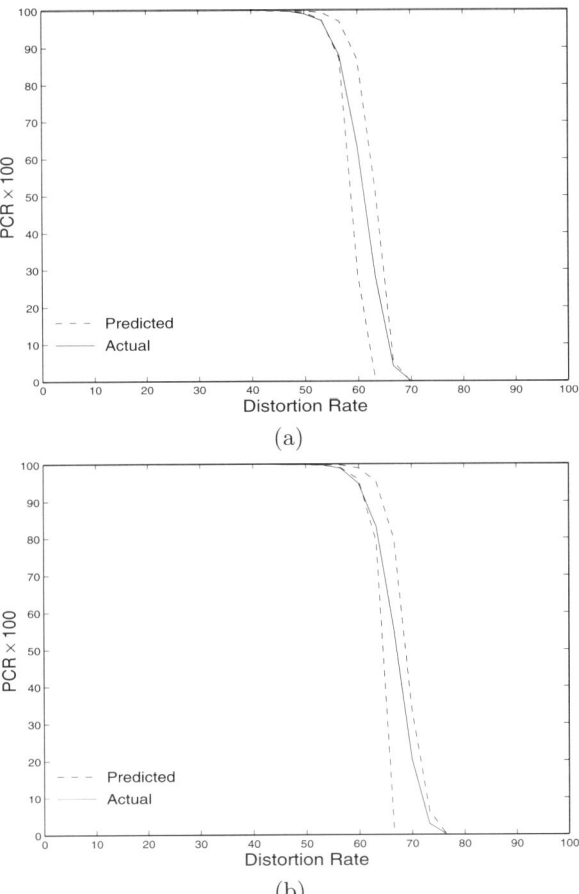

**Figure 1.15.** Actual and predicted PCR plots for synthetic-distortion sets involving different clutter regions: (a) $\mathcal{TD}_4$, (b) $\mathcal{TD}_5$.

proportional to $D$. At low distortion levels, the size of $\mathcal{H}_i(D)$ is small, and the probability that it has no peak hypothesis is generally high. Accordingly, the upper bound defined in (1.15) becomes a less tight one (recall that this upper bound represents the probability that no peak hypothesis beats the true one). Note that we can quantitatively measure the tightness of the upper bound using the ratio

$$F = \frac{1 - \text{predicted\_upper\_bound}}{1 - \text{actual\_PCR}},$$

where $F \in [0, 1]$. As distortion increases, the size of $\mathcal{H}_i(D)$ increases as well, and the probability it has no peak hypothesis decreases. This increases the tightness of the upper bound, until it eventually coincides with the actual PCR, at which case $F$ becomes equal to one.

We now turn to analyzing the real-distortion group. Actual and predicted PCR plots for real-distortion test sets $\mathcal{TD}_6$ and $\mathcal{TD}_7$ are shown in Figures 1.16(a) and 1.16(b), respectively. These figures show that our prediction method succeeds in predicting reasonably accurate bounds on actual performance. For example, the breakpoint in performance is predicted accurately in both cases. However, we also observe that the predicted bounds are over-optimistic in the knee section of the plot. The reason is obviously the existence of some differences between the actual distortion models and the assumed uniform ones. One of the important differences is that the assumed uniform occlusion model does not account for the spatial correlation among occluded/unoccluded features.

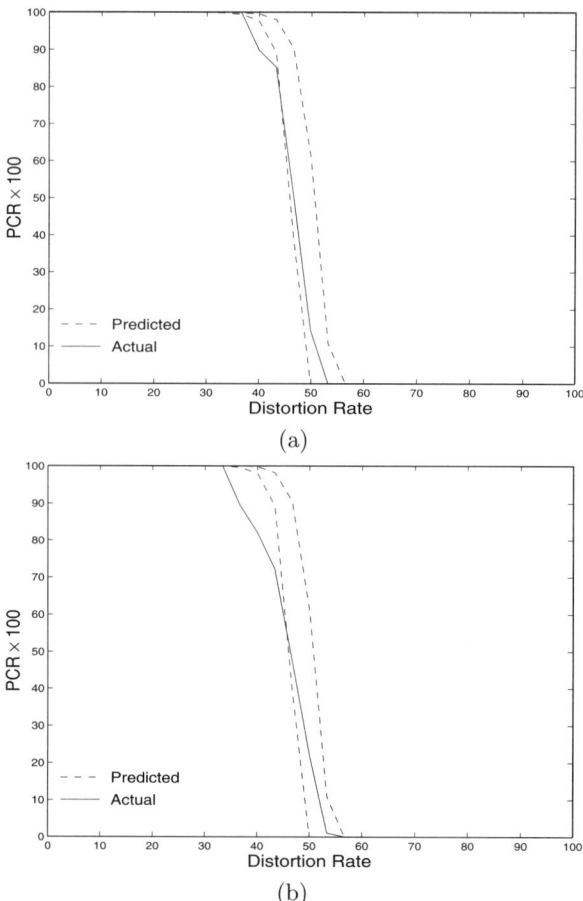

**Figure 1.16.** Actual and predicted PCR plots for real-distortion group: (a) $\mathcal{TD}_6$, (b) $\mathcal{TD}_7$.

A goal of our research is to develop statistical techniques for learning uncertainty, occlusion, and clutter models. The topic of learning distortion models has received some attention in the literature, particularly in learning distributions of feature positional uncertainty [29]. We present here an initial method for learning the statistical model of occlusion, for the purpose of performance prediction. A possible direct approach is to learn the spatially correlating occlusion model from training data (e.g., using Markov random fields [26]), and then use it to determine the prediction bounds. In particular, the occlusion model can be used to determine the PDF of the number of unoccluded similar features (PDF of $N_o$), which is subsequently used to determine the PDF of votes for an erroneous hypothesis (see Section 1.6.2). The difficulty with this approach is that it involves not only a spatially correlating occlusion model, but also a spatially correlating similarity model, one that considers the spatial correlation among similar/dissimilar feature pairs. Fortunately, in the context of our problem, there is a significantly simpler approach: Instead of learning the spatially correlating occlusion and similarity models, and then estimating their combined effect on the PDF of $N_o$, we directly learn the PDF of $N_o$. The learning process can be outlined as follows:

1. We use half of the test set for the learning process (odd-numbered views).
2. For each selected test view, $\widehat{\mathcal{M}}_i$, we search for the best matching model view, $\mathcal{M}_i$, as described earlier when estimating the distortion rate for a test view (see Section 1.7.2). From this match, we eliminate unmatched features of $\mathcal{M}_i$. The resulting view, $\overline{\mathcal{M}}_i$, corresponds to model view $\mathcal{M}_i$ after occlusion, but without either uncertainty or clutter.
3. For each erroneous hypothesis corresponding to $\mathcal{M}_i$, $\mathcal{M}_j^{T_i} \in \mathcal{H}_i$, we calculate the number of similar feature pairs with each of $\mathcal{M}_i$ and $\overline{\mathcal{M}}_i$, denoted by $N_j^{T_i}$ and $\bar{N}_j^{T_i}$, respectively.
4. For each tuple $(\mid \mathcal{M}_i \mid, O =\mid \mathcal{M}_i \mid - \mid \overline{\mathcal{M}}_i \mid, N_j^{T_i})$, we histogram corresponding values of $\bar{N}_j^{T_i}$ to estimate the conditional PDF of $N_o$.

We have replaced the PDF of $N_o$ defined in (1.6) by the learned PDF. The resulting new predicted bounds are shown in Figures 1.17(a) and 1.17(b) for test sets $\mathcal{TD}_6$ and $\mathcal{TD}_7$, respectively. Comparing these figures to the corresponding ones in Figure 1.16, we observe that the predicted bounds obtained using the learned PDF of $N_o$ are considerably more accurate than the ones obtained by assuming the uniform distortion models. In particular, the extent of over-optimism between actual and predicted plots almost disappears for set $\mathcal{TD}_6$, and decreases by about 65% for set $\mathcal{TD}_7$. This is assuming that the extent of over-optimism is measured by the area between the actual plot and the predicted lower-bound plot at the knee section. More accurate prediction can be obtained through comprehensive learning of the PDFs of all the distortion parameters involved.

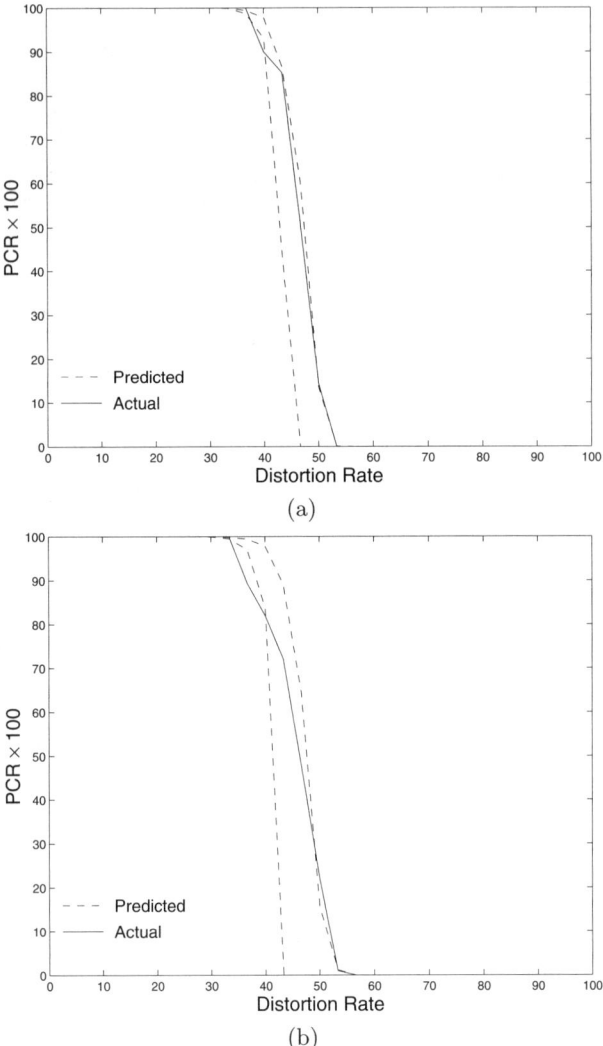

**Figure 1.17.** Actual and predicted PCR plots for real-distortion group using learned PDF of $N_o$: (a) $\mathcal{TD}_6$, (b) $\mathcal{TD}_7$.

## 1.8 Conclusions

Object recognition performance is typically evaluated empirically. A major limitation of empirical evaluation is that it does not give us an insight into the recognition process. Such an insight is fundamental for transforming the field of object recognition from an art to a science. Most efforts for formal analysis of object recognition performance focus on the problem of object/clutter discrimination. This work extends those efforts by also considering the problem of object/object discrimination. An integrated approach is presented for predicting lower and upper bounds on recognition performance. Such an approach simultaneously considers data distortion factors such as uncertainty, occlusion, and clutter, in addition to model similarity. This is in contrast to the other few object/object discrimination approaches, which consider only a subset of these factors. The method is validated in the context of a recognition task involving real SAR data, and point features for matching. Validation is done by comparing predicted lower and upper PCR bounds with actual PCR plots determined experimentally. The selected test sets involve both synthetic and real distortion. In all cases, the results demonstrate that the prediction method consistently succeeds in predicting reasonably tight bounds on actual performance. The predicted bounds, however, are slightly over-optimistic in the case of test sets involving real distortion. More accurate bounds can be obtained through comprehensive learning of all the distortion models involved in the recognition process. As an initial step towards achieving this goal, we have presented a method for learning the statistical distribution of the parameter that encodes combined effect of occlusion and similarity on performance. The impact of using such a learned distribution on the accuracy of the predicted bounds has been demonstrated. To the best of our knowledge, this work is the first that validates a prediction method for object recognition using real data.

We finally discuss possible ways to extend the work presented here. The recognition task considered in this work involves discretized 2D point features for both data and model objects, and a 2D translation space. The proposed method, however, can be extended in a number of ways. We explore some of them below:

- Consideration of more general data/model transformations (e.g., rigid, affine) is straightforward. It would mainly involve extending the set of erroneous hypotheses for a model object, to reflect the extra degrees of freedom (see Sect. 1.6.3).
- Replacement of discretized feature sets by non discretized sets is a more involved process. This is because, in such a case, the erroneous-hypothesis set for a model object would be of infinite size. One possible approach is sampling the erroneous hypotheses in the transformation space, and then "dilating" the consistency regions associated with model features to account for non-sampled hypotheses (e.g., [21]). We note that consistency-region dilation would be needed only when computing lower bounds on performance.
- It is also possible to incorporate feature attributes [5], and/or increase dimensionality of feature locations. Both cases would mainly involve increasing the dimensionality of consistency and clutter regions to be equal to the sum of location dimensionality and number of attributes. Otherwise, the method remains basically the same.

- Another interesting possibility is to describe the positional uncertainty by a Gaussian distribution, instead of a uniform one. This would involve replacing the uniform vote-based criterion by a weighted one, which depends on the distance between corresponding data and model features, as well as the standard deviation of the Gaussian distribution [18]. This would require making modifications to both the measure of object/hypothesis similarity, and the method used to calculate the vote distributions for both true and erroneous hypotheses. The prediction method, however, remains conceptually the same.
- The performance theory presented here can be integrated with adaptive algorithms to optimize object recognition performance in practical scenarios (e.g., with respect to a desired trade-off between probability of false alarm and probability of correct recognition [30]).

In conclusion, the work presented here is believed to lay a theoretical/conceptual foundation that is important for developing a general theory for performance prediction of object recognition.

## Appendix

We present an approximate method for estimating the conditional PDF of clutter votes $V_c$, considering the constraint that no two features can be eight neighbors (imposed by the feature extraction process). We can represent the feature-adjacency constraint by a separation region, $R_s(\cdot)$. In our case, $R_s(\cdot)$ is a $3 \times 3$ window centered at the feature's location. The method can be outlined as follows:

- It can be seen that the feature-adjacency constraint reduces effective clutter and clutter-vote regions, $R'_c$ and $R'_{V_c}$, to smaller ones, $R''_c$ and $R''_{V_c}$, respectively (see Figure 1.18). We can express the area of $R''_c$ as

$$\text{AREA}(R''_c) \approx \text{AREA}(R'_c) - \text{AREA}(R_s(\cdot))(|\mathcal{M}_i| - O).$$

For clutter-vote region $R''_{V_c}$, let us first consider features in $\mathcal{M}_i$ that are similar to those in $\mathcal{M}_j^{T_i}$, but are lying *outside* the overlapping regions (e.g., upper right feature of $\mathcal{M}_i$ in Figure 1.18). Generally, the separation region corresponding to one of these features partially covers the consistency region corresponding to the similar feature in $\mathcal{M}_j^{T_i}$. We can approximate the area of the covered section, denoted by $E_d$, as follows. (1) Both the consistency and separation regions are modeled as squares, having the same areas as their original respective regions. (2) The relationship between square consistency regions corresponding to similar feature pairs is modeled by movement of one square with respect to the other parallel to one of the edges, until the area of the overlapping region is the same as the original one (which corresponds to feature/feature similarity). (3) Assuming the square modeling of consistency regions, we can easily estimate expected location of a feature of $\mathcal{M}_i$, assuming that it is outside the overlapping region. Given the expected location, we can easily estimate the area of the covered section, $E_d$. We can then estimate the area of $R''_{V_c}$ as

$$\text{AREA}(R''_{V_c}) = \text{AREA}(R'_{V_c}) - (n_o - v_s)E_d.$$

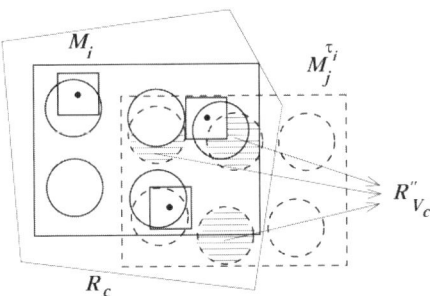

**Figure 1.18.** An illustration of clutter vote region $R''_{V_c}$, in the presence of separation regions, shown as squares.

- The above discussion, as well as that in Section 1.6.2, implies that features of $\mathcal{M}_j^{\tau_i}$ that can result in votes for $\mathcal{M}_j^{\tau_i}$ are effectively associated with consistency regions of various sizes. In order to simplify the calculations, we assume that we have an "effective" number of these features, $n_e$, such that each feature is associated with a full consistency region. It can be easily shown that $n_e = \lfloor \text{AREA}(R''_{V_c})/\text{AREA}(R_u(\cdot)) + 0.5 \rfloor$. In such a case, the effective area of the consistency region, $E_u$, is $\text{AREA}(R''_{V_c})/n_e$, which is approximately the same as $\text{AREA}(R_u(\cdot))$.
- Next, we need to estimate the average *effective* area of a separation region, $E_s$. This area is simply the ratio between the area corresponding to the union of the separation regions of the $C$ clutter features, to $C$. If the clutter features are spread apart from each other, then $E_s$ is close to $\text{AREA}(R_s(\cdot))$. However, if their density is high, which is typical in our case, then there can be significant overlap between their separation regions, thus making $E_s$ considerably smaller than $\text{AREA}(R_s(\cdot))$. In our work, we have

$$E_s \approx \min(\text{AREA}(R_s(\cdot)), \text{AREA}(R''_c)/C). \qquad (1.18)$$

- Now, we are finally at a position to approximate the conditional PDF of $V_c$:

$$P_{V_c}(v_c; n_o, v_s)$$
$$\approx K(C, n_o) \times$$
$$\frac{L(\text{AREA}(R''_{V_c}), E_u, v_c) L(\text{AREA}(R''_c - R''_{V_c}) - v_c(\max(0, E_s - E_u)), E_s, C - v_c)}{L(\text{AREA}(R''_c), E_s, C)}$$

where $L(X, Y, n) = \prod_{i=0}^{n-1}(X - iY)$.

## Acknowledgments

This work was supported in part by AFOSR grant F49620-02-1-0315. The contents and information do not reflect positions or policies of the U.S. Government.

## References

[1] Arman, F., Aggarwal, J.: Model-based object recognition in dense-range images: A review. ACM Comput. Surveys **25** (1993) 5–43
[2] Besl, P., Jain, R.: Three-dimensional object recognition. ACM Comput. Surveys **17** (1985) 75–145
[3] Chin, R., Dyer, C.: Model-based recognition in robot vision. ACM Comput. Surveys **18** (1986) 67–108
[4] Suetens, P., Fua, P., Hanson, A.: Computational strategies for object recognition. ACM Comput. Surveys **24** (1992) 5–61
[5] Bhanu, B., III, G.J.: Recognizing target variants and articulations in synthetic aperture radar images. Optical Engineering **39** (2000) 712–723
[6] Boshra, M., Zhang, H.: A constraint-satisfaction approach for 3-D object recognition by integrating 2-D and 3-D data. Comput. Vision Image Understand. **73** (1999) 200–214
[7] Wells, W.: Statistical approaches to feature-based object recognition. Int. J. of Computer Vision **21** (1997) 63–98
[8] Alter, T., Grimson, W.: Verifying model-based alignments in the presence of uncertainty. In: Proc. IEEE Conf. Comput. Vision and Patt. Recogn., San Juan, Puerto Rico (1997) 344–349
[9] Boykov, Y., Huttenlocher, D.: A new Bayesian framework for object recognition. In: Proc. IEEE Conf. Comput. Vision and Patt. Recogn. Volume 2., Fort Collins, Colorado (1999) 517–523
[10] III, G.J., Bhanu, B.: Recognition of articulated and occluded objects. IEEE Transactions on Pattern Anal. and Mach. Intell. **21** (1999) 603–613
[11] Dhome, M., Richetin, M., Lapreste, J., Rives, G.: Determination of the attitude of 3-D objects from a single perspective view. IEEE Transactions on Pattern Anal. and Mach. Intell. **11** (1989) 1265–1278
[12] Huttenlocher, D., Ullman, S.: Recognizing solid objects by alignment with an image. Int. J. of Computer Vision **5** (1990) 195–212
[13] Dhome, M., Kasvand, T.: Polyhedra recognition by hypothesis accumulation. IEEE Trans. on Pattern Anal. and Mach. Intell. **9** (1987) 429–439
[14] Stockman, G.: Object recognition and localization via pose clustering. Comput. Vision Graphics Image Process. **40** (1987) 361–387

[15] Faugeras, O., Hebert, M.: The representation, recognition and locating of 3-D objects. Int. J. of Robotic Res. **5** (1986) 27–52
[16] Grimson, W., Lozano-Perez, T.: Localizing overlapping parts by searching the interpretation tree. IEEE Trans. on Pattern Anal. and Mach. Intell. **9** (1987) 469–482
[17] Grimson, W., Huttenlocher, D.: On the verification of hypothesized matches in model-based recognition. IEEE Transactions on Pattern Anal. and Mach. Intell. **13** (1991) 1201–1213
[18] Sarachik, K.: The effect of Gaussian error in object recognition. IEEE Transactions on Pattern Anal. and Mach. Intell. **19** (1997) 289–301
[19] Lindenbaum, M.: An integrated model for evaluating the amount of data required for reliable recognition. IEEE Transactions on Pattern Anal. and Mach. Intell. **19** (1997) 1251–1264
[20] Irving, W., Washburn, R., Grimson, W.: Bounding performance of peak-based target detectors. In: Proc. SPIE Conference on Algorithms for Synthetic Aperture Radar Imagery IV. Volume 3070. (1997) 245–257
[21] Lindenbaum, M.: Bounds on shape recognition performance. IEEE Transactions on Pattern Anal. and Mach. Intell. **17** (1995) 665–680
[22] Grenander, U., Miller, M., Srivastava, A.: Hilbert-Schmidt lower bounds for estimators on matrix lie groups for ATR. IEEE Transactions on Pattern Anal. and Mach. Intell. **20** (1998) 790–802
[23] Boshra, M., Bhanu, B.: Predicting performance of object recognition. IEEE Trans. on Pattern Anal. and Mach. Intell. **22** (2000) 956–969
[24] Boshra, M., Bhanu, B.: Validation of SAR ATR performance prediction using learned distortion models. In: Proc. SPIE Conference on Algorithms for Synthetic Aperture Radar Imagery VII. Volume 4053., Orlando, Florida (2000) 558–566
[25] Boshra, M., Bhanu, B.: Predicting an upper bound on SAR ATR performance. IEEE Trans. on Aerospace and Electronic Syst. **37** (2001) 876–888
[26] Li, S.: Markov Random Field Modeling in Image Analysis. Springer-Verlag, New York (2001)
[27] Ying, Z., Castanon, D.: Feature-based object recognition using statistical occlusion models with one-to-one correspondence. In: Proc. Int. Conf. on Comput. Vision. Volume 1., Vancouver, Canada (2001) 621–627
[28] Ross, T., Worrell, S., Velten, V., Mossing, J., Bryant, M.: Standard SAR ATR evaluation experiments using the MSTAR public release data set. In: Proc. SPIE Conference on Algorithms for Synthetic Aperture Radar Imagery V. Volume 3370., Orlando, Florida (1998) 566–573
[29] Pope, A., Lowe, D.: Probabilistic models of appearance for 3-D object recognition. Int. J. of Computer Vision **40** (2000) 149–167
[30] Bhanu, B., Lin, Y., Jones, G., Peng, J.: Adaptive target recognition. Machine Vision and Applications **11** (2000) 289–299

# Chapter 2

# Methods for Improving the Performance of an SAR Recognition System

Bir Bhanu[1] and Grinnell Jones III[2]

[1] Center for Research in Intelligent Systems, University of California, Riverside, California 92521, bhanu@cris.ucr.edu
[2] Center for Research in Intelligent Systems, University of California, Riverside, California 92521, grinnell@cris.ucr.edu

**Summary.** The focus of this chapter is on methods for improving the performance of a model-based system for recognizing vehicles in synthetic aperture radar (SAR) imagery under the extended operating conditions of object articulation, occlusion, and configuration variants. The fundamental approach uses recognition models based on quasi-invariant local features, radar scattering center locations, and magnitudes. Three basic extensions to this approach are discussed: (1) incorporation of additional features; (2) exploitation of a priori knowledge of object similarity represented and stored in the model-base; and (3) integration of multiple recognizers at different look angles. Extensive experimental recognition results are presented in terms of receiver operating characteristic (ROC) curves to show the effects of these extensions on SAR recognition performance for real vehicle targets with articulation, configuration variants, and occlusion.

## 2.1 Introduction

In this chapter we are concerned with methods for improving the recognition of vehicles in Synthetic Aperture Radar (SAR) imagery. The recognition systems start with real SAR chips (at one foot resolution) of actual military vehicles from the MSTAR public data [1] and end with the identification of a specific vehicle type (e.g., a T72 tank). Several major challenges for identifying the vehicles are the presence of significant external configuration variants (fuel barrels, searchlights, etc.), articulated configurations (such as a tank with its turret rotated), and partial occlusion. The detection theory [2, 3], pattern recognition [4, 5, 6], and neural network [7] approaches to SAR recognition all tend to use global features that are optimized for standard, non-articulated, non-occluded configurations. Approaches that rely on global features are not appropriate for recognizing occluded or articulated objects because occlusion and articulation change global features like the object outline and major axis [8]. Our previous work [9, 10, 11, 12], relied on local features to successfully recognize articulated and highly occluded objects. SAR recognition results for our basic approach are compared (in [9]) to other different approaches using real SAR images from the MSTAR public data.

In our research on SAR automatic target recognition (ATR), we initially started using invariant locations of SAR scattering centers as features [12] and later developed our basic recognition approach based on using quasi-invariant locations and magnitudes of the scattering centers [9, 10, 11]. This followed the traditional approach to improving recognition performance by finding additional features that can help to distinguish between the objects. This is the first method of improvement discussed in this chapter. The second method of exploiting model similarity was inspired by the research on predicting the performance of recognition systems by Boshra and Bhanu [13] that introduced the idea that recognition performance depends on the distortion in the test data and the inherent similarity of the object models. Related to the third method for improving recognition performance, we had previously shown that a significant number of SAR scattering center locations do not typically persist over a few degrees of rotation [10]. However, this had been viewed as a problem for modeling, rather than a potential opportunity for independent observations at different look angles. In this chapter, we integrate results from multiple look angles to improve recognition performance.

This chapter discusses three basic approaches to improve the performance of an SAR recognition system:

1. Incorporation of additional features.
2. Exploitation of a priori knowledge of object similarity.
3. Integration of multiple recognizers at different look angles.

The next section discusses SAR target characteristics; Section 2.3 gives a description of the basic SAR recognition system; Section 2.4 introduces the additional feature of peak shape factor and shows performance improvements for additional features; Section 2.5 describes techniques to measure and utilize model similarity to improve recognition performance; Section 2.6 demonstrates the independence of multiple look angle SAR recognition and the results for performance improvements; and the final Section 2.7 provides the conclusions of the chapter.

## 2.2 SAR Target Characteristics

The typical detailed edge and straight line features of man-made objects in the visual world do not have good counterparts in SAR images for subcomponents of vehicle-sized objects at one-foot resolution. However, there is a wealth of peaks corresponding to scattering centers. The relative locations of SAR scattering centers, determined from local peaks in the radar return, are related to the aspect and physical geometry of the object, independent of translation and serve as distinguishing features. Target regions-of-interest (ROI) are found in the MSTAR SAR chips by reducing speckle noise using the Crimmins algorithm (see [14]), thresholding at the mean plus two standard deviations, dilating to fill small gaps among regions, eroding to have one large ROI and little regions, discarding the small regions with a size filter and dilating to expand the extracted ROI. The scattering centers are extracted from the SAR magnitude data (within the boundary contour of the ROI) by finding local eight-neighbor maxima. The parameters used in extracting ROIs are held constant for all the results reported.

Objects from the MSTAR public data used in this research include: BMP2 armored personnel carriers (APCs), a BTR70 APC, T72 tanks, a ZSU23/4 anti-aircraft gun, and a BRDM2 APC. Photo images of the MSTAR articulated objects used in this paper, T72 tank serial number (#) a64 and ZSU 23/4 anti-aircraft gun #d08, are shown in Figures 2.1 and 2.2. Example SAR images and the ROI, with the locations of the scattering centers superimposed are shown in Figure 2.3 for baseline and articulated versions of the T72 and ZSU (at 30° radar depression angle, 66° target azimuth).

(a) turret straight.  (b) turret articulated.

**Figure 2.1.** T72 tank #a64.

(a) turret straight.  (b) turret articulated.

**Figure 2.2.** ZSU 23/4 anti-aircraft gun #d08.

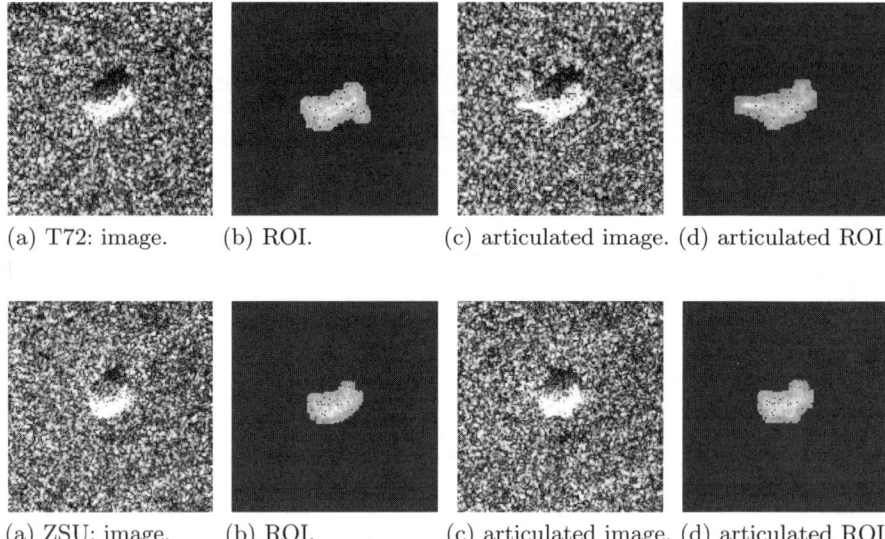

(a) T72: image.  (b) ROI.  (c) articulated image.  (d) articulated ROI.

(a) ZSU: image.  (b) ROI.  (c) articulated image.  (d) articulated ROI.

**Figure 2.3.** MSTAR SAR images and ROIs (with peaks) for T72 tank #a64 and ZSU 23/4 #d08 at 66° azimuth.

### 2.2.1 Azimuthal Variance of Scatterer Locations

The typical rigid body rotational transformations for viewing objects in the visual world do not apply much for the specular radar reflections of SAR images. This is because a *significant* number of features *do not* typically persist over a few degrees of rotation. Since the radar depression angle is generally known, the significant unknown target rotation is (360°) in azimuth. Azimuth persistence or invariance can be expressed in terms of the percentage of scattering center locations that are unchanged over a certain span of azimuth angles. It can be measured (for some base azimuth $\theta_o$) by rotating the pixel locations of the scattering centers from an image at azimuth $\theta_o$ by an angle $\Delta\theta$ and comparing the resulting range and cross-range locations with the scatterer locations from an image of the same object at azimuth $\theta_0 + \Delta\theta$. More precisely, because the images are in the radar slant plane, we actually project from the slant plane to the ground plane, rotate in the ground plane, and project back to the slant plane. Since the objects in the chips are not registered, we calculate the azimuth invariance as the maximum number of corresponding scattering centers (whose locations match within a given tolerance) for the optimum integer pixel translation. This method of registration by finding the translation that yields the maximum number of correspondences has the limitation that for very small or no actual invariance it may find some false correspondences and report a slightly higher invariance than in fact exists. To determine scattering center locations that persist over a span of angles, there is an additional constraint that for a matching scattering center to "persist" at the $k$th span $\Delta\theta_k$, it must have been a persistent scattering center at all smaller spans $\Delta\theta_j$, where $0 \le j < k$. Averaging the results of these persistent scattering center locations over 360 base azimuths gives the mean azimuth invariance of the object.

Figure 2.4 shows an example of the mean azimuth invariance (for the 40 strongest scatterers) as a function of azimuth angle span using T72 tank #132, with various definitions of persistence. In the "exact match" cases the center of the rotated scatterer pixel from the image at $\theta_o$ azimuth is within the pixel boundaries of a corresponding scatterer in the image at $\theta_0 + \Delta\theta$. In the "within 1 pixel" cases, the scatterer location is allowed to move into one of the 8 adjacent pixel locations. Note that for a 1° azimuth span, while only 20% of the scatterer locations are invariant for an "exact match," 65% of the scatterer locations are invariant "within 1 pixel." The cases labeled "persists" in Figure 2.4 enforce the constraint that the scatterer exist for the entire span of angles and very few scatterers continuously persist for even 5°. In the upper two cases (not labeled "persists") scintillation is allowed and the location invariance declines slowly with azimuth span. The "within 1 pixel" results (that allow scintillation) are consistent with the one-foot ISAR results of Dudgeon et al. [15], whose definition of persistence allowed scintillation. Because of the higher scatterer location invariance with 1° azimuth span, in our research we use azimuth models at 1° increments for each target, in contrast to others who have used 5° [16], 10° [17], and 12 models covering specific azimuth ranges [6].

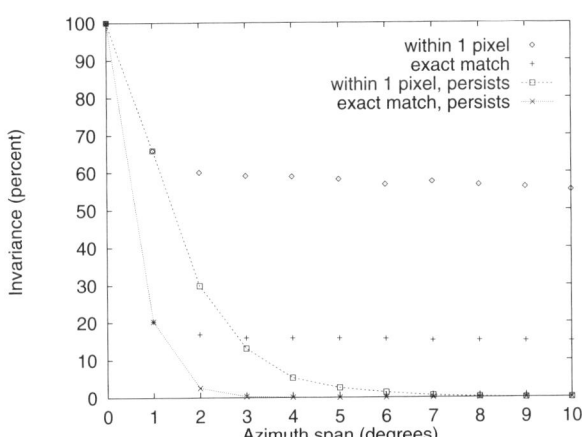

**Figure 2.4.** Scatterer location persistence, T72 #132.

The fact that the SAR scatterer locations *do not* persist over a span of even a few degrees, demonstrated in Figure 2.4, strongly indicates that observations at different azimuth angles are independent. Thus, what had previously been viewed as a "problem" for modeling, now presents a significant opportunity to improve recognition performance by integrating the results of SAR observations at multiple look angles. This is the basis for the approach which will be discussed later in Section 2.6.

### 2.2.2 Scatterer Location and Magnitude Invariance

Many of the scatterer locations and magnitudes are invariant to target conditions such as articulation or configuration variants. Because the object and ROI are not registered, we express the scattering center location invariance with respect to articulation or configuration differences as the maximum number of corresponding scattering centers (whose locations match within a stated tolerance) for the optimum integer pixel translation.

Figure 2.5 shows the location invariance of the strongest 40 scattering centers with articulation for T72 tank #a64 and also for ZSU 23/4 anti-aircraft gun #d08 (at a 30° depression angle) as a function of the hull azimuth. The combined average invariance for both articulated vehicles is 16.5% for an exact match of scattering centers and 56.5% for a location match within one pixel (3x3 neighborhood) tolerance. (Note that not all 360 degrees are present in the MSTAR data, so the missing azimuth angles are ignored in this research.) Similarly, Figure 2.6 shows the percent of the strongest 40 scattering center locations that are invariant for two example configuration variants, T72 #812 vs. #132 and BMP2 vehicle #9563 vs. #C21, at a 15° depression angle. (While the MSTAR data has 11 configurations of the T72, it only has three configurations of the BMP2, so to avoid biasing the experiments toward the T72, we use only three of the T72 configurations.) The combined average invariance for configuration variants of the T72 (#812 and #s7 vs. #132) and BMP2 (#9563 and #9566 vs. #c21) is 15.3% for an exact match of scatterer locations and 57.15% for a location match within one pixel. Since less than 20% of the SAR scattering center locations exactly match for object articulation and configuration variants, while over 50% of these locations are quasi-invariant within a 3∗3 pixel tolerance, in our research we accommodate this 3∗3 tolerance for scattering center locations in the recognition system.

Because of the very large dynamic range for scatterer magnitudes we use a scaled scatterer amplitude (S), expressed as a radar cross-section in square meters, given by S = $100 + 10\log_{10}(i^2 + q^2)$, where $i$ and $q$ are the components of the complex radar return, and we define a percent amplitude change $(A_{jk})$ as: $A_{jk} = 100(S_j - S_k)/S_j$. Note that this form allows a larger variation for the stronger signal returns. Figure 2.7 shows the probability mass functions (PMFs) for percent amplitude change for the strongest 40 articulated vs. nonarticulated scattering centers of T72 tank #a64 and ZSU 23/4 gun #d08. Curves are shown both for the cases where the scattering center locations correspond within one pixel tolerance and for all the combinations of scatterers whose locations do not match. Similarly, Figure 2.8 shows the PMFs for percent amplitude change for the strongest 40 scattering centers with configuration variants, T72 #812 vs. #132 and BMP2 #9563 vs. #C21,

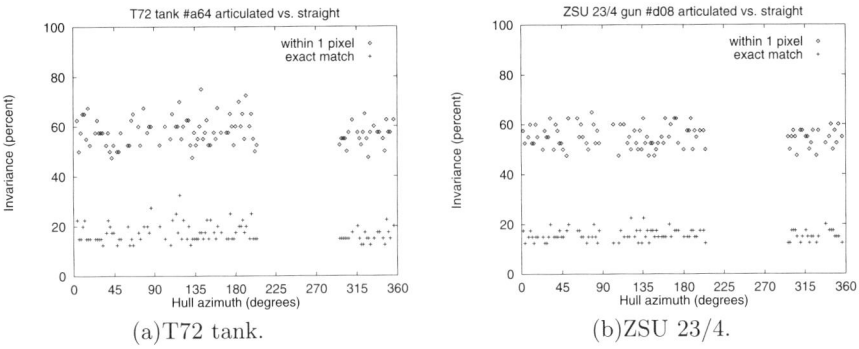

**Figure 2.5.** Scatterer location invariance with articulation.

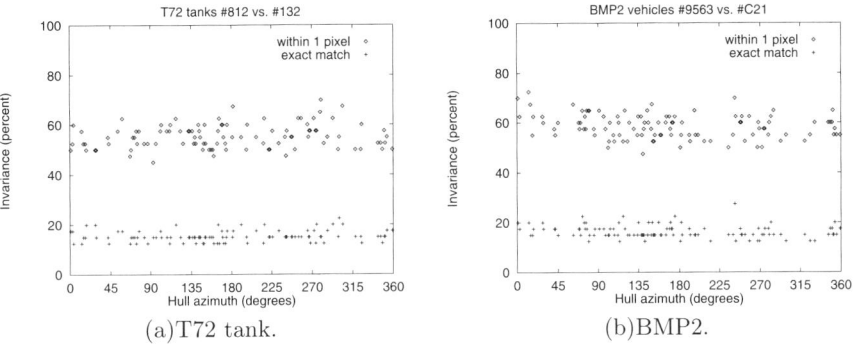

**Figure 2.6.** Scatterer location invariance with configuration.

at a 15° depression angle. If we define scatterer *magnitude invariance* as the number of scatterers with corresponding locations whose magnitudes match (within a stated percent amplitude change tolerance), then the overall average scatterer magnitude invariance (within ± one pixel location and ± nine percent amplitude change) is 50.8 percent for the articulation cases and 51.7 percent for the configuration variant cases.

### 2.2.3 Target Occlusion

There is no real SAR data with occluded objects available to the general public (limited data on vehicles in revetments [18] and partially hidden behind walls [19] has been reported to exist, but it has not yet been released for unrestricted use). In addition, there is no standard, accepted method for characterizing or simulating occluded targets. Typically occlusion occurs when a tank backs up into a tree line, for example, so that the back end is covered by trees and only the front portion of the tank is visible to the radar. Thus, the "bright target" becomes a much smaller sized object to the recognition

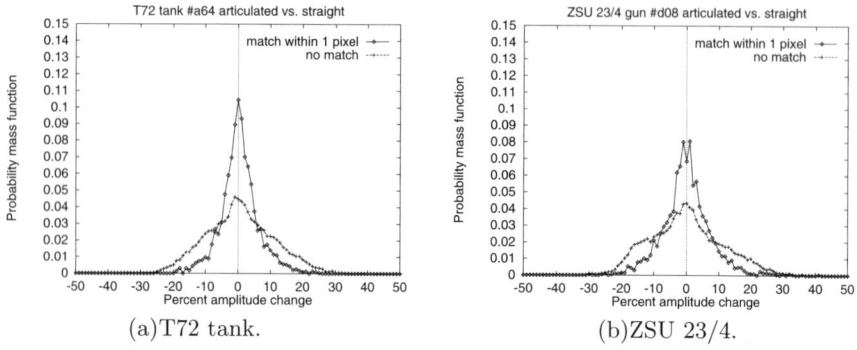

**Figure 2.7.** Scatterer magnitude invariance with articulation.

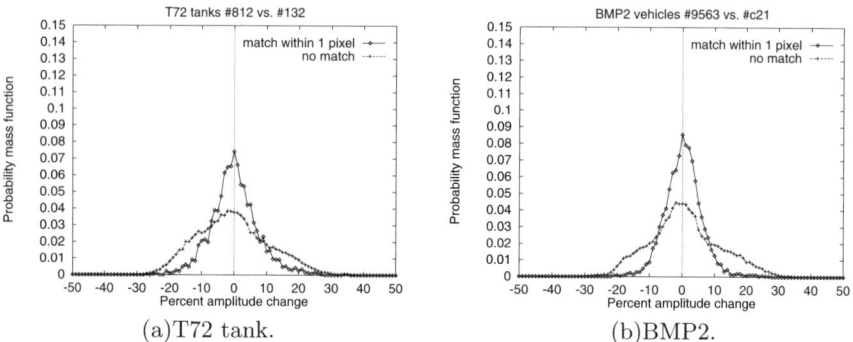

**Figure 2.8.** Scatterer magnitude invariance with configuration.

system. In addition, the tree tops can produce "bright" peaks that are similar to the strength of target peaks at many azimuths.

The occluded test data in our research is simulated by starting with a given number of the strongest scattering centers and then removing the appropriate number of scattering centers encountered in order, starting in one of four perpendicular directions $d_i$ (where $d_1$ and $d_3$ are the cross-range directions, along and opposite the flight path, respectively, and $d_2$ and $d_4$ are the up range and down range directions). Then the same number of scattering centers (with random magnitudes) are added back at *random locations* within the original bounding box of the chip. This keeps the number of scatterers constant and acts as a surrogate for some potential occluding object. Our approach, using simulated occlusion provides an enormous amount of data with varying known amounts of occlusion for carefully controlled experiments.

## 2.3 Basic SAR Recognition System

The basic SAR recognition algorithm is an off-line model construction process and a similar on-line recognition process. The approach is designed for SAR and is specifically intended to accommodate recognition of occluded and articulated objects. Standard nonarticulated models of the objects are used to recognize these same objects in nonstandard, articulated, and occluded configurations. The models are a look-up table and the recognition process is an efficient search for positive evidence, using relative locations of the scattering centers in the test image to access the look-up table and generate votes for the appropriate object (and azimuth pose).

Establishing an appropriate local coordinate reference frame is critical to reliably identifying objects (based on locations of features) in SAR images of articulated and occluded objects. These problems require the use of a local coordinate system; global coordinates and global constraints do not work, as illustrated in Figure 2.3, where the center of mass and the principal axes change with articulation. In an SAR image the radar range and cross-range directions are known and choosing any local reference point, such as a scattering center location, establishes a reference coordinate system. The relative distance and direction of the other scattering centers can be expressed in radar range and cross-range coordinates, and naturally tessellated into integer buckets that correspond to the radar range/cross-range bins. The scale is determined by the bin size, which is a function of the frequency of the radar (e.g., X-band for the one foot resolution used in this research). The recognition system takes advantage of this natural system for SAR, where a single basis point performs the translational transformation and fixes the coordinate system to a "local" origin.

The model construction algorithm for the recognition system is outlined in Figure 2.9 (where the "origin" is the stronger of a pair of scatterers and a "point" is another weaker scatterer). Because of the specular radar reflections in SAR images, a significant number of features do not typically persist over a few degrees of rotation (as shown in Figure 2.4). Consequently, we model each object at $1°$ azimuth increments. The relative locations and magnitudes of the $N$ strongest SAR scattering centers are used as characteristic features (where $N$, the number of scattering centers used, is a design parameter). Any local reference point, such as a scattering center location, could be chosen as a basis point ("origin") to establish a reference coordinate system for building a model of an object at a specific azimuth angle pose. For ideal data, picking the location of the strongest scattering center as the origin is sufficient. However, for potentially corrupted data where any scattering center could be spurious or missing (due to the effects of noise, target articulation, occlusion, nonstandard target configurations, etc.), we use all $N$ strongest scattering centers in turn as origins to ensure that a valid origin is obtained. Thus, to handle occlusion and articulation, the size of the look-up table models (and also the number of relative distances that are considered in the test image during recognition) are increased from $N$ to $N(N-1)/2$. Using a technique like geometric hashing [20], the models are constructed using the relative positions of the scattering centers in the range (R) and cross-range (C) directions as the initial indices to a look-up table of labels that give the associated target type, target pose, "origin" range and cross-range positions, and the magnitudes (S) of the two scatterers. Since the relative distances are not unique, there can be several of these labels (with different target, pose, etc., values) at each lookup table entry.

---

1. For each model Object do 2
   2. For each model Azimuth do 3, 4, 5
      3. Obtain the location $(R,C)$ and magnitude $(S)$ of the strongest $N$ scatterers.
      4. Order $(R, C, S)$ triples by descending $S$.
      5. For each origin O from 1 to $N$ do 6
         6. For each point P from O+1 to $N$ do 7, 8
            7. $dR = R_P - R_O$; $dC = C_P - C_O$.
            8. At look-up table location $dR, dC$ append to list entry with: Object, Azimuth, $R_O$, $C_O$,$S_O$, $S_P$.

---

**Figure 2.9.** Model construction algorithm.

The recognition algorithm is outlined in Figure 2.10. The recognition process uses the relative locations of the $N$ strongest scattering centers in the test image to access the look-up table and generate votes for the appropriate object, azimuth, range, and cross-range translation. (In contrast to many model-based approaches to recognition [21], we are not "searching" all the models.) Further comparison of each test data pair of scatterers with the model look-up table result(s) provides information on the magnitude changes (between the data and the model) for the two scatterers. Limits on allowable values for translations and magnitude changes are used as constraints to reduce the number of false matches. The number of scattering centers used and the various constraint limits are design parameters that are optimized, based on experiments, to produce the best recognition results. (Another approach to optimizing these tuning parameters, based on reinforcement learning is presented in [22].) Given the MSTAR targets are "centered" in the chips, a ±5 pixel limit on allowable translations is imposed for computational efficiency. The experimentally determined optimum limit on the allowable percent difference in the magnitudes of the data and model scattering centers was ±9%, which is consistent with the measured probability mass functions of scatterer magnitude invariance with target configuration variants and articulations (previously shown in Figures 2.8 and 2.7). To accommodate some uncertainty in the scattering center locations, the eight-neighbors of the nominal range and cross-range relative location are also probed in the look-up table, and the translation results are accumulated for a 3∗3 neighborhood in the translation subspace. A city-block weighted voting method reduces the impact of the more common small relative distances. The recognition process is repeated with different scattering centers as basis points, providing multiple "looks" at the model database to handle spurious scatterers that arise due to articulation, occlusion, or configuration differences. The recognition algorithm actually makes a total of $9N(N-1)/2$ queries of the look-up table to accumulate evidence for the appropriate target type, azimuth angle, and translation. The models (labels with object, azimuth, etc.) associated with a specific look-up table entry are the "real" model and other models that happen by coincidence, to have a scatterer pair with the same (range, cross-range) relative distance. The constraints on magnitude differences filter out many of these false matches. In addition, while these collisions may occur at one relative location, the same random object-azimuth pair doesn't keep showing up at other relative locations with appropriate scatterer magnitudes and mapping to a consistent 3*3 neighborhood in translation space, while the "correct" object does.

---

1. Obtain from test image the location $(R, C)$ and magnitude $(S)$ of $N$ strongest scatterers.
2. Order $(R, C, S)$ triples by descending $S$.
3. For each origin O from 1 to $N$ do 4
   4. For each point P from O+1 to $N$ do 5, 6
      5. $dR = R_P - R_O$; $dC = C_P - C_O$.
      6. For DR from $dR - 1$ to $dR + 1$ do 7
         7. For DC from $dC - 1$ to $dC + 1$ do 8, 9, 10
            8. weighted_vote = |DR| + |DC|.
            9. Look up list of model entries at DR, DC.
            10. For each model entry $E$ in the list do 11
               11. IF $|\text{tr} = R_O - R_E| <$ translation_limit
                   and $|\text{tc} = C_O - C_E| <$ translation_limit
                   and $|1 - S_{EO}/S_O| <$ magnitude_limit
                   and $|1 - S_{EP}/S_P| <$ magnitude_limit
                   THEN increment accumulator array [Object, Azimuth, tr, tc]
                      by weighted_vote.
12. Query accumulator array for each Object, Azimuth, tr and tc, summing the votes in a 3x3 neighborhood in translation subspace about tr, tc; record the maximum vote_sum and the corresponding Object.
13. IF maximum vote_sum > threshold
    THEN result is Object ELSE result is "unknown."

---

**Figure 2.10.** Recognition algorithm

The basic decision rule used in the recognition is to select the object-azimuth pair (and associated "best" translation) with the highest accumulated vote total. To handle identification with "unknown" objects, we introduce a criteria for the quality of the recognition result that the votes for the potential winning object exceed some minimum threshold $v_{\min}$. By varying the decision rule threshold we obtain a form of receiver operating characteristic (ROC) curve with probability of correct identification, PCI = $P\{$decide correct object|object is true$\}$, vs. probability of false alarm, $P_f = \{$decide any object|unknown is true$\}$. We call the algorithm a 6D recognition algorithm since, in effect, we use the range and cross-range positions and the magnitudes of pairs of scattering centers. (When using 40 scatterers, this 6D algorithm takes an average of 2.5 seconds to process a test chip on a Sun Ultra2 without any optimizations.)

More formally, a radar image of object $c$ at azimuth pose $a$ consists of $N$ (or more) scatterers, each scatterer $k$ with a magnitude $S_k$ and range and cross-range locations $R_k$ and $C_k$, which (for consistency) are ordered by decreasing magnitude such that $S_k \geq S_{k+1}$ where $k = 1, \ldots, N$. A *model* $M$ of object $c$ at azimuth $a$ is given by

$$M(c,a) = \{V_1(c,a), V_2(c,a), \ldots, V_{N(N-1)/2}(c,a)\}, \qquad (2.1)$$

which is comprised of the set of all pairwise *observations*, $V_i$,

$$V_i(c,a) = \{f_1, f_2, \ldots, f_6\}_i, \qquad (2.2)$$

where $i = 1, 2, \ldots, N(N-1)/2$, $f_1 = R_P - R_O$, $f_2 = C_P - C_O$, $f_3 = R_O$, $f_4 = C_O$, $f_5 = S_O$, $f_6 = S_P$, and with the individual scatterers in each pair labeled $O$ and $P$ so that $S_O \geq S_P$ for consistency (see Figure 2.11).

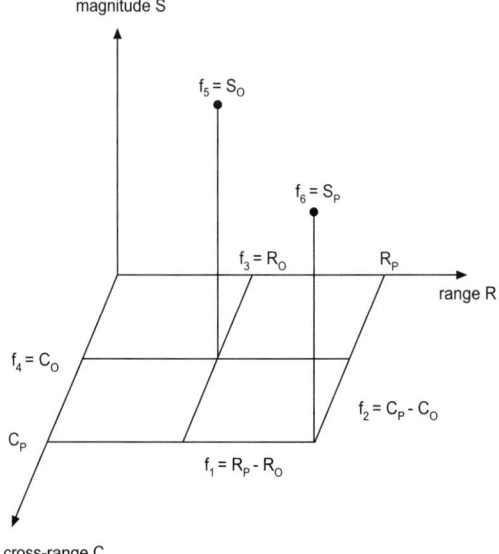

**Figure 2.11.** Observation for a pair of scatterers $O$ and $P$.

We define a *match*, $H$, as

$$H(V_i, V_j) = \begin{cases} w \text{ if } |(f_b)_i - (f_b)_j| \leq \delta_b, \forall \, b = 1, \ldots, 6, \\ 0 \text{ otherwise} \end{cases} \quad (2.3)$$

where the weight $w = |(f_1)_i| + |(f_2)_i|$ and the match constraints are $\delta_1 = \delta_2 = 0$ pixels, $\delta_3 = \delta_4 = 5$ pixels and $\delta_5 = \delta_6 = L$ percent.

The recognition result, $T$, for some test image (with a test class, $x$, and test azimuth, $y$, to be determined) is a maximal match that is greater than a threshold, $D$, given by

$$T = \begin{cases} [c, a] \text{ , if arg max}_{c,a,t} \left( \sum_{l=1}^{9} \sum_{k=1}^{N(N-1)/2} \sum_{n=1}^{9} H_t^l(V_k^n(x,y), V_m(c,a)) \right) > D, \\ \text{"unknown," otherwise} \end{cases} \quad (2.4)$$

where $V_m \in M(c, a) \, \forall m$ such that $|(f_1)_{V_k^n} - (f_1)_{V_m}| = 0$ and $|(f_2)_{V_k^n} - (f_2)_{V_m}| = 0$, and the subscript $t$ applied to a match denotes that the match, $H_t$, is associated with the relative translation $t(R, C) = (\Delta f_3, \Delta f_4)$ of the stronger scatterers in the two observations. Note that in this formulation the constraint on $m$ avoids exhaustive search of all the models and can be implemented as a look-up table. The nine observations (denoted by the superscript $n$ in $V_k^n$) are made to account for location uncertainty by taking the 3*3 neighbors about the nominal values for the relative locations $f_1$ and $f_2$ of scatterer pair $k$ in the test image. Similarly, the nine matches (denoted by the superscript $l$ in $H_t^l$) are computed at the 3*3 neighbors located ±1 pixel about the resulting nominal value for translation, $t(R, C)$, of the scatterers in the test image from the model.

## 2.4 Incorporation of Additional Features

A traditional approach to improving recognition performance is to find additional features that distinguish between the objects. The basic SAR recognition system, the 6D system described in Section 2.3, successfully evolved from an earlier simpler 2D version, [12] that used only relative distances and the "exact match" scatterer locations. The 6D version adds: (1) the scatterer magnitudes as additional features; 2) the within-one-pixel quasi-invariance of scatterer locations; and (3) the consistent translation constraint.

In addition to the scattering center locations and magnitude features, a "shape factor" can be used as a measure of the sharpness of the local peak in the radar return associated with a scattering center. We define the shape factor $F = S_k / \sum_{i=1}^{8} S_i$, where $S_k$ is the amplitude of the peak and the $S_i$'s are the amplitudes of the eight neighbors. Figure 2.12 shows the PMFs for percent shape factor change for the strongest 40 scattering centers of T72 #812 vs. #132 (at 15° depression angle). Curves are shown both for cases where the scattering center locations correspond within one pixel tolerance and for all the combinations of scatterers whose locations do not match. For the cases with locations that match within one pixel, the percent shape factor change

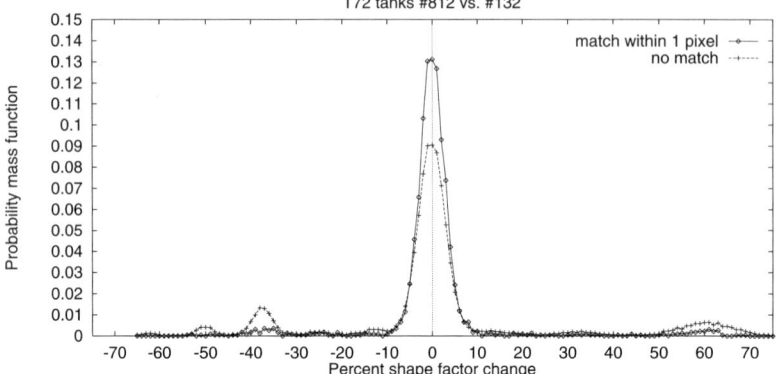

**Figure 2.12.** Shape factor change with configuration.

mean and standard deviation are 1.3 and 15.7, while for the nonmatching cases they are 5.3 and 31.3, respectively.

The peak shape factors can be used as additional features to create an 8D version of the basic recognition system in a manner similar to the way that the magnitudes of the peaks are handled in the 6D system. Figure 2.13 shows ROC curves for the MSTAR T72 and BMP2 configuration variants with the BTR confuser using the 2D, 6D, and 8D recognition systems. Each of the systems was optimized for the forced recognition configuration variant case: the 2D system at 20 scatterers; the 6D system at 36 scatterers (with ±5 pixel translation and ±9% amplitude change limits); the 8D system at 50 scatterers (with a ±30% shape factor change limit). Both the 6D and 8D system results are a substantial improvement over the earlier 2D system results. While Figure 2.13 shows that the 8D system gave worse results than the 6D system in the region below 0.1 $P_f$, reoptimizing the operating parameters (e.g., using 45 scatterers) gives the 8D system better results in the region below 0.1 $P_f$ at the cost of a slightly reduced forced recognition rate.

## 2.5 Exploitation of Model Similarity

### 2.5.1 Similarity Measurement

Model similarity can be measured in terms of collisions *collisions*, where a collision is an instance when observations of two different objects map into the same location (within some specified region of uncertainty) in feature space, i.e., if $H(V_i, V_j) = 1$ and $c_i \neq c_j$. The recognition system described in the preceding section has a 6-dimensional (6D) feature space based on the range and cross-range positions and magnitudes of pairs of scatterers (see equation (2.2)). As noted before (in equation (2.1)), the model of an object at some azimuth, with $N$ scatterers, is represented by $N(N-1)/2$ observations

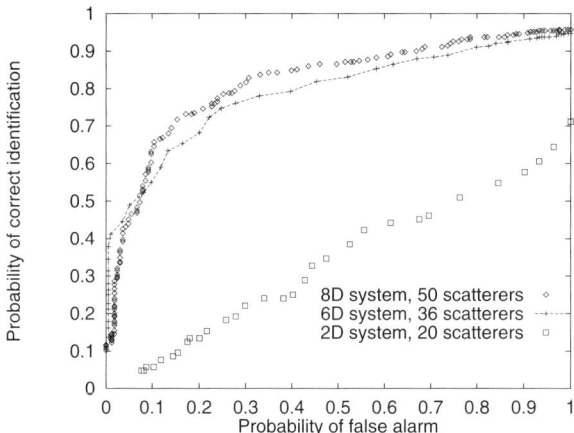

**Figure 2.13.** ROC curves for configuration variants with 2D, 6D and 8D systems.

using pairs of scatterers with each pair mapped into the 6D feature space. While the 6D feature space could be represented by a simple 6D array in concept, the large range of potential feature values and high dimensionality make other implementations more practical. The nature of the SAR problem, with discrete pixel values for distances and a large dynamic range for scatterer magnitudes, leads to a natural model implementation, shown previously in Figure 2.9, where the relative range and cross-range locations of a scatterer pair are direct indices to a physical 2D array of lists that contain another 4D of information and the label with the object and pose. Thus, the model construction algorithm of Figure 2.9 does not directly provide collisions in all six dimensions of feature space. In order to determine if two objects map to the same location in feature space we need to apply the same constraints as are used in the recognition algorithm (see step 10 of Figure 2.10 and equation (2.3)), because the constraints dictate the size of the region or bucket in feature space that is considered the same location.

The general approach to measure the similarity of one model object with respect to several other objects is to first build the look up table models of the other objects using the normal model construction algorithm of Figure 2.9, and then use a modified version of the recognition algorithm of Figure 2.10 with the subject model object (at all the modeled azimuths) as the test conditions to obtain a histogram of the number scatterer pairs that have various numbers of collisions. Basically the modified algorithm uses the first ten steps of Figure 2.10, with the consideration of each pair of scatterers as a separate occurrence (starting a new count of collisions at step 5) and if the constraints are satisfied (at step 10) then a collision is counted. The total number of observations is equal to $AN(N-1)/2$, where $A$ is the number of azimuths modeled (some of the MSTAR data was sequestered, so not all 360° were available).

Figure 2.14 shows example model collision histograms (at $N = 39$ and $L = 9$) for four MSTAR vehicles (at 15° depression angle): BMP2 armored personnel carrier (APC) serial number (#) c21; BTR70 APC #c71; T72 tank #132 and ZSU23/4 anti-aircraft gun #d08. Note that the ZSU23/4 has significantly fewer collisions with the other vehicles, because the ZSU23/4 SAR scatterers cover a larger area than the other objects, and thus, have fewer collisions.

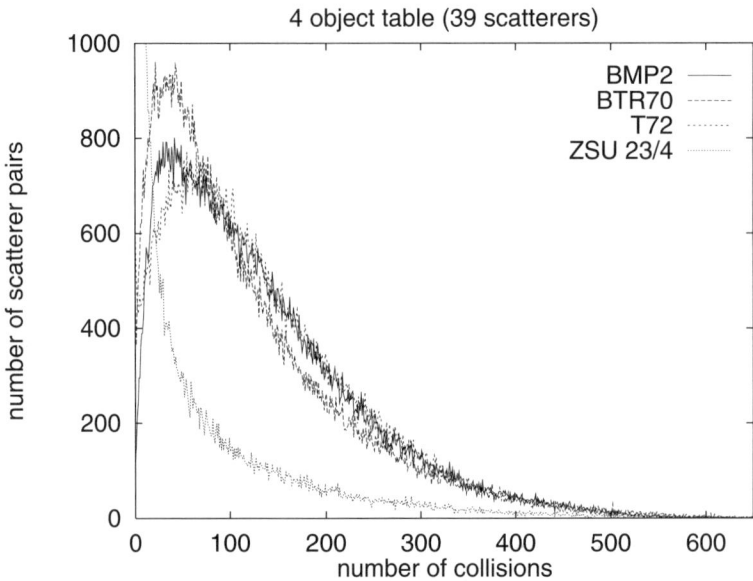

**Figure 2.14.** Example recognition model look-up table collision histograms.

The similarity of a pair of scatterers of given object (at a given azimuth) to the other objects modeled can be measured by the number of collisions with other objects in the look-up table. This can be expressed as a relative measure by using the collision histogram. For convenience, the population of collisions for a particular object is mapped into equal partitions (each with 10% of the total number of collisions). As an example, for the collision histograms in Figure 2.14 we obtain the results in Table 2.1, which shows the number of collisions for a given percent of the population. For the BMP2, for example, 27 collisions or less is in the 10% of the population that is the least similar to the other three models (whereas 90% of the BMP2 scatterer pairs have 274 or less collisions).

### 2.5.2 Weighted Voting

The a priori knowledge of the similarities between object models, expressed as the number of collisions for a given percent of the population, can be captured

**Table 2.1.** Number of collisions for a given percent of the population (example for $N = 39$, $L = 9$).

| Object | Number of collisions | | | | | | | | | |
|---|---|---|---|---|---|---|---|---|---|---|
| BMP2 | 27 | 46 | 66 | 87 | 110 | 136 | 167 | 209 | 274 | 676 |
| BTR70 | 21 | 37 | 53 | 70 | 91 | 116 | 148 | 192 | 266 | 712 |
| T72 | 27 | 48 | 68 | 89 | 111 | 137 | 168 | 209 | 271 | 667 |
| ZSU 23/4 | 0 | 0 | 0 | 0 | 0 | 1 | 3 | 18 | 78 | 760 |
| Population percent | 10 | 20 | 30 | 40 | 50 | 60 | 70 | 80 | 90 | 100 |

by assigning weighted votes to model entries in the look-up table, based on collisions with other objects. This is accomplished off-line by again using a version of the recognition algorithm to obtain the number of look-up table collisions for a particular observation with a pair of scatterers from a subject model and azimuth, as before, and then based on the number of collisions determine the population partition (e.g., using Table 2.1) and finally a given weight function is used to assign a weight label to that instance of the particular model observation entry in the look-up table. Thus, in this approach the model similarities, collisions and associated weightings are all precomputed and appropriate weightings are stored in the look-up table during the off-line modeling process. The similarity-weighted models are obtained using the weighted version of an observation (similar to equation (2.2)) given by

$$\hat{V}_i(c, a) = \{w, f_1, f_2, \ldots, f_6\}_i, \tag{2.5}$$

where $w$ is the weight. The similarity weighted version of a match (similar to equation (2.3)), given by

$$\hat{H}(V_i, \hat{V}_j) = \begin{cases} w & \text{if } |(f_b)_i - (f_b)_j| \leq \delta_b, \ \forall \, b = 1, \ldots, 6 \\ 0 & \text{otherwise} \end{cases} \tag{2.6}$$

can be substituted in equation (2.4) to obtain weighted recognition results. The various weight functions, used in this research to specify $w$, are shown in Figure 2.15 which plots the weight value assigned vs. percent of collision population. Function 1 (Figure 2.15(a)) applies equal weight to all the values and is later referred to as unweighted. Functions 2–4 (Figures 2.15(b)–(d)), the convex weight functions, penalize the most similar features (in the right tail of the histogram). Function 5 (Figure 2.15(e)), with equal steps is linear. While functions 6–7 (Figures 2.15(f)–(g)), which reward uniqueness (the left tail of the histogram) are concave. These weight functions illustrate a range of possibilities from function 2, which penalizes only the most similar 10% of the population, to function 7, which rewards only the most dissimilar 10%. These seven weight functions are used and comparative performance results are obtained in experiments described in the next subsection.

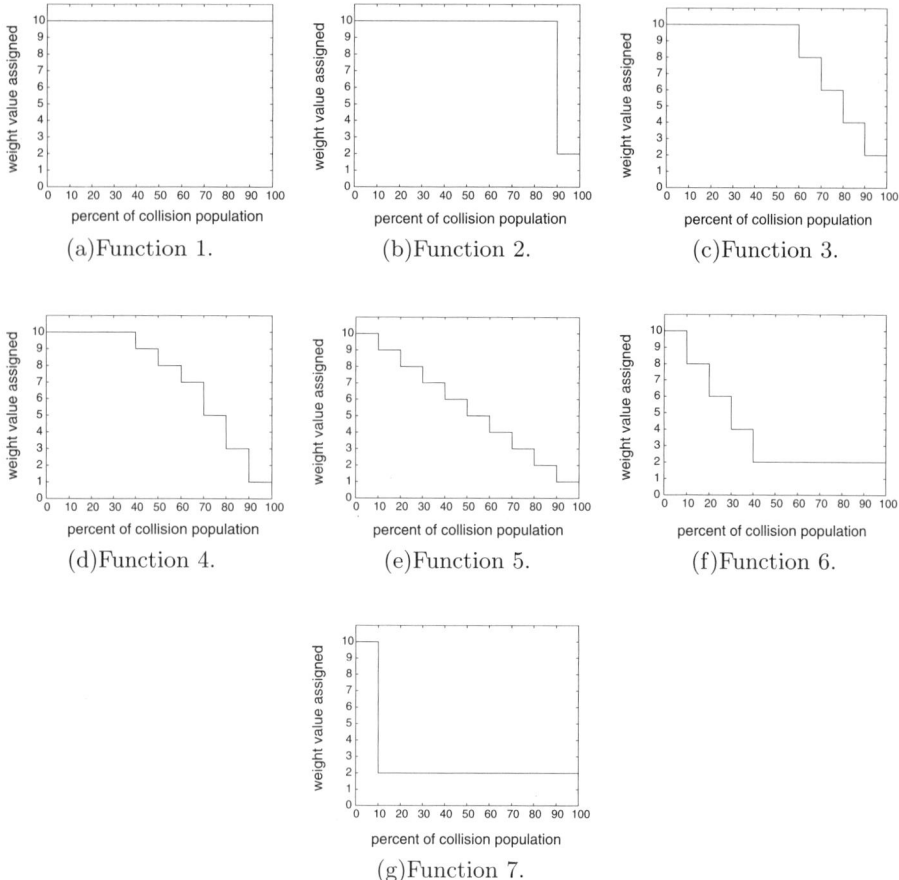

**Figure 2.15.** Table weighting functions.

### 2.5.3 Configuration Variant Experiments

Our previous results [9] (using a distance-weighted voting technique where the weight was proportional to the sum of the absolute values of the relative range and cross-range distances between the scatterer pair) showed that for the real vehicles used in the MSTAR data, the differences of configurations for an object type are a more significant challenge for recognition than articulation (where the model and the test data are the same physical object under different conditions). Similarly, the previous results [11] on occluded objects (using an unweighted voting technique) demonstrated significantly better recognition results than the configuration variant cases. For these reasons, in this research we follow a similar approach and optimize the recognition system for the difficult configuration variant cases and then utilize the same system parameters for the articulation and occlusion cases. In these (15° depression angle) configuration variant experiments, the two object model cases use T72 tank #132 and BMP2 APC #C21 as models, while the four object model cases add BTR70 APC #c71 and ZSU23/4 gun #d08. The test data are two other variants of the T72 (#812, #s7) and two variants of the BMP (#9563, #9566). In addition, BRDM2 APC #e71 is used as an unknown confuser vehicle.

The forced recognition results for MSTAR configuration variants are shown in Figure 2.16 for both two object and four object look up table models using various weight functions (defined earlier in Figure 2.15). These results use the optimal parameters ($N,L$) for each weight function and table size. For the two object cases, function 3 gives the best results, a recognition rate of 95.81%, compared to the unweighted case of 95.17%. For the four object cases, the convex and linear weighting functions all provide better forced recognition performance than the unweighted case. The concave weighting functions result in worse performance than the unweighted case. The best four object result is 94.17% for function 2, compared to the unweighted case of 92.27%. Thus, increasing the number of objects modeled from two to four, reduces the forced recognition rate by 2.9% (95.17 - 92.27) for the unweighted case, while using model similarity information in the optimum weight function reduces that loss to 1% (95.17 - 94.17).

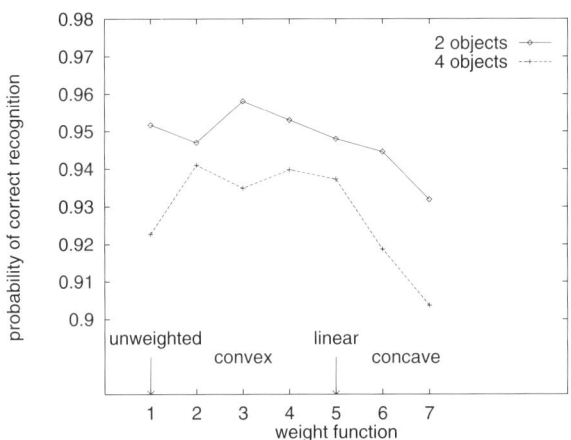

**Figure 2.16.** Effect of table size and weighting function on forced recognition of MSTAR configuration variants.

Table 2.2 shows example confusion matrices that illustrate the effect of going from a two object recognition system to a four object model recognition system for the MSTAR configuration variant data. In both cases the system parameters ($N,L$) are optimized for forced recognition (2 objects at (38,11) and 4 at (38,12)), both are unweighted cases (constant weight of 10), and both are for $d = 1700$. (At least 1700 votes, with a weight of 10, is equivalent to 19 or more scatterers that "matched".) Comparing the two object results on the left of Table 2.2 with the four object results on the right, we observe that basically a large number of confusers and a few targets move from the Unknown column to the additional models. Thus, while the recognition results are similar for 2 and 4 models (PCI = 0.773 and 0.790 respectively) there are increased false alarms ($P_f$ = 0.13 and 0.32 respectively) which would move the knee of the ROC curve to the right.

**Table 2.2.** Effect of 2 and 4 models on MSTAR configuration variant confusion matrices (unweighted, $d = 1700$).

| test targets [serial number] | Identification results (config. modeled) | | | Identification results (configuration modeled) | | | | |
|---|---|---|---|---|---|---|---|---|
| | BMP2 (#C21) | T72 (#132) | Unk. | BMP2 (#C21) | T72 (#132) | BTR70 (#C71) | ZSU23/4 (#d08) | Unk. |
| BMP2 [#9563,9566] | 189 | 3 | 25 | 189 | 2 | 8 | 0 | 18 |
| T72 [#812,s7] | 8 | 131 | 58 | 11 | 138 | 1 | 0 | 47 |
| BRDM2 (confuser) | 28 | 4 | 214 | 27 | 5 | 47 | 0 | 167 |

Table 2.3 shows an example MSTAR configuration variant four object confusion matrix for weight function 4. The system parameters (37,9) are optimized for forced recognition with weight function 4 and a $d$ of 1100 is chosen to yield a PCI of 0.776, which is similar to the results shown in Table 2.2. (At least 1100 votes, with an average weight for function 4 of 7.3, is equivalent to 18 or more scatterers matched.) Comparing the earlier four-object unweighted results, shown on the right of Table 2.2, with the weighted results of Table 2.3, we observe that half the misidentifications (11 of 22) are moved to the unknown column. This reduction in misidentifications shows that the model-weighting approach is increasing the distinguishability of the modeled objects. This reduction in misidentifications does not show up directly in the ROC curve results, which treat the off-diagonal target misidentifications the same as the misses where a target is called unknown (i.e., both are cases where the target was not correctly identified). However, the weight function (which effectively reduces the average weighting) allows a similar PCI to be achieved with a lower vote threshold (1100 votes vs. 1700 votes) and results in fewer false alarms. Thus, the lower $P_f$ of 0.276 for the weighted case, vs. 0.321 for the unweighted case, would move the ROC curve to the left.

ROC curves are generated for the four-object configuration variant cases by using the optimum parameters for the forced recognition case and varying the vote threshold. Figure 2.17 shows that the ROC curves for the convex and linear weight functions provide generally better performance than the unweighted case. In addition, Figure 2.18 shows that the concave weight functions give worse performance than the unweighted case (except for the region

**Table 2.3.** Example MSTAR configuration variant confusion matrix for weight function 4 ($d = 1100$).

| test targets [serial number] | Identification results (configuration modeled) | | | | |
|---|---|---|---|---|---|
| | BMP2 (#C21) | T72 (#132) | BTR70 (#C71) | ZSU23/4 (#d08) | Unknown |
| BMP2 [#9563,9566] | 179 | 6 | 1 | 0 | 32 |
| T72 [#812,s7] | 4 | 143 | 0 | 0 | 50 |
| BRDM2 (confuser) | 30 | 6 | 32 | 0 | 178 |

where PCI $< 0.5$, $P_f < 0.05$). The convex weight functions penalize the most common features and so are not much affected by noise (due to configuration differences or other confuser vehicles). On the other hand, the concave weight functions reward (very strongly reward in function 7) the relatively unique features, which makes them susceptible to conditions where noise is strongly rewarded.

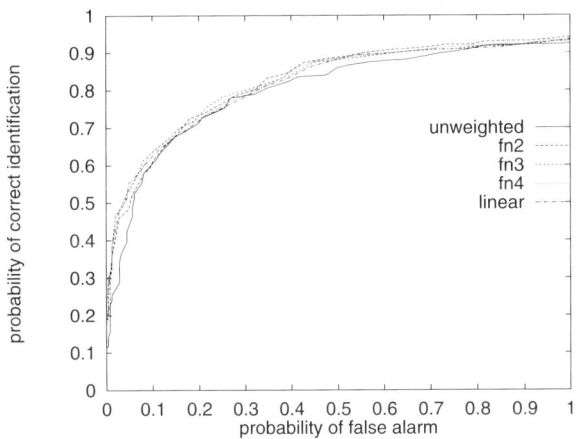

**Figure 2.17.** MSTAR configuration variant ROCs for beneficial weight functions (four objects).

### 2.5.4 Articulation Experiments

In the articulation experiments the models are nonarticulated versions of T72 #a64 and ZSU23/4 #d08 and the test data are the articulated versions of these same serial number objects and BRDM2 #e71 as a confuser vehicle (all at $30°$ depression angle).

Figure 2.19 shows the ROC curves, with excellent articulated object recognition results for both the weight function 2 and the unweighted cases. Since

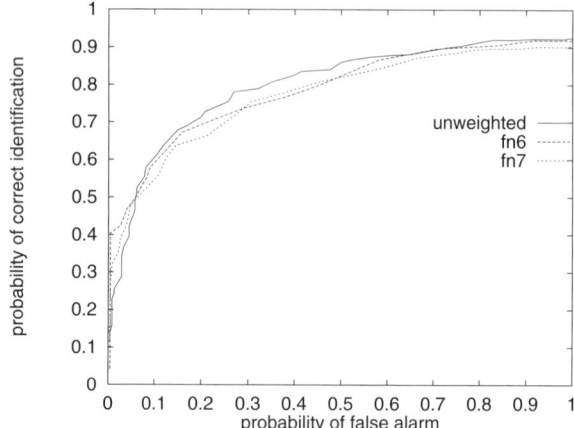

**Figure 2.18.** MSTAR configuration variant ROCs for concave weight functions (four objects).

weight function 2, with $N = 39$ and $L = 9$, gives the optimum ROC results for the two object (T72, BMP2) configuration experiments and the optimum unweighted parameters are $N = 38$ and $L = 11$, these same parameters are used for the articulation experiments.

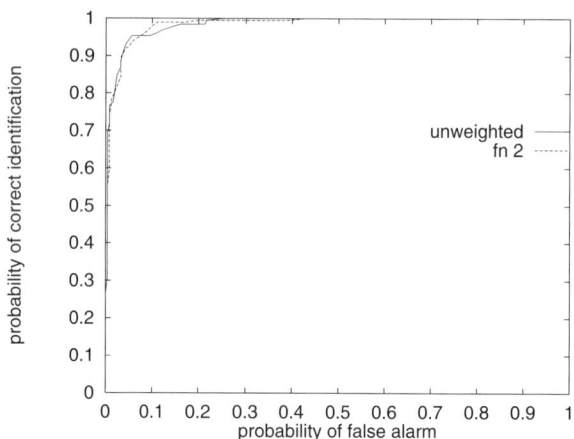

**Figure 2.19.** Articulation recognition results.

### 2.5.5 Occlusion Experiments

The occlusion experiments use the same four models as the configuration variant experiments: T72 tank #132, BMP2 APC #C21, BTR70 APC #c71 and ZSU23/4 gun #d08 (all at 15° depression angle). The occluded test data is generated using the technique described previously in Subsection 2.2.3. In our previous work on occluded objects [11], the confuser vehicle was occluded. However, while the target may be occluded, the confuser vehicle may not necessarily be occluded in practical situations. Hence, in this case the BRDM2 APC (#e71) is an *unoccluded confuser* vehicle, which is a more difficult problem.

Figure 2.20 shows the effect of occlusion on ROC curves for weight function 2, with $N = 40$ and $L = 9$ (while $N = 40$ is not optimum, it yields occlusion in 5% increments). Here with the unoccluded confuser, excellent recognition results are achieved for less than 45 percent occlusion, compared with the prior 70 percent occlusion with an occluded confuser [11].

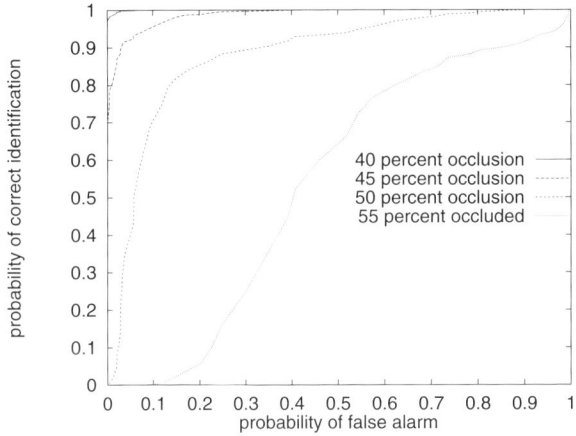

**Figure 2.20.** Effect of occlusion on receiver operating characteristics.

## 2.6 Multiple Recognizers at Different Look Angles

### 2.6.1 Independence of Multiple Look Angle SAR Recognizers

The azimuthal variance of SAR scatterer locations, previously demonstrated in Figure 2.4, strongly indicates that observations at different azimuth angles are independent. Given the probability of one SAR recognizer failing, $F1$, where $F1 = 1 - PCI$; then if two recognizers are independent, the probability that both recognizers are wrong, $F2$, is simply $F2 = F1^2$. In order to obtain the most failures, we pick the object configuration variant case, which is the most difficult case for our SAR recognition approach, compared to the depression angle change and object articulation cases.

In the configuration variant experiments a single configuration of the T72 (#132) and BMP2 (#C21) vehicles are used as the models and the test data are two other variants of each vehicle type (T72 #812, #s7 and BMP2 #9563, #9566), all at 15° depression angle. These results are obtained using the optimum parameters of 38 scattering centers and a percent magnitude change of less than ±11%. These parameter settings are slightly different from the corresponding (36 and 9%) values used previously in Section 2.4, because here we use unweighted voting, as opposed to the previous distance-weighted voting method. Figure 2.21 takes the experimental forced recognition results for the configuration variant cases with all available combinations of two different azimuths and plots the probability that two recognizers are both wrong as a function of the difference in azimuth angle of the object from the two recognizers. (While Figure 2.21 emphasizes the small angles by only showing up to ±60 degrees, the results out to ±180 degrees are similar.) The single recognizer result, 19 failures in 414 trials, is an F1 failure rate of 0.0459, which is plotted for reference as point "a" in Figure 2.21. For an F1 of 0.0459 the predicted value of F2 is 0.0021, which is very close to the overall experimental average F2 of 0.0025. The other interesting observation from Figure 2.21 is that the results are independent of the angle difference, even for very small values like 1 degree. This demonstration that multiple-look-angle SAR recognition results are independent, even for small angles down to 1 degree, provides the scientific basis for both measuring and improving the quality of recognition results.

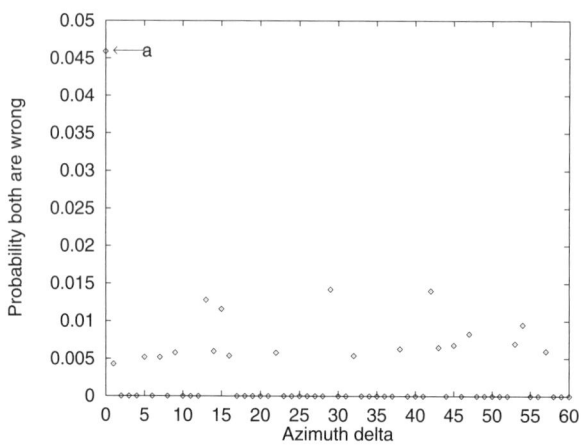

**Figure 2.21.** Probability that two recognizers are both wrong.

### 2.6.2 Multiple-Look-Angle Configuration Variant Results

In contrast to the forced recognition case described in the previous subsection, in these configuration variant experiments the BTR70 armored personnel carrier (#c71) is used as an unmodeled confuser vehicle to test the recognition system and the vote threshold parameter is used to generate "unknown" results. The other test conditions and parameters are the same as the forced recognition case (most significantly, the models are one configuration of the T72 and BMP2 vehicles and the test vehicles are two different configurations).

Figure 2.22 shows the effect of multiple look angle recognizers on the probability of false alarm using the BTR70 as a confuser. (The BTR70 is a more difficult case than other confusers such as the BRDM2 armored personnel carrier or the ZSU 23/4 anti-aircraft gun [9].) In the cases with two recognizers, the decision rule is that if either gives results above the vote threshold, the result is declared a target (which, for these BTR70 confusers, would be a false alarm). Thus, with this "target bias" decision rule the multiple recognizer cases have higher false alarms than a single recognizer. It is important to note that the penalty in increased false alarms is small for the left tail of the curve. Figure 2.22 also shows that the false alarm rates are similar for all the two-look angle recognizer cases and that agreement on the "target" is basically irrelevant for a false alarm.

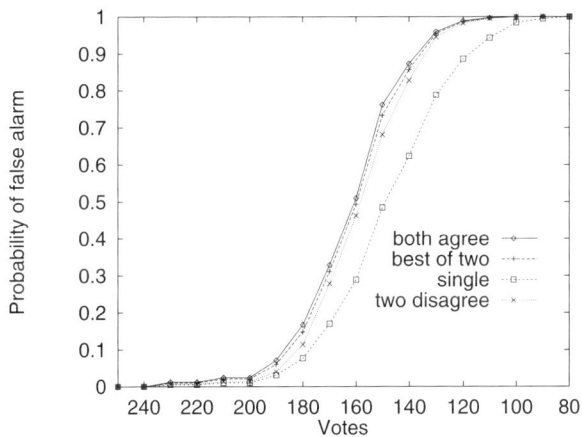

**Figure 2.22.** Effect of multi-look on probability of false alarm.

Figure 2.23 shows the effect of multiple-look-angle recognizers on the probability of correct recognition for the configuration variant case where the test data are different configurations of the T72 and BMP2. The top curve shows the results for the 91.7 percent of the time when two recognizers at different look angles agree on the result. The bottom curve is for the remaining 8.3 percent of the time when the two recognizers disagree and the answer that gets the most votes is chosen. The second curve, labeled "best of two," uses a decision level fusion rule that simply picks the target based on which of the two recognizers got the most votes. (This case is also the weighted average of the agree and disagree cases). In Figure 2.23 the probability of correct recognition decreases as the vote threshold increases (to the left in Figure 2.23), because the higher threshold causes more targets to be classified as "unknown." The recognition results for using the best of two recognizers at different look angles are substantially better than the results for a single recognizer. This is basically the result of fewer "misses," where a target object is classified as an "unknown"; because there are two opportunities to get above the vote threshold and declare a "target."

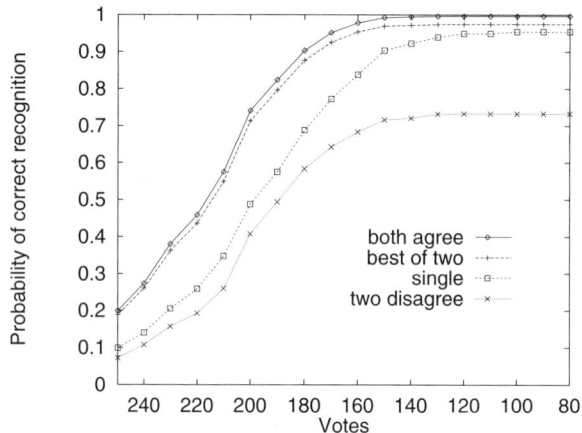

**Figure 2.23.** Effect of multi-look on probability of correct recognition.

Figure 2.24 combines the results of Figures 2.22 and 2.23 and shows the effect of using multiple-look-angle recognizers on the ROC curve for the configuration variant cases. These recognition results, using the best-of-two recognizers at different look angles, are substantially better than the results for a single recognizer. For example, at a 0.10 $P_f$ the PCI for the best of two look angles is 0.8324, compared to 0.7091 for a single recognizer. The performance improvement is because the cost in increased false alarms (in Figure 2.22) is low compared to the benefits in increased recognition (in Figures 2.23), due to fewer targets being classified as "unknown."

### 2.6.3 Multiple-Look-Angle Articulation Results

In the articulated object experiments the models are nonarticulated versions of T72 tank #a64 and ZSU 23/4 anti-aircraft gun #d08 (with the gun turret straight forward) and the test data are articulated versions of these same serial number objects (with the turret rotated) and BRDM2 armored personnel carrier #e71 as a confuser vehicle (all at 30° depression angle). The results of applying the same techniques (and all the same recognition system parameters) in these articulated object experiments are shown as ROC curves in Figure 2.25. Again the results for using two recognizers at different look angles and picking the answer with the largest number of votes are better than the single recognizer results.

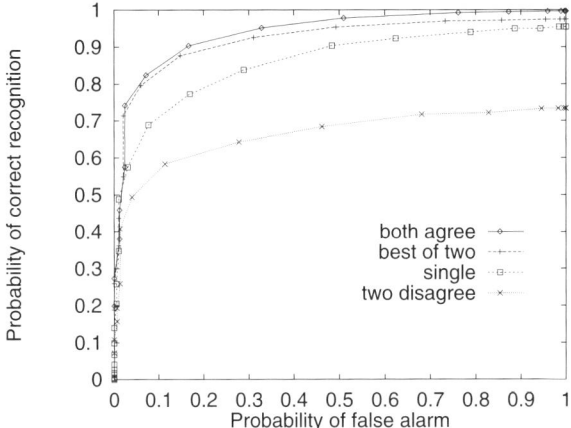

**Figure 2.24.** Effect of multilook on configuration variant ROC curve.

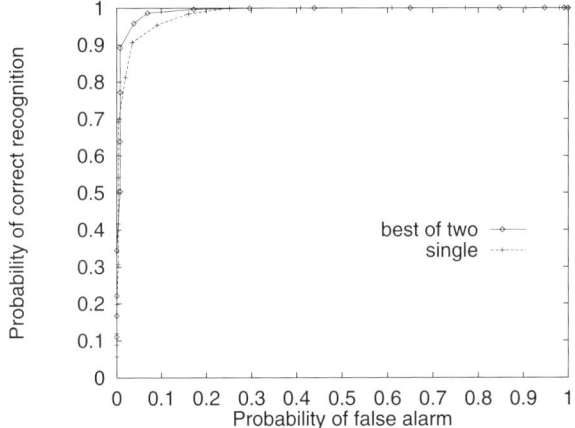

**Figure 2.25.** Effect of multilook on articulated object ROC curve.

### 2.6.4 Multiple-Look-Angle Occlusion Results

The occlusion experiments use four models: T72 tank #132, BMP2 APC #C21, BTR70 APC #c71 and ZSU23/4 gun #d08 and the unmodeled confuser vehicle is BRDM2 APC (#e71) (all at 15° depression angle). The occluded test data is generated using the technique described previously in Subsection 2.2.3, while the BRDM2 APC (#e71) is the more difficult *unoccluded confuser* vehicle.

Figure 2.26 shows the effect of multiple-look-angle recognizers on the probability of correct recognition for the case of 50% occluded targets and an unoccluded confuser. The same techniques (and all the same recognition system parameters) used in the prior configuration variant and articulation experiments are applied to these occluded object experiments. Here again, using two recognizers at different look angles and a decision-level fusion rule of picking the answer with the largest number of votes gives better results than a single recognizer.

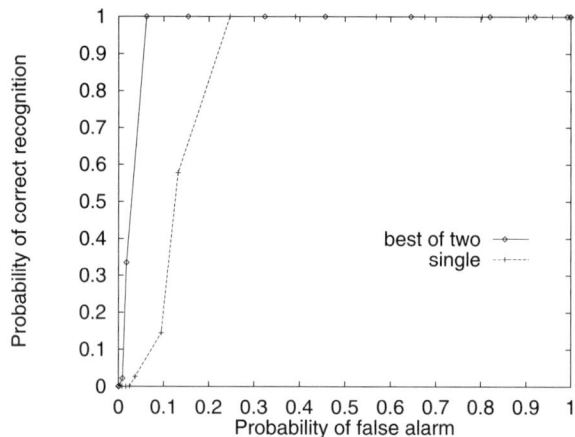

**Figure 2.26.** Effect of multi-look on 50% occluded object ROC curve.

## 2.7 Conclusions

The locations and magnitudes of a significant number of SAR scatterers are quasi-invariant with target configuration variations and articulations. A model-based recognition system, using inexact match of local features can successfully handle difficult conditions with object configuration variants, articulation, and occlusion. A comparison of experimental recognition results for MSTAR configuration variants is shown in Table 2.4 for different methods of improving performance of SAR recognition systems involving incorporation of additional features; exploitation of knowledge of model similarity; and integration of multiple recognizers at different look angles. Recognition rates are shown for forced recognition cases and for cases with a 0.10 confuser false alarm rate.

The use of additional scatterer magnitude features, the consistent translation constraint, and the ability to handle inexact matches of these local features are fundamental to the success of the basic 6D recognition system, compared to the earlier 2D system [12] which only used relative scatterer locations and required an exact location match. Additional quasi-invariant features, such as peak shape factor in the 8D system, can provide some increase in recognition performance.

The similarities between object models can be effectively quantified using histograms of collisions in feature space. This a priori knowledge of object similarity can be successfully used to improve the performance of SAR target

Chapter 2 Improving an SAR Recognition System 67

**Table 2.4.** Comparison of results for MSTAR T72 and BMP2 configuration variants (6D systems, except as noted by *).

|  | recognition rate | |
|---|---|---|
|  | forced | at 0.10 $P_f$ |
| Additional features [Figure 2.13]: | | |
| *   2D system | 0.712 | 0.058 |
|     6D system | 0.947 | 0.554 |
| *   8D system | 0.957 | 0.636 |
| Use model similarity: | | |
| function 2 [Figure 2.17], 4 models | 0.942 | 0.609 |
| 2 models | 0.949 | 0.744 |
| Multiple look angles: | | |
| best of two [Figure 2.24] | 0.975 | 0.832 |

recognition. The approach can increase the distinguishability of the modeled objects, reduce misidentifications, and result in decreased false alarms. This is especially beneficial in cases where the number of modeled objects is large. One set of results shown in Table 2.4 for "model similarity" is for a more difficult case with four models, while all the other cases in the table are for two models. In the most difficult configuration variant cases, the convex and linear weight functions, which penalize the most common features, give better performance than the concave weight functions, which strongly reward relatively unique features.

The fundamental azimuthal variance of SAR scatterer locations can be successfully used as the basis for a principled and effective multiple-look-angle SAR recognition approach. The experiments demonstrate that SAR recognition results at different azimuths are independent, even for small azimuths, such as one degree. In addition, using decision level fusion of two observations at different look angles can substantially increase SAR recognition performance for target configuration variants.

These techniques have also been successfully applied to recognition of articulated objects and occluded objects. In experiments using MSTAR SAR data with a BRDM2 APC confuser and articulated versions of a T72 tank and ZSU 23/4 anti-aircraft gun, the basic 6D recognition system with non-articulated T72 and ZSU 23/4 models achieved a 95.6–95.7% recognition rate (depending on system parameters) with a 0.10 $P_f$. Comparable rates using the model similarity and multiple-look-angle approaches increased to 98.0% and 99.0% respectively (see Figures 2.19 and 2.25). Similarly, for experiments with 50 percent occluded objects (with an unoccluded confuser), using the model similarity approach achieved a 70.1% recognition rate at $0.10 P_f$ and multiple look angles achieved 100%, while the basic 6D system only achieved a 20.7% recognition rate (see Figures 2.20 and 2.26).

Higher-resolution SAR data, instead of the one-foot resolution now commonly available, would greatly increase the discriminating power of the current scatterer location and magnitude features. Additional features such as valleys or ridges could also be used, especially with higher-resolution data. The approach of exploiting knowledge of model similarity will become more critical in scaling the recognition problem from 2 to 4 objects to a more realistic 20 to 40 objects. The multiple-look-angle approach could be readily extended to use more than two look angles and more formal methods, like Dempster–Shafer and Bayesian techniques could be used for decision-level fusion. In addition, while the research described in this chapter independently addresses three basic approaches to improve recognition performance: (1) incorporation of additional features; (2) exploitation of a priori knowledge of model similarity; and (3) integration of multiple recognizers at different look angles, these techniques could be combined in practice to address difficult operating conditions involving large numbers of different objects with the combined effects of non-standard configurations, articulation and occlusion.

## Acknowledgements

This work is supported by AFOSR grant F49620-02-1-0315; the contents and information do not necessarily reflect the position or policy of the U.S. Government.

## References

[1] Ross, T., Worrell, S., Velten, V., Mossing, J., Bryant, M.: Standard SAR ATR evaluation experiments using the MSTAR public release data set. In: SPIE Proceedings: Algorithms for Synthetic Aperture Radar Imagery V. Volume 3370. (1998) 566–573
[2] Carlson, D., Kumar, B., Mitchell, R., Hoffelder, M.: Optimal trade-off distance classifier correlation filters (OTDCCFs) for synthetic aperture radar automatic target recognition. In: SPIE Proceedings: Algorithms for Synthetic Aperture Radar Imagery IV. Volume 3070. (1997) 110–120
[3] Casasent, D., Shenoy, R.: Synthetic aperture radar detection and clutter rejection MINACE filters. Pattern Recognition **30** (1997) 151–162
[4] Meth, R., Chellappa, R.: Automatic classification of targets in synthetic aperture radar imagery using topographic features. In: SPIE Proceedings: Algorithms for SAR Imagery III. Volume 2757. (1996) 186–193
[5] Ryan, T., Egaas, B.: SAR target indexing with hierarchical distance transforms. In: SPIE Proceedings: Algorithms for Synthetic Aperture Radar Imagery III. Volume 2757. (1996) 243–252
[6] Verly, J., Delanoy, R., Lazott, C.: Principles and evaluation of an automatic target recognition system for synthetic aperture radar imagery

based on the use of functional templates. In: SPIE Proceedings: Automatic Target Recognition III. Volume 1960., Orlando, FL (1993) 57–71
[7] Casasent, D., Shenoy, R.: Feature space trajectory for distorted-object classification and pose estimation in SAR. Optical Engineering **36** (1997) 2719–2728
[8] Yi, J.H., Bhanu, B., Li, M.: Target indexing in SAR images using scattering centers and the Hausdorff distance. Pattern Recognition Letters **17** (1996) 1191–1198
[9] Bhanu, B., Jones, G.: Recognizing target variations and articulations in synthetic aperture radar images. Optical Engineering **39** (2000) 712–723
[10] Jones, G., Bhanu, B.: Recognizing articulated targets in SAR images. Pattern Recognition **34** (2001) 469–485
[11] Jones, G., Bhanu, B.: Recognizing occluded objects in SAR images. IEEE Transactions on Aerospace and Electronic Systems **37** (2001) 316–328
[12] III, G.J., Bhanu, B.: Recognition of articulated and occluded objects. IEEE Transactions on Pattern Analysis and Machine Intelligence **21** (1999) 603–613
[13] Boshra, M., Bhanu, B.: Predicting an upper bound on performance of target recognition in SAR images. IEEE Transactions on Aerospace and Electronic Systems **37** (2001) 876–888
[14] : Khoros Pro v2.2 User's Guide. Addison-Wesley-Longman (1998)
[15] Dudgeon, D., Lacoss, R., Lazott, C., Verly, J.: Use of persistent scatterers for model-based recognition. In: SPIE Proceedings: Algorithms for Synthetic Aperture Radar Imagery. Volume 2230., Orlando, FL (1994) 356–368
[16] Novak, L., Halversen, S., Owirka, G., Hiett, M.: Effects of polarization and resolution on SAR ATR. IEEE Transactions on Aerospace and Electronic Systems **33** (1997) 102–115
[17] Ikeuchi, K., Shakungaa, T., Wheelera, M., Yamazaki, T.: Invariant histograms and deformable template matching for SAR target recognition. In: Proc. IEEE Conf. on Comp. Vision and Pattern Rec. (1996) 100–105
[18] Mossing, J., Ross, T.: An evaluation of SAR ATR algorithm performance sensitivity to MSTAR extended operating conditions. In: SPIE Proceedings: Algorithms for SAR Imagery V. Volume 3370. (1998) 554–565
[19] Wissinger, J., Ristroph, R., Diemunsch, J., Severson, W., Freudenthal, E.: MSTAR's extensible search engine and inferencing toolkit. In: SPIE Proceedings: Algorithms for SAR Imagery VI. Volume 3721. (1999) 554–570
[20] Lamden, Y., Wolfson, H.: Geometric hashing: A general and efficient model-based recognition scheme. In: Proc. International Conference on Computer Vision. (1988) 238–249
[21] Grimson, W.E.L.: Object Recognition by Computer: The Role of Geometric Constraints. The MIT Press, Cambridge, MA (1990)
[22] Bhanu, B., Lin, Y., Jones, G., Peng, J.: Adaptive target recognition. Machine Vision and Applications **11** (2000) 289–299

# Chapter 3

# Three-Dimensional Laser Radar Recognition Approaches

Gregory Arnold[1], Timothy J. Klausutis[2], and Kirk Sturtz[3]

[1] Air Force Research Lab, AFRL/SNAT, Bldg. 620, 2241 Avionics Circle, Dayton, Ohio 45433 `Gregory.Arnold@wpafb.af.mil`
[2] Air Force Research Lab, AFRL/MNGI, Bldg. 13, 101 West Eglin Blvd., Eglin AFB, FL 32542 `Timothy.Klausutis@eglin.af.mil`
[3] Veridian Incorporated, 5200 Springfield Pike, Suite 200, Dayton, Ohio 45431 `ksturtz@mbvlab.wpafb.af.mil`

**Summary.** Three-dimensional laser radars measure the geometric shape of objects. The shape of an object is a geometric quality that is more intuitively understood than intensity-based sensors, and consequently laser radars are easier to interpret. While the shape contains more salient (and less variable) information, the computational difficulties are similar to those of other common sensor systems. A discussion of common approaches to 3D object recognition, and the technical issues (called operating conditions), are presented. A novel method that provides a straightforward approach to handling articulating object components and multiscale decomposition of complex objects is also presented. Invariants (or more precisely covariants) are a key element of this method. The presented approach is appealing since detection and segmentation processes need not be done beforehand, the object recognition system is robust to articulation and obscuration, and it is conducive to incorporating shape metrics.

## 3.1 Introduction

Lasers provide many advantages for the object recognition problem, especially when compared to passive electro-optical (video) sensors. For robust object recognition, it is desirable for the sensor to provide measurements of the object that are stable under many viewing and environmental conditions. Furthermore, these sensor measurements, or signatures, should be easily exploitable and provide enough richness to allow object-to-object separability. Specifically, 3D imaging laser radar (ladar) greatly simplifies the object recognition problem by accurately measuring the geometric shape of an object in 3D (preserving scale). In contrast many other sensing techniques inherently suffer a loss of information by projecting 3D objects onto 2D or 1D images. Another advantage of ladar shape measurements for object recognition is that the object signature is far less variable than other sensing modalities (e.g., the shape of the object does not vary due to lighting, diurnal affects, thermal loading, range, etc.). However, there is still difficulty since the ladar effectively samples an object's surface slightly differently each time. Another complicating factor is that ladar does not sample a scene on a uniform sampling lattice. Furthermore, a general complication for recognition is the lack of a general theory of discrimination (i.e., how to tell objects apart). This chapter will present the fundamentals of generic ladar systems, and detail a method for handling the differences in images due to articulation and viewpoint changes.

Generically, ladars can be thought of as an orthographic projection of the world onto the sensor (see Figure 3.1). Many 3D imaging ladars provide a range and intensity measurement at every point in the sampling lattice. Direct detection and coherent detection are two common ladar detection techniques. A complete treatise of ladar detection techniques is beyond the scope of this chapter [1]. The intensity value is a measurement of the amount of energy reflected from the appropriate region of the sampling lattice. This measurement is directly related to the monostatic bidirectional reflectance distribution function (mBRDF) of the material illuminated by the laser pulse. The intensity image is effectively a narrow-band, actively illuminated 2D image. This chapter focuses on the range measurement. The inherent data coordinate system for 3D ladar is {angle, angle, range}=$\{\theta, \phi, \rho\}$, where $\theta$ is the depression angle and $\phi$ is the azimuth angle from which the transmitted laser energy propagates from the sensor for each point in the sampling lattice. This is a polar coordinate system that can be transformed into a rectilinear $\{x, y, z\}$ coordinate system. Many different types of ladars exist, but for simplicity a flash ladar constructed with a focal plane array (FPA) of detectors will be assumed as the standard in this chapter. The term flash implies that the whole range and intensity image is measured at one time by spotlight illuminating the entire scene with one laser pulse. Alternately, a scanning 3D ladar images a scene by scanning one or several Laser beams over the entire sampling lattice. Although scanning ladars will be briefly described, the assumption is that appropriate motion compensation for platform motion has been done such that the resulting range image from the scanning ladar is equivalent to a (3-D) flash ladar. This assumes that the scanning mechanism operates in a linear fashion.

Operating conditions [2] have been discussed in many papers since their inauguration during the DARPA Moving and Stationary Target Acquisition and Recognition (MSTAR) program. Essentially the operating conditions are an attempt to describe everything that can affect the sensed image. They include sensor, target, and environmental parameters. Understanding all the standard and extended operating conditions is a first step to the development of an object recognition system. For the purposes of this chapter, they are categorized as (i) conditions whose effect on the image can be modeled (i.e., by a group action), (ii) conditions that obscure the image of the object (but are not easily modeled), or (iii) conditions that do not affect the part of the image corresponding to the object (i.e., changes in background). The goal is to

Chapter 3 Three-Dimensional Laser Radar Recognition Approaches 73

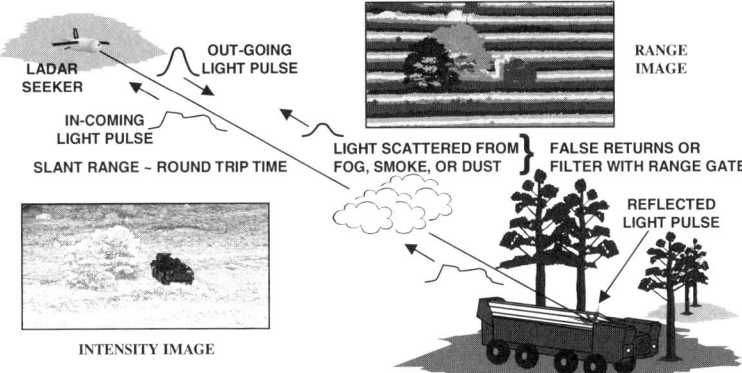

**Figure 3.1.** General ladar scene. This conceptual diagram illustrates the typical scenario and primary drivers of the sensed imagery. Most ladars return both a range and intensity at each pixel.

model and mitigate sufficient geometric effects to create a robust recognition system. The approach is model–based and object-centric.

A 3D ladar object recognition system can be thought of as a "simple" data fusion technique. It is simpler than more general fusion techniques because (typically) only one sensor and one image is involved. Even for this simple case the algorithm must efficiently fuse the information collected from each image pixel (or voxel, short for volumetric picture element, since it contains a third dimension). Enough pixels must be considered simultaneously to remove the unknown nuisance parameters of the model. For example, a single range return (without any other knowledge) cannot contain any information about the shape of the object. For the purposes of this chapter, "shape" is defined as what is left after translation and rotation have been removed. At least four points are needed (for 3D) to extract the local shape. Noise and small changes in the sampling grid can seriously affect the measured shape if the four points are in close proximity. Close proximity is preferred to mitigate obscuration and articulation, but dispersed locations are preferred to mitigate noise and achieve separability. Thus, the algorithm must make intelligent and efficient use of multiscale information to balance these issues.

The goal of 3D ladar object recognition systems is to classify images based on how similar or dissimilar the shape from the image is as compared to the shape of each object in a database. A metric is required to quantitatively compare and sort objects based on shape. The triangle inequality property of a metric enables efficient sifting through a large database by removing whole classes of objects that are different enough to be immediately removed from consideration. A modified form of the Procrustes metric is used for this object recognition system. This metric is constructed by considering the quotient space that is invariant to translation and rotation.

While 2D and 1D ladar systems exist and have many useful applications, for the purposes of this chapter ladar will imply 3D imaging ladar. A 3D model will be referred to as an *object*, and an {angle, angle, range} projection of that object (with a ladar at an arbitrary orientation) as an *image* or *range image*.

## 3.2 Ladar Sensors

### 3.2.1 Hardware

ladar sensors are active devices that avoid numerous issues inherent in passive systems and in stereo 3D reconstruction approaches. The two primary types of active range sensors are direct detection (also known as time-of-flight) and coherent detection.

Most direct-detection sensors periodically emit a short intermittent pulse and the distance is calculated by measuring the round-trip time (given the speed of light). This eliminates one type of shadow region that occurs with standard stereo 3D reconstruction approaches when a portion of the scene is not visible to both sensors [3]. However, shadow regions will still occur due to self-occlusion and occlusion of the background by objects in the foreground. This is one of the major differences in 3D image formation using ladar and the sensors used in the medical community. ladar is a reflective sensor while many 3D medical imaging sensors are partially transmissive.

Coherent detection techniques measure distance with a continuous (or multipulse) laser beam by measuring the phase shift of the reflected signal with respect to the original signal. Continuous-beam lasers require more power and are less covert. Furthermore, coherent detection techniques require more complex hardware than direct detection techniques.

Many ladar sensors have a single emitter and detector that are scanned across the scene at a constant angular resolution (see Figure 3.2). Although this is (currently) less expensive than the flash arrays that follow, there are numerous disadvantages. The primary disadvantage is that it takes longer to image a scene by scanning. A secondary problem is the nonlinearities induced by the scanning motion and by scene and object motion during the scanning process.

### 3.2.2 Projection

The appropriate choice of the projection model for ladar is muddled since the projection model will be different depending upon which coordinate system is utilized! In the native {angle, angle, range} coordinate system the projection model is a pinhole camera (full perspective model). Consider two parallel line segments in a plane orthogonal to the line of sight of the ladar. As the lines are moved farther away from the sensor, the angle subtended with respect

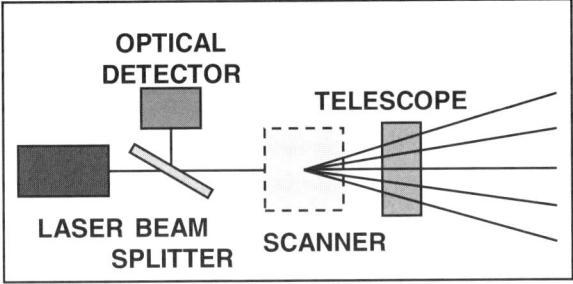

**Figure 3.2.** Typical ladar sensor hardware. A simplified diagram of the most common scanning, direct–detect system.

to the sensor decreases as $\frac{1}{\text{range}}$ (in each dimension). Therefore the angular extent of the lines within the image decreases as a function of their range.

The farther an object is from the sensor, the fewer pixels that will be on the object's surface (commonly called pixels-on-target). Lines that are parallel in 3D (but not parallel to the FPA) converge to a vanishing point in the {angle, angle, range} image. This is known as the "train tracks phenomenon." Standard video cameras are also perspective projection. A fundamental difference is that video cameras cannot recover the absolute size of the object. In contrast, since a ladar measures the range, the size of the object can be calculated (up to the pixelation error).

An orthographic projection model is appropriate for a rectangular coordinate system. It is not possible to perform this coordinate transformation directly with video cameras. The addition of range information enables the conversion for ladar from polar to rectangular coordinates. This is not precisely correct since the rectangular voxels should grow larger as the depth increases analogous to the change in size of the polar voxels. Alternately, the $x$–$y$ precision decreases for voxels that are farther from the sensor. For those familiar with radar systems, which are also orthographic, this is analogous to the fact that the signal-to-noise ratio decreases as the object moves farther away from the sensor.

### 3.2.3 Angle–Angle

A flash ladar's FPA is analogous to the CCD arrays used in digital cameras and video cameras where each array element contains a tiny receiver. Unlike CCD cameras, ladar is an active sensor and a single laser is used to spotlight illuminate the entire scene with each laser pulse. The laser is co-located with the receiver to form a monostatic system. Therefore, each element in the focal plane array can be viewed as containing both a receiver and a transmitter.

The FPA structure is inherently different from scanning systems as illustrated in Figure 3.3. Lenses can be used to transform to either array format, but FPAs are easier to build with a constant array size. A constant angular size has a modest advantage during the polar-to-rectangular transformation that is commonly done in software. Neither system will obtain a uniform sampling of an observed surface in general. In Figure 3.3 a constant array size would obtain a uniform sampling of a flat surface that is parallel to the FPA, but as the surface rotates out of plane it will be sampled non uniformly. The only surface that would be sampled uniformly in general is a spherical surface with its origin placed at the focal point of a constant angular size ladar system. In summary, most ladar systems are natively {angle, angle, range} (i.e., polar coordinates). Algorithms that are conducive to polar coordinate systems will have an inherent computational and noise advantage, because it is not possible to transform from the (discretized) polar coordinate system to the (discretized) rectangular coordinate system without some interpolation scheme.

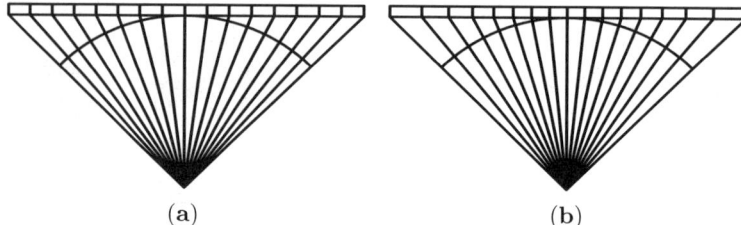

(a)            (b)

**Figure 3.3.** ladar focal plane arrays. A notional one-dimensional focal plane array is portrayed with a constant array size (**a**), and with a constant angular size (**b**). The choice of coordinate system will affect sensor design and the corresponding algorithm development. For FPAs, uniformly sized arrays, (**a**), are easier to construct and can achieve a constant angular size using a lens. For scanning ladar systems, it is easier to use constant angular step scan mirrors creating a constant angular array (**b**). Caution is still necessary since neither a constant angular or constant array size imply a uniform sampling of the object surface.

### 3.2.4 Range

Each pixel of the ladar measures the energy returned as a function of time. This is called the "range profile," and a notional example of a range profile for one pixel is illustrated in Figure 3.4. The majority of systems return the location and intensity of the $n$th-peak, where $n$ is typically the first or last peak. This is commonly called "first bounce" or "last bounce." Some newer systems return the entire range profile or the first $m$ detected pulse returns. This has advantages for reasoning about obscuration and validating edges and surfaces more complex than flat plates.

The returned pulse is the convolution of the transmitted signal with the object surface. Thus, the determination of the actual range is accomplished

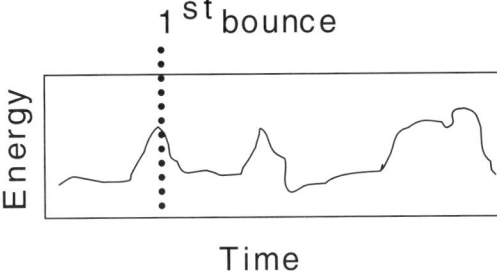

**Figure 3.4.** ladar range profile for one pixel. The range profile is the returned signal as a function of time (or distance). When multiple objects are within the field of view, more than one peak will occur. Current ladar systems typically return the first or last bounce peak of the range profile. Newer systems return multiple peaks or the entire profile.

by deconvolving the transmitted signal from the returned signal. This is generally an ill-posed problem. The common solution is to assume that the object surface is flat and therefore a simple peak detector is used to determine the estimated range. Hardware designers employ several common pulse detection strategies and these directly affect the location of the sensed return pulse. Furthermore, the selection of pulse detection technique will change the noise statistics. When multiple surfaces are within the same resolution cell, multiple peaks occur within the returned signal such as are illustrated in Figure 3.4. The $n$th-peak approach is a trade-off between accuracy and speed.

### 3.2.5 Transmission

Atmospheric effects such as scintillation (atmospheric transmission) and semitransparent obscuration (clouds and smoke) limit the useful range of ladar systems. Platform vibration mitigation can also be difficult to achieve for high-resolution systems. Currently, ranges beyond 5 km are typically relegated to 1D or 2D systems unless advanced techniques, like pulse-doublets, is used. Experimental high-resolution 3D systems have greater operational ranges.

The reflectance properties of the object can severely affect the returned signal, or cause a complete dropout (i.e., where no return is recorded for the given pixel). Dropouts can occur if the reflectance is too high (shiny metal reflects the emitted energy away from the detector) or too low (the material attenuates the reflected energy to a peak value below the threshold of the detector).

Range gating is a technique that is commonly applied to mitigate the affects of obscuration and semitransparent obscurants. The concept is to only examine the part of the signal that could have come from something near the estimated object location. In other words, if some portion of the energy is returned too quickly, then it is probably the result of reflecting off nearby clouds or smoke. Multiple reflections can cause a portion of the energy to return late. The finer the ability to accurately range gate, the better the signal-to-noise ratio will be. The associated peak detector or more advanced algorithms are more robust by applying this simple geometric constraint to the data.

The wavelength of the ladar is an interesting question from the point of view of computer vision beyond the visible spectrum. Lasers in the visible wavelengths are obviously very common; however, most sensor platforms would prefer to be covert. Atmospheric transmission and lack of generic tunable band-gap materials also limit the choice of wavelength. A common wavelength today is $1.06\mu m$. However, this wavelength is not eye-safe and therefore the potential applications are limited. Longer wavelengths are being investigated not only for eye safety, but also for better weather and aerosol penetration.

### 3.2.6 Synopsis

Many other topics with regard to ladar system design could be addressed here. For example, polarization, noise statistics, and hardware limitations are all very important considerations for the design and exploitation of ladar data. However, this section is intended to be a basic initiation into key features of ladar hardware and phenomenology.

The primary purpose of this chapter is to discuss object recognition capabilities given a ladar sensor. ladar sensor engineers often want to know how to optimize their sensor design for a particular application. However, a discrimination theory does not currently exist and therefore sensor optimization for object recognition is often based on standard pattern recognition techniques applied to very limited data sets. Approaches that are beginning to address this shortcoming are discussed after the presentation of existing algorithm approaches.

## 3.3 Is 3D Ladar Object Recognition a Solved Problem?

ladar object recognition is not a solved problem, primarily due to computational complexity issues. It has been argued that computer vision is generally an ill-posed problem that will never be solved. From an information-theoretic point of view, 3D sensors contain more information than 2D sensors because 2D sensors suffer a loss of information from the projection of $\mathbb{R}^3 \mapsto \mathbb{R}^2$. Still, the computational problems are no easier for 3D than they are for 2D or 1D.

### 3.3.1 Technical Challenges

The computation complexity issues are primarily due to the operating conditions briefly mentioned in Section 3.1 and further illustrated in Table 3.3

Chapter 3  Three-Dimensional Laser Radar Recognition Approaches    79

on p. 92. Ross [2] is an excellent paper on defining operating conditions. The following list represents the subset of the operating conditions that are key to the 3D ladar object recognition problem:

- Translation
- Rotation
- 2.5D projection
- Surface resampling
- Number of pixels-on-target
- Point correspondence (labeling, registration)
- Obscuration
- Articulation
- Fidelity
- Unknown objects

An algorithm that can efficiently and effectively handle all of these problems would be a major advancement in the fields of object recognition and computer vision. Note that "noise" is not explicitly listed. This is intentional since it is the belief of the authors that noise is not the fundamental limitation of current object recognition approaches.

Added to all these difficulties, the development process is seldom conducive to solving these hard problems. Typically, most of the money is spent on the hardware and data collections. The exploitation efforts are not begun until the final stages of the development. This leads to an additional set of issues:

- There is never enough data.
- The sensor parameters are undefined or incorrect.
- The ground truth will be incomplete or incorrect.
- The operational "requirements" will be way beyond the state-of-the-art.
- The money will be limited.
- Time will be limited.
- Expectations will be too high.
- The capabilities will be oversold.
- The sensor models will not be understood.
- Mother Nature is enigmatic.

Each of the technical challenges will now be addressed in more detail.

**Translation**

The first challenge refers to the fact that the coordinate system is sensor-centered, not object-centered. Therefore, the $\{x, y, z\}$ or $\{\theta, \phi, \rho\}$ pixel locations for a particular point on the object can change arbitrarily. In other words, the algorithm should recognize the object whether it appears at the top of the image or the bottom of the image or anywhere in between. Translation is the easiest challenge to solve, but caution is necessary. The standard approach is to move the centroid of a region or volume of interest to the origin.

If the object is consistently segmented from the background every time then this simple method will suffice. However, approaches that are more complicated are needed if the algorithm is to successfully match partially obscured or articulated objects.

**Rotation**

The above method of removing translation can be interpreted as putting the data into a standard position. A proof exists [4] that the equivalent approach cannot be done to handle rotation. In general, however, the eigenvectors of the inner product of the (rectangular) coordinates are invariant to rotation. The primary difficulty is with degenerate objects or completely symmetric objects such as a sphere. The difficulty manifests itself as the need for heuristic algorithms to choose the sign and ordering of the eigenvectors. Mathematically, quotienting out 3D rotation from the ladar data does not result in a smooth manifold.

**2.5-D Projection**

The term 2.5D is used to convey the fact that ladar sensors cannot see through an object [1]. As noted in Section 3.2.1, ladar is a reflective sensor, not a transmissive sensor like most medical 3D sensors. A "true" 3D sensor would return a computer-aided design (CAD) model of the object (i.e., front, back, and all the information in between). A ladar only returns the range to the first dispersive surface. Thus, there is a requirement to match an "image" to a model in the database. Ideally, the ladar data can be compared directly to the appropriate portion of the CAD models that are used to describe the objects of interest.

**Surface Resampling**

The sampling grid's relative location on the surface is slightly different each time the surface of the object is sampled. Sometimes these differences are negligible, but a shift of half a pixel can be significant, especially for low resolutions or near, sharp edges. Typically, algorithm developers consider each point in a point cloud as the distance to the object at that particular grid point. This assumption would be correct if the instantaneous field of view (IFOV) of the detectors in the FPA were infinitesimally small. However, this is never the case and the detectors see a portion of the target larger than an infinitesimally small point. Therefore, the range measurement returned is the average distance of the surface(s) within the pixel's extent (depending upon the peak extraction algorithm). This simplification is harder to make when working with the complete range profile. The implication is that approaches (like graph matching) based on vertex locations will fail without compensating for this affect. Point correspondence algorithms must carefully handle this affect too.

## Number of Pixels-on-Target

The absolute range from the sensor to the surface of the object is measured and therefore size (scale) is a known quantity. However, the farther the object is from the sensor, the fewer the pixels per unit surface area. For example, for a fixed angular resolution ladar, halving the range to an object quadruples the number pixels on the object. It is possible to calculate the area of the pixel at the object's surface. However, at the initial stages of the object recognition system it is often more computationally efficient to normalize out the number of pixels-on-target. This is a trade-off between discrimination capability and computational cost.

## Point Correspondence

Even ignoring the issues about resampling the surface, registration is a computationally complex problem—potentially factorial in the number of points to be corresponded. There are some nice algorithms available from the operational research and pattern recognition literature (called the bipartite matching or Hungarian algorithm)[5, 6, 7]. A nice summary and approach is provided in [8] and a powerful new approach is presented in [9]. However, even these have complexity that is polynomial in the number of points to be imaged.

An additional consideration is that "point correspondence" typically implies that there are equal numbers of points to register. The problem becomes even more complex when the goal is to match as many points as possible from two different point clouds. The above references have various approaches to solve this problem.

An elegant solution to this problem is required before any algorithm is computationally feasible. Two (previously mentioned) complications are that corresponding points will never be precisely the same due to noise, and the points may not physically represent precisely the same region from the scene or CAD model, i.e., the image grid shifted. The first complication necessitates the use of a metric, and the second complication implies that the ultimate goal is to register the measured point cloud directly to a CAD model. In other words, there may *not* be an advantage to converting the CAD model to a point cloud for the purposes of matching.

## Obscuration

Can an object still be recognized if it is not completely visible? Of course one can never see *all* of an object simultaneously, and algorithms must work even if everything that is expected to be visible in an image is not. Some parts of the object are more salient than others, so a true understanding of an algorithm's capability to handle obscuration can only be made with an understanding of the saliency of the visible surfaces. Jones [10] shows relatively good results against obscuration and articulation in 2D. Hetzel [11] presents an approach for 3D data that has an additional advantage as segmentation is not required.

## Articulation

Articulation and obscuration each force a tradeoff between local and global features for object recognition algorithms. Articulation refers to the fact that objects have specific ways of changing. For example, cars have doors that can open, and people can move their arms, legs, and head. It can also be something that may or may not appear on an object, such as a spoiler on a sports car. The ability to recognize an object is typically limited to one instantiation without the ability to recognize each of these variations as allowable changes of the same object.

## Fidelity

A physical and geometric model of the world, the sensor, and everything in between may be conceivable, but it is computationally unachievable. Each step in the processing chain, from the ladar probing the world to the output of the probability of a matching database model, makes simplifications to achieve computability and potentially compromising fidelity. Ideally these simplifications would be achieved while preserving the fundamental discrimination capability or be constructed in such a way as to provide incremental steps back to achieving the ideal discrimination capability. The goal is to be able to recognize all the objects (and corresponding poses) in the image with a finite number of computations.

In general, increased fidelity requires greater computation. The link between complexity and database indexing (search) is very strong. The indexing step is an $n$-class recognition problem, whereas the final validation is a one-class hypothesis verification. Most of the computations are consumed by the search (throwing out the models that do not match). The fundamental question is how to minimize the required fidelity and still guarantee that potential matches cannot be incorrectly pruned.

## Unknown Objects

This problem has been addressed the least in current literature. The question is how to realize that something is not represented in the database. Ultimately, it boils down to drawing a threshold in a one-class recognition problem, but that threshold should make sense geometrically and with respect to the appropriate noise model. All that can typically be done with current systems is to decide which database object the image looks most like. Thus, the final threshold is drawn based on experiments with limited data sets. As mentioned in the previous technical challenge, the indexing step must be dependent upon the objects in the database, but the final step of verifying the hypothesized identity should be independent of the other objects in the database. Only a controlled environment, where unknown objects cannot occur, would allow the algorithm to bypass the final verification step.

## 3.3.2 Shape Representations

Object representation is a critical first issue in the construction of an object recognition system. From an information-theoretic point of view, the algorithm should use the raw data directly from the sensor since each transformation potentially introduces additional noise into the system or loses relevant information.

A rough hierarchy can be imposed upon the different representations that have been proposed by various authors. The following list of shape representations is ordered in an ascending level of abstractness:

1. Point cloud.
2. Features (points, edges, corners, normals).
3. Triangular mesh.
4. NURBS (biquadratic surfaces).
5. Superquadrics/generalized cylinders (geons).

Note that the number of parameters necessary to describe the shape of an object decreases as the level of abstractness increases. Consequently, object matching based on the more abstract representation is less complicated. A useful survey of techniques for data representation can be found in [12, 13, 14]. Each representation is briefly discussed subsequently.

**Point Cloud**

A point cloud generally denotes the raw data that is available directly from the sensor.

**Features**

Features are a first-level abstraction of point clouds. Feature points could be as simple as pruning the point cloud to only include points at "interesting" places, such as along edges or corners. The features could also be an augmentation of the point cloud with a local surface normal.

**Triangular Mesh**

A triangular mesh is the most common form of mesh that is used. Triangular meshes can be interpreted as a linear approximation (and interpolation) of the point cloud data. Ideally, the choice of vertices for the triangles is made to minimize the difference between the measured point cloud and the corresponding point cloud that would be generated from the mesh. A mesh is typically constrained to be continuous; however this is generally not sufficient to enforce a unique mesh representation. Additional constraints are often applied in order to achieve a unique representation, such as attempting to make the all of the triangles nearly equilateral. However, this is not possible in general, so corresponding recognition algorithms must be able to handle different representations for the same object.

## NURBS

Nonuniform rational B-splines (NURBS) [15] are a generalization of biquadratic surfaces. These representations typically model a 2D surface embedded in 3D [16]. Patches of NURBS can be thought of as the generalization of meshes. The patches form a complete covering of the object's (measured) surface. Then, the surface within the bounds of each of these patches is represented by a NURBS surface. An immediate complication is defining a robust and unique decomposition of data into model patches, similar to meshes. Thus, the recognition algorithm cannot rely on the same object being modeled the same way for every instantiation of the object.

## Superquadrics/Generalized Cylinders

Superquadrics [17] and generalized cylinders [18] (or geons) model a volume of data. They define specific mathematical forms to model the data. For example, generalize cylinders estimate a cross-section and then sweep the cross-section along the measured data. At each step along the swept out path the centroid of the cross-section is located, and the size of the cross-section is estimated. Objects are constructed by assembling multiple generalized cylinders together. This is clearly an ill–posed problem for obscured or unseen portions of the object. Generally, a symmetry assumption is made, or efforts are made to demarcate what portions of the surface were constructed from measured data.

## Synopsis

All of these higher-level object representations have great promise; however, none of them have lived up to that promise to date. While a successful higher-level representation would greatly simplify the object recognition process, in practice these representations have simply traded simplicity in one portion of the overall system for added complexity in another with zero or negative gain.

The "negative gain" comes from the fact that these representations are exactly that — representations. Object recognition is based on discrimination. While one can argue that there is a functional dependence between representation and discrimination, it is easy to generate examples such that any given representation is the worst possible choice to differentiate the exemplars. However, this is not meant to encourage the other common extreme, which is to collect *enough* data to distinguish the classes (i.e., data-driven approaches). The goal is to find the right representation to optimize the discrimination capability with respect to the storage or speed requirements. This can only be discovered by modeling the entire object recognition system and using the scientific method such that the data is simply an experiment to validate or refute a hypothesized system model.

### 3.3.3 Shape Recognition/Indexing Approaches

Two obvious approaches to recognition are image-based and model-based. Jain [19] provides an excellent overview of statistical pattern analysis techniques. The most common image-based approach is template matching, and correlation is the most common similarity measure used in template matching. Some common approaches are briefly described below. Every approach has three common problems:

1. How to detect and segment the object from the background.
2. How to build and search a large database (avoiding local minima).
3. How to interpret results when the match is not perfect.

The following list and detailed descriptions represents many common approaches to shape recognition that have been developed and tested. References [12] and [13] are good survey papers.

1. Image-based matching (matched filter, template match).
2. Model-based matching.
3. Geometric hashing.
4. Hough transform.
5. Evidence accrual.
6. Learning approaches (neural networks, genetic algorithms, etc.).
7. Tree search (graph matching).
8. Principal components (eigenspace, eigenface, appearance-based).
9. Invariance (Fourier, moments, spherical harmonics, spin images).

Table 3.1. A summary of some typical characteristics of various approaches to shape recognition. A more detailed description of each method follows.

| Approach | Model | Empirical | Voting | Alignment |
|---|---|---|---|---|
| Image-based |  | ✓ | ✓ | ✓ |
| Model-based | ✓ |  | ✓ | ✓ |
| Hashing | ✓ | ✓ | ✓ |  |
| Hough |  | ✓ | ✓ |  |
| Evidence accrual |  | ✓ | ✓ |  |
| Learning |  | ✓ |  | ✓ |
| Tree | ✓ | ✓ |  | ✓ |
| Principal Components | ✓ | ✓ |  | ✓ |
| Invariance | ✓ | ✓ | ✓ | ✓ |

**Image-Based Matching**

Image-based or pixel-level matching generally implies the most common matched-filter-type comparison. This could be a comparison between two different measurements, but more generally is a comparison between a synthesized signature and the measured image using a "hypothesize and verify" type approach.

Intuitively, a matched filter is the average of the squared error between each measured pixel and the corresponding template pixel, or a nearest-neighbor type algorithm in a very high-dimensional space. A plethora of variations on this approach is available for purposes such as robustness to outliers and improving relative separability. "Image-based" refers to the fact that the algorithm is trained on measured or synthetic image data.

Image-based approaches suffer from both the complexity of the correspondence problem (finding the template in the image) as well as the complexity of building the template database. Moreover, it has been shown that "every consistent recognition scheme for recognizing 3D objects must in general be model based" [20]. Although this was referring to 2D imagery, the pretext still stands and thus pixel-level matching is best suited as a final validation step.

Pixel-level validation promises the maximum discrimination capability that is achievable (again from an information theoretic point of view). Several additional considerations are implied by pixel-level validation that are not commonly lumped into standard matched filtering techniques. First, a template is generated from a modeling and simulation capability since precise information is required which cannot be created apriori by data collections (due to cost and complexity). Second, the projection of the model into the image is used to determine the best estimated segmentation. This information should then be used to (a) perform a pixel-level validation for the visible parts of the object, (b) ignore the portions of the object that are obscured, and (c) provide a consistency check that the background looks like background (as opposed to unmodeled portions of the object).

A background consistency check is crucial to avoid the "box-in-a-box" problem. This problem appears when trying to recognize different sized objects. For example, a small box looks exactly like a larger box, except (from the sensor point of view) there is additional information that would be ignored without the consistency check. In other words, it would be easy to "recognize" the smaller box within an image of the larger box (however, it would be hard to recognize the larger box given an image of the smaller box). Alternately, is half a car still a car? It is possibly still a car if the other half of the car is obscured, but not likely if the measurements indicate the other half of the car is missing (as evidenced by measurements coming from behind where the other half of the car should be). Even if the algorithm was intelligent enough to realize two known objects are easily confused for each other, a background consistency check is still required so that unknown objects will successfully be rejected.

**Model-Based Matching**

The fundamental premise of model-based approaches is that the desired object and few other things will obey the constraints defined by the model. The system's ultimate performance will be bounded by how well the models can predict the real world and all of its subtleties. As with other methods, inverse methods do not currently exist for efficiently matching measured data back to these models. Therefore, current versions of model-based matching are very similar to template matching. The basic approach is "hypothesize and verify," where the geometry and physics models are used to synthesize an image corresponding to the hypothesis and then a verification technique is applied to compare the measured and synthesized data. Most of these approaches can be classified as

1. voting-based (geometric hashing, pose clustering, or generalized Hough transforms): voting in a parameter space for potential matches, or
2. alignment-based: searching for additional model-to-image matches based on the transformation computed from a small number of hypothesized correspondences.

The selectivity and an error analysis based on these approaches has been reported [21, 22, 23]. These techniques are discussed in more detail subsequently.

The different voting-based techniques primarily differ in their choice of a transformation space in which to tabulate votes to elect potential matches [24, 10]. These techniques are computationally efficient, accurate, and robust theoretically. The difficulty is in applying these techniques with a noise model. Most of the demonstrated systems assume that the robustness of the system will handle the noise. However, Grimson [23] demonstrates that the affect of the noise is dependent on location. This violates a fundamental assumption of the geometric hashing technique that the noise is independent of where the selected basis was located. Grimson suggests a modified voting technique, but the computational expense and loss in selectivity is clear in comparison to the alignment-based approach, especially as the noise increases.

Huttenlocher and Ullman [25] presents an alignment-based approach. Jacob and Alter [21] demonstrates that an alignment-based approach has better error characteristics than voting-based techniques, primarily due to the error approximations being much more accurate in the alignment-based approach (the transformation into the voting space generally make the error very difficult to estimate). Grimson [23] demonstrates, based on the selectivity of a matching set, that a large number of matches are necessary to reduce the probability of false match toward zero. Although Jacobs [21] developed a linear approach for matching points, this still does not handle the combinatorics associated with determining the initial minimal match between the image and object.

To conclude, the model-based approach has significant advantages such as limited dependence on measured data (so it is extendible to unsampled data regimes), and orders of magnitude reduction in online storage requirements. The typical disadvantage to model–based approaches is the online cost of synthesizing images corresponding to the hypothesis, especially when complex interactions occur between objects and backgrounds.

## Geometric Hashing

Geometric hashing is essentially an index tabulating an exhaustive enumeration of feature values. The goal is to build an index offline so that the occurrence of a feature in the image is linked to the occurrence of those features for a particular model or models in the database. For example, given a set of features with integer labels between one and five, a hash table would record which models have which features. In addition, it may record the relative frequency of occurrence of the objects for each features value (using some apriori information about the relative frequency of occurrence of the models).

Given the following occurrence data:

- Model 1: Features 2,4
- Model 2: Features 1,5
- Model 3: Features 5,5,4,2

the hash table in Table 3.2 would be constructed.

**Table 3.2.** An example hash table. The table contains the number of times a feature is present for each model. The occurrence of a feature in the image can be directly indexed into potential matches from the database.

| Features | Model 1 | Model 2 | Model 3 |
|---|---|---|---|
| 1 | 0 | 1 | 0 |
| 2 | 1 | 0 | 1 |
| 3 | 0 | 0 | 0 |
| 4 | 1 | 0 | 1 |
| 5 | 0 | 1 | 2 |

Now, given an image containing a feature of type 1, it is obvious that Model 2 is the only possible match in the database. Furthermore, if a feature of type 3 is found, no models match. If feature type 5 is found, Model 3 is more likely than Model 2.

Additional information, such as the look angle from which the feature was visible, may also be stored in the hash table. This is useful for further refining the hypothesis to include pose, especially when there is significant overlap between the features and models.

This is a simplistic model of geometric hashing. The principal difficulties are: (a) enumerating and binning all the images so that they can be recorded in the hash table, (b) the separability of the hash table once extensive enumeration has been accomplished, and (c) noise forces the discrete features to be treated probabilistically. This makes the matching much more complex, and once again highlights the need for metrics. References [26] and [27] implement hash tables to achieve efficient indexing.

## Hough Transform

The Hough transform converts images into the parameter space of lines. This technique is presented in most pattern recognition and computer vision books [28]. The generalized Hough transform extends this technique to parameter spaces for other object descriptors (generally geometric curves and surfaces). The concept is that an edge extraction technique has been applied to the image, and now the goal is to combine the evidence to identify lines despite noise and missing segments.

Lines in a 2D image could be parameterized by an angle and the location of an intercept along an axis. However, this parameterization does not uniquely identify lines that are parallel to the axis of intercept. Many different parameterizations are possible, but for demonstration a simple angle–angle representation will be used. A notable benefit of this particular representation is that both parameters have the same domain and range. Thus, the bin size and number of bins should be the same for each parameter. The relative proportion of bins is not as obvious if, for example, slope and intercept are used as the parameters. The general goals in choosing the parameterization include minimizing the number of parameters, nondegeneracy of the parameters, computational complexity, boundedness of the parameters, and uniformity of the parameter distributions. Figure 3.5 illustrates a binning scheme that might be used for the angle–angle parameterization. Notice that the middle bins are effectively more forgiving (coarser quantization) than the bins near the edges. Also, some bins cannot possibly have any support from the image. Thus, an improvement upon this scheme would make the bins a uniform size in the image.

**Figure 3.5.** Hough transform binning. Here are three examples of the 30 × 30 different bins corresponding to sampling the circle every 6 degrees. The square represents the image and the circle is shown to demonstrate the construction of the bins. The bins correspond to approximately equivalent lines in the image. Any pixels falling within a bin gives support to that particular line.

Lines in a 2D image intersect a circle circumscribing the image in two locations (the circle must be around the outside of the image in order to avoid lines tangent to the circle). These intersections uniquely define the line, and provide a minimal global parameterization of lines in the image. The circle is discretized into 30 × 30 bins, and a two-dimensional Hough space is created, each axis corresponding to the bin locations along the circle. Each feature in the image votes for all the bins that contain that appropriate point in the image. The relation between the image locations and Hough bins can be pre-computed offline so that the online computation is minimal. Bins containing multiple features will always have more votes than bins containing few or no features. Thus, the bins with the most votes are the most likely to contain a line. The presence of a line still needs to be verified.

**Evidence Accrual**

Evidence accrual is most commonly another name for Bayesian statistical inference methods [29, 26], although Dempster–Schafer is also common. The approaches and applications are well beyond the scope of this chapter. In general, they are similar to Hough transforms in that ultimately a voting space is established (appropriately normalized to produce a valid probability). Both positive and negative evidence can be accrued. "Negative evidence" refers to evidence that contradicts the hypothesis. Evidence accrual can be used both for determining what models are most likely given an image, and for deciding whether to accept or reject a particular hypothesis given an image. Dempster–Schafer's advantage in this respect is that it has a framework to incorporate the *unknown* class and it can generate confidence levels with so-called belief functions [30].

**Learning Approaches**

Neural networks are another area that are well beyond the scope of this chapter [31]. Many different types of networks (generally biologically inspired) have been designed, all with the basic idea of taking the available input data as well as truth information and producing the desired output data. A learning (training) process is used along with feedback (when available) to generate the desired output for a given input. The networks are generally constructed by multiple levels of massively interconnected computational nodes and thresholds. Neural networks are an excellent tool for a quick assessment of the computability of a desired process. However, the inability to guarantee robust decisions and handle unknown objects limits their applicability to object recognition. The reliance of the training process on available data is its fundamental weakness, both because a model is required [20] and because it is typically difficult to predetermine how the decision boundaries generalize to new data. In particular, it is necessary to avoid overtraining, a condition in which the network memorizes the training data and recalls it perfectly but does not generate the true class boundaries.

Genetic algorithms are another biologically inspired approach to object recognition. Genetic algorithms are essentially an optimization technique akin to simulated annealing and gradient descent. The "training" is an iterative process of creation and destruction of potential solutions. Creation involves taking existing solutions and randomly mutating or hybridizing to create new solutions. All the solutions are evaluated based on a measure of success, and those that fall below some threshold are destroyed. This very useful optimization technique is appropriate when other methods of modeling and simplifying the problem are not possible.

**Tree Search**

Tree search algorithms are easily as numerous as pixel-level validation variations. The most common approach is to assume the data can be efficiently and correctly segmented into chunks based on local similarity constraints. Then a tree structure is used to describe the relative geometric relation between the chunks. Labels are added to the nodes of the tree that correspond to information extracted from the individual chunks. Finally, trees may be replaced by linked graphs so that missing information will not be detrimental to the matching process [16]. The primary difficulty is how to handle the nonuniqueness of the various partitionings of the image without making the search computationally intractable. "Decision trees" are an automated approach to building trees [32].

**Principal Components**

Principal components, also known as eigenspaces, Hotelling, and Karhunen–Loeve transforms, minimize the mean-squared error in the reconstructed data as elements are dropped from a linear basis. The concept is developed in most pattern recognition and computer vision books [32]. Variations on the basic concept have been suggested that are invariant to translation, rotation, and other useful group actions. It is both an advantage and a disadvantage that the optimal (linear) transformation is data dependent. This enables the data to be transformed such that the principal components are linearly independent. However, the ability to determine and remove any linear dependence between the coefficients representing the objects is limited to the available data. Furthermore, any functional dependence between the components cannot be addressed. Finally, the standard technique is not appropriate for object recognition, since it is optimized to minimize representational error, not discrimination. Section 3.4.2 will detail how Procrustes analysis successfully uses the eigenspace to determine the distance between objects. Reference [33] is the seminal paper on eigenfaces, and [34] develops appearance-based techniques.

## Invariance

Some constraints are required so that not everything can be matched to everything. Data-driven approaches infer the constraints from the available data, whereas invariance-based approaches use a model (group action) of how the objects transform. The invariance approach is to equivalence sets of images that differ only by some (predefined) group actions. The concept is best expressed in the statistical shape analysis literature, shape is what is left after translation, rotation, and scale are removed [35]. Therefore, if the goal is to measure changes in shape, then invariance is a natural tool.

Invariance to a particular group action can be achieved many different ways. This includes both variations in the actual invariant function, but also in methods of calculating the function. In particular, invariants can be explicit or implicit. Implicit invariants are functions that are independent of the group parameter. For example, $x_1 - x_2$ is invariant to translation along the $x$-axis. Explicit functions achieve invariance by fixing the group parameter. This is also known as "standard position." The best example is moving the centroid of the data to the origin. No matter how the data is translated, by moving the centroid back to the origin, the compensated data is explicitly invariant to translation.

It only takes a quick example to demonstrate why invariant approaches are important. Consider a standard hypothesis-and-verify technique. Even if the operating condition parameters can be safely quantized as in Table 3.3, the total number of hypothesis is beyond exhaustive computational capabilities.

**Table 3.3.** Enumeration of quantized operating conditions. The total number of hypothesis is beyond exhaustive computational capabilities.

| Parameter | Quantized Bins |
|---|---|
| Object type | 20 |
| Object aspect | 72 |
| Depression angle | 5 |
| Articulation (1 DoF) | 36 |
| Configuration (4 binary) | 16 |
| Obscuration | 400 |
| Correspondence | 20 |
| Netting | 5 |
| Total Hypotheses | $165,888,000,000 = 1.6 \times 10^{11}$ |

The most common problem with invariant-based techniques is a lack of understanding of the affect of the transformations into an invariant space. It is trivially true that a constant function is invariant to any group action, so it is only illustrative to point out that many objects are mapped to the same equivalence class by this function. An invariant basis can be derived such that it only equivalences objects that are the same up to the desired transformation; however, the affect of noise can be substantially different for each invariant basis. Therefore the choice of invariant basis must be carefully considered. Furthermore, it is not trivial to determine the distribution of the objects in invariant space given a distribution of objects in parameter space.

Invariants provide the most promising theoretical approach, but current approaches have not achieved their potential for *solving* the object recognition problem. However, progress is being made using several different techniques. Johnson's thesis [36] is the seminal paper on spin images and their application to ladar object recognition. Funkhouser et al. [37] present spherical harmonics. Lo and Don [38] present an excellent summary of invariants formed using moments, and [39] makes significant advances.

## 3.4 Shape Metrics

### 3.4.1 Background

The motivation for metrics is illustrated by some common questions, "What is the potential efficacy of this sensor to my application?" and "How close is my algorithm to achieving the maximum achievable separability?" However, no theory of discrimination exists and therefore it has not been possible to answer these questions in a general context. Object recognition is fundamentally driven by the ability to differentiate objects. Alternately, a method is needed to measure the difference between objects.

The metrics of interest for object recognition are invariant to the modeled group action (e.g., rotation and translation). This is intuitively obvious as one would not expect the distance (measurement of shape difference) between two objects to change when those objects have been rotated.

Shape metrics provide a basis for answering these questions. However, metrics for object recognition have been notoriously difficult to find and very little work has been done in developing methods for constructing them. An exception comes from the field called statistical shape analysis or morphometrics [4, 35, 40]. Specifically, these books summarize and extend the research (primarily for biological applications) that although founded in work developed throughout the century, really accelerated with papers by Kendall [41] and Bookstein [42].

The goal in this field is to develop a shape metric based on "landmark" features. From the object recognition point of view, these are simply pixel locations of extracted features. Since shape is of fundamental interest, two objects are considered similar independent of translation, rotation, and scale. In other words, two objects are considered equivalent if they can be brought into correspondence by translating, rotating, and scaling. This is called the similarity group.

The metric developed for statistical shape analysis is commonly called the Procrustes, Procrustean, or (in one specific case) the Fubini–Study metric.

The proof that it is a metric is developed in Section 3.4.2. The Procrustes metric is a quotient metric. Intuitively, it is easy to conceptualize considering the space of objects modulo the similarity group. Quotienting the group action out of objects can be very difficult both analytically and numerically. Despite these difficulties, this does provide a constructive approach to developing metrics invariant to other group actions.

Although an object recognition system must eventually make a hard decision (i.e., make a binary decision as to whether an image is consistent with a particular hypothesized object), the concept is that this decision will occur in the final stages of the algorithm when a pixel-level validation and background consistency is performed. This chapter is applicable to both this final validation stage and the weeding or indexing problem where the goal is to order the most likely candidates first. The indexer should not predetermine how many candidates the algorithm will examine before making a final decision. The metric will naturally rank order the models based on their similarity to the image. Therefore the system will validate the best matches first (based on the extracted data or features) and continue until either the validation is highly confident in a match, it rejects all the potential matches, or a time constraint is exceeded.

A preliminary study on the fundamental separability of objects (sets of landmarks) based on this metric can be found in [43]. The need for appropriate metrics is a major theme of this chapter. There have been numerous metrics, measures, and distances proposed throughout the history of object recognition, computer vision, and pattern recognition. Alternative 3D pseudometrics are presented in [3] and [36]. More general metrics can be found in [44] which provides a nice summary of different metrics. Csiszár argues for a unique choice of metrics in [45]. Finally, [46] questions whether human perception satisfies any of the axioms of a metric, and [47] contains an excellent summary of what is known about biological vision systems.

### 3.4.2 Procrustean Metrics

In this section the partial Procrustes metric is derived and compared relative to other choices of a metric. The three metrics commonly used in statistical shape analysis are the *full* Procrustes, the *partial* Procrustes, and the Procrustes metric. Each metric corresponds to a different model and the appropriate selection of a given metric, in the context of the object recognition problem, depends upon the sensor being employed.

Let $G$ be the similarity group — the group generated by the rotation, translation, and scale (dilation)— and consider the componentwise group action $\mathbb{R}^m \times \cdots \times \mathbb{R}^m \times G \to \mathbb{R}^m \times \cdots \times \mathbb{R}^m$. The associated quotient space, $\Sigma_m^k$, is called the shape-space of order $(m, k)$

$$\Sigma_m^k \equiv \mathbb{R}^m \times \cdots \times \mathbb{R}^m / G,$$

where $k$ corresponds to both the number of features and the number of copies of $\mathbb{R}^m$. Note that $m = 2$ represents a two-dimensional "image" space, $m = 3$ represents three-dimensional "volume" space, and higher values of $m$ represent higher-dimensional feature spaces. The procedure to calculate the (representative) equivalence classes of this quotient space as well as a metric on this space follows.

The first step in determining the shape of an object, $\mathbf{X}$, is to quotient out translation by moving the centroid along each axis to a standard position (the origin)

$$\mathbf{X} \to \mathbf{X} - \bar{\mathbf{X}},$$

where $\bar{\mathbf{X}}$ represents the centroid of $\mathbf{X}$. This can be rewritten as a matrix product

$$\mathbf{X} \to -\frac{1}{k} \underbrace{\begin{bmatrix} 1-k & 1 & 1 & \cdots & 1 \\ 1 & 1-k & 1 & \cdots & 1 \\ 1 & 1 & 1-k & \cdots & 1 \\ \vdots & \vdots & \vdots & \ddots & \vdots \\ 1 & 1 & 1 & \cdots & 1-k \end{bmatrix}}_{\mathbf{C}} \underbrace{\begin{bmatrix} x_1 & y_1 & z_1 \\ x_2 & y_2 & z_2 \\ x_3 & y_3 & z_3 \\ \vdots & \vdots & \vdots \\ x_k & y_k & z_k \end{bmatrix}}_{\mathbf{X}}$$

where $\mathbf{C}$ is the $k \times k$ centering matrix (the matrix that takes the centroid to the origin), which is symmetric and idempotent (so $\mathbf{C} = \mathbf{C}^T = \mathbf{CC}$).

The centering matrix has rank $k-1$ and taking the Cholesky decomposition of $\mathbf{C}$ results in the first row consisting entirely of zeros. Therefore, the Cholesky decomposition is a method for removing a linearly dependent row from $\mathbf{C}$. In particular, the Helmert submatrix, $\mathbf{H}$, is defined by

$$\mathbf{H}^T \mathbf{H} = \mathbf{C},$$

where $\mathbf{C}$ is the centering matrix presented above, and $\mathbf{H}$ is the Cholesky decomposition into a lower triangular matrix followed by the removal of the row of zeros. Thus $\mathbf{H}$ removes translation and reduces the dimension of the object (analogous to dropping one point). This is exactly the number of parameters that were removed from the group action (one parameter for the translation of each axis). Therefore, the Helmertized object, $\hat{\mathbf{X}} = \mathbf{HX}$, is of dimension $(k-1) \times m$ and is a representative of the original object $\mathbf{X}$ in the quotient space for translation. Note that $\mathbf{H}$ can be computed directly [35].

The scale is then removed by the scaling operation $\mathbf{x} \mapsto \mathbf{x}/\|\mathbf{x}\|$, where $\|\cdot\|$ is the $l_2$-norm as this norm is invariant to rotation. The resulting space, $\mathbb{R}^m \times \cdots \times \mathbb{R}^m$ modulo translation and scale, is called the preshape space and the equivalence class of $\mathbf{X}$ is $\mathbf{W} = \mathbf{HX}/\|\mathbf{HX}\|$. Elements of the preshape space are denoted by $\mathbf{W}$ and will also be of dimension $(k-1) \times m$.

A continuous global quotient space does *not* exist for rotation [4]. However, the distance between the equivalence classes of $\mathbf{W}$ modulo $SO(m)$ can be defined and does exist,

$$d_P(\mathbf{X}_1, \mathbf{X}_2) = \inf_{\mathbf{R}_1, \mathbf{R}_2 \in \mathrm{SO}(m)} \|\mathbf{W}_1 \mathbf{R}_1 - \mathbf{W}_2 \mathbf{R}_2\|, \tag{3.1}$$

where the matrix norm $\|\mathbf{W}\| = \sqrt{\mathrm{Tr}(\mathbf{W}^T \mathbf{W})}$ is once again the $l_2$-norm, and $\mathrm{SO}(m)$ is the set of $m \times m$ rotation matrices. This function can be simplified by defining $\mathbf{R} = \mathbf{R}_1 \mathbf{R}_2^{-1}$ and using $\mathrm{Tr}(\mathbf{BAB}^{-1}) = \mathrm{Tr}(\mathbf{A})$ to obtain

$$\begin{aligned}
d_P(\mathbf{X}_1, \mathbf{X}_2) &= \inf_{\mathbf{R}_1, \mathbf{R}_2 \in \mathrm{SO}(m)} \|\mathbf{W}_1 \mathbf{R}_1 - \mathbf{W}_2 \mathbf{R}_2\| \\
&= \inf_{\mathbf{R}_1, \mathbf{R}_2 \in \mathrm{SO}(m)} \|(\mathbf{W}_1 \mathbf{R}_1 \mathbf{R}_2^{-1} - \mathbf{W}_2) \mathbf{R}_2\| \\
&= \inf_{\mathbf{R}, \mathbf{R}_2 \in \mathrm{SO}(m)} \|(\mathbf{W}_1 \mathbf{R} - \mathbf{W}_2) \mathbf{R}_2\| \\
&= \inf_{\mathbf{R} \in \mathrm{SO}(m)} \|(\mathbf{W}_1 \mathbf{R} - \mathbf{W}_2)\|.
\end{aligned}$$

The properties of a metric can be found in any standard math reference book [48]. The properties of the Procrustes metric are ultimately induced by the properties of the norm. In particular, the triangle inequality property of the metric follows from the triangle inequality for the norm

$$\begin{aligned}
d_P(\mathbf{X}_1, \mathbf{X}_3) &= \inf_{\mathbf{R} \in \mathrm{SO}(m)} \|\mathbf{W}_1 \mathbf{R} - \mathbf{W}_3\| \\
&= \inf_{\mathbf{R}, \mathbf{R}_2 \in \mathrm{SO}(m)} \|\mathbf{W}_1 \mathbf{R} - \mathbf{W}_2 \mathbf{R}_2 + \mathbf{W}_2 \mathbf{R}_2 - \mathbf{W}_3\| \\
&\leq \inf_{\mathbf{R}, \mathbf{R}_2 \in \mathrm{SO}(m)} (\|\mathbf{W}_1 \mathbf{R} - \mathbf{W}_2 \mathbf{R}_2\| + \|\mathbf{W}_2 \mathbf{R}_2 - \mathbf{W}_3\|)
\end{aligned}$$

because by defining $\overline{\mathbf{R}} = \mathbf{R} \mathbf{R}_2^{-1}$

$$\begin{aligned}
&\inf_{\mathbf{R}, \mathbf{R}_2 \in \mathrm{SO}(m)} (\|\mathbf{W}_1 \mathbf{R} - \mathbf{W}_2 \mathbf{R}_2\| + \|\mathbf{W}_2 \mathbf{R}_2 - \mathbf{W}_3\|) \\
&= \inf_{\overline{\mathbf{R}} \in \mathrm{SO}(m)} \|\mathbf{W}_1 \overline{\mathbf{R}} - \mathbf{W}_2\| + \inf_{\mathbf{R}_2 \in \mathrm{SO}(m)} \|\mathbf{W}_2 \mathbf{R}_2 - \mathbf{W}_3\| \\
&= d_P(\mathbf{X}_1, \mathbf{X}_2) + d_P(\mathbf{X}_2, \mathbf{X}_3).
\end{aligned}$$

Thus this function defines a metric on the shape space. Expanded out in terms of the $l_2$-norm, the distance is

$$\begin{aligned}
d_P(\mathbf{X}_1, \mathbf{X}_2)^2 &= \inf_{\mathbf{R} \in \mathrm{SO}(m)} \|\mathbf{W}_1 \mathbf{R} - \mathbf{W}_2\|^2 \\
&= \inf_{\mathbf{R} \in \mathrm{SO}(m)} \mathrm{Tr}((\mathbf{W}_1 \mathbf{R} - \mathbf{W}_2)^T (\mathbf{W}_1 \mathbf{R} - \mathbf{W}_2)) \\
&= \inf_{\mathbf{R} \in \mathrm{SO}(m)} \mathrm{Tr}(\mathbf{W}_1^T \mathbf{W}_1 + \mathbf{W}_2^T \mathbf{W}_2 - 2 \mathbf{R}^T \mathbf{W}_1^T \mathbf{W}_2) \\
&= \mathrm{Tr}(\|\mathbf{W}_1\|^2 + \|\mathbf{W}_2\|^2 - 2 \sup_{\mathbf{R} \in \mathrm{SO}(m)} \mathbf{R}^T \mathbf{W}_1^T \mathbf{W}_2) \\
&= 2(1 - \sup_{\mathbf{R} \in \mathrm{SO}(m)} \mathrm{Tr}(\mathbf{R}^T \mathbf{W}_1^T \mathbf{W}_2)).
\end{aligned} \tag{3.2}$$

## Globally Optimal Solution

Finding the argument of the supremum in (3.2) is equivalent to finding the argument of the infimum in (3.1). An analytic globally optimal solution to this optimization problem follows using singular value decomposition

$$\mathbf{W}_1^T \mathbf{W}_2 = \mathbf{U}^T \mathbf{\Lambda} \mathbf{V},$$

where $\mathbf{U}$ is orthonormal, $\mathbf{V} \in SO(m)$, and $\mathbf{\Lambda}$ is zero with a diagonal of singular values. The singular values are positive except the smallest, $\lambda_m$, is the negative if and only if $\det(\mathbf{W}_1^T \mathbf{W}_2) < 0$ [35]. Note that the singular values, $\{\lambda_i\}_{i=1}^m$, are the square roots of the eigenvalues of the matrix $\mathbf{W}_1^T \mathbf{W}_2 \mathbf{W}_2^T \mathbf{W}_1$ and has a maximum $\sum_{i=1}^m \lambda_i = 1$ (which corresponds to $\mathbf{W}_1$ and $\mathbf{W}_2$ matching).

Now the metric can be written as

$$d_P(\mathbf{X}_1, \mathbf{X}_2)^2 = 2(1 - \sup_{\mathbf{R} \in SO(m)} \text{Tr}(\mathbf{R}^T \mathbf{U}^T \mathbf{\Lambda} \mathbf{V})).$$

Since $\text{Tr}(\mathbf{R}^T \mathbf{U}^T \mathbf{\Lambda} \mathbf{V}) \leq \text{Tr}(\mathbf{\Lambda})$ with equality occurring when $\mathbf{R} = \mathbf{V}\mathbf{U}^T$, the supremum occurs at $\mathbf{R} = \mathbf{V}\mathbf{U}^T$. Thus

$$d_P(\mathbf{X}_1, \mathbf{X}_2) = \sqrt{2\left(1 - \sum_{i=1}^m \lambda_i\right)}. \tag{3.3}$$

This metric is called the partial Procrustes metric in the statistical shape literature [35]. The full Procrustes metric between two objects $\mathbf{X}_1$ and $\mathbf{X}_2$ can be written with respect to their corresponding preshapes $\mathbf{W}_1$ and $\mathbf{W}_2$ as

$$d_F(\mathbf{X}_1, \mathbf{X}_2) = \inf_{\mathbf{\Gamma} \in SO(m), \beta \in \mathbb{R}} \|\mathbf{W}_2 - \beta \mathbf{W}_1 \mathbf{\Gamma}\|.$$

Expressing this metric in terms of the eigenvalues of $\mathbf{W}_1^T \mathbf{W}_2 \mathbf{W}_2^T \mathbf{W}_1$, as done above for the partial Procrustes metric, gives

$$d_F(\mathbf{X}_1, \mathbf{X}_2) = \sqrt{1 - \left(\sum_{i=1}^m \lambda_i\right)^2}. \tag{3.4}$$

The Procrustes metric is defined as the closest great circle distance between $\mathbf{W}_1$ and $\mathbf{W}_2$ on the preshape sphere. Kendall [41] shows that the preshape space is indeed a sphere. In fact, the Procrustes distance on $\Sigma_2^k$ is equivalent to the Fubini-Study metric on $\mathbf{CP}^{k-2}(4)$. From trigonometry it follows

$$d(\mathbf{X}_1, \mathbf{X}_2) = \arccos\left(\sum_{i=1}^m \lambda_i\right). \tag{3.5}$$

Finally, another distance measure is the full generalized Procrustes metric defined by

$$d_g(\mathbf{X}_1, \mathbf{X}_2) = \inf_{\substack{\mathbf{\Gamma}_1, \mathbf{\Gamma}_2 \in SO(m) \\ \beta_1, \beta_2 \in \mathbb{R} \\ \gamma_1, \gamma_2 \in \mathbb{R}^m}} \|(\beta_1 \mathbf{X}_1 \mathbf{\Gamma}_1 + \mathbf{1}_k \gamma_1) - (\beta_2 \mathbf{X}_2 \mathbf{\Gamma}_2 + \mathbf{1}_k \gamma_2)\|$$

$$= d_F(\mathbf{X}_1, \mathbf{X}_2).$$

where an additional constraint such as $\beta_1 \beta_2 = 1$ is necessary to prevent degenerate solutions. The full generalized form is typically used in statistical shape analysis for estimating the mean shape of a set of objects. This development demonstrates it for comparing just two objects. Therefore, the only difference between the full generalized form and the full ordinary form (as it is called in [35]) is the inclusion of the translation in the optimization. For two objects, the full generalized form is equivalent to the full ordinary Procrustes metric (by using the fact that the objects can be centered, i.e., have their centroids moved to the origin, without loss of generality). This follows since the cross terms in the norm due to the translation will disappear because $\mathbf{C}\mathbf{1}_k = \mathbf{0}$ for any centering matrix $\mathbf{C}$ such as the Helmert matrix ($\mathbf{C} = \mathbf{H}^T \mathbf{H}$).

The three metrics can be interpreted as a path in shape space corresponding to the chord length (partial Procrustes), arc distance (standard Procrustes), and orthogonal length (full Procrustes). It should be noted that these three metrics are order preserving relative to each other, i.e.,

$$d_F(\mathbf{X}_1, \mathbf{X}_2) < d_F(\mathbf{X}_1, \mathbf{X}_3)$$
$$\updownarrow$$
$$d_P(\mathbf{X}_1, \mathbf{X}_2) < d_P(\mathbf{X}_1, \mathbf{X}_3)$$
$$\updownarrow$$
$$d(\mathbf{X}_1, \mathbf{X}_2) < d(\mathbf{X}_1, \mathbf{X}_3).$$

Therefore, any of these metrics can be used if all that matters is the relative ordering of the objects. Dryden and Mardia [35] argues that the full Procrustes distance is the natural choice as it optimizes over the full set of similarity parameters and because it "appears exponentiated in the density for many simple probability distributions for shape" (p.44).

**Point Correspondence**

The existing literature on shape metrics presumes the point correspondence between the objects is known or is handled elsewhere. This is also known as the labeling problem. Conceptually, this is the internal labeling problem (matching $k$ image features to $k$ model features) as opposed to the external labeling problem (choosing which subset of the image features and which subset of the model features to compare). An exhaustive computation of the external problem requires $\binom{P}{k}$ computations (where $P$ is the number of image features) for each of the $\binom{K}{k}$ feature sets (where $K$ is the number of model features) of each model in the database. The internal problem requires $k!$ computations (exhaustive). See [49] for more information. Algorithms for point correspondence that are polynomial in complexity were referenced in Section 3.3.1, however the ultimate goal is to develop an analytic solution to this problem so that theoretical analysis of the label-invariant metric is possible.

An implicit assumption has been made that the same subset of points can be found on the image and the object to compare. Zhang [50] presents a technique for consistently extracting the same number of points using Legendre polynomials. The internal labeling problem for Procrustes is explored in [51].

## 3.5 Current Approach

The subsequently described algorithm handles most of the technical challenges previously discussed for a successful 3D object recognition system. The conclusions and future research sections will discuss desired improvements.

The method is based on looking at "spheres" of data extracted from the image and comparing these to "spheres" of data extracted from the models. This method specifically avoids feature detection for improved robustness, and this approach specifically handles articulated objects. Recognizing such an object can involve a search in a high-dimensional space that involves all the articulating degrees of freedom, in addition to the usual unknown viewpoint. This algorithm uses invariants to reduce the search space to a manageable size.

The input is a range image (in rectangular coordinates) of an unknown object in an unknown articulation position, such as a backhoe with each link in an arbitrary position. The desired output of the system is the identity of the object, and the articulation and viewpoint parameters. The object is identified using a database of known models. Since an object, like the backhoe, could have ten degrees of freedom (DoF) it is infeasible to store all images of each articulated object. Subsampling and storing 10 images of each DoF for just this one object would require $10^{10}$ images, which is prohibitive. Consequently an efficient procedure for representing and searching a database of objects is described.

Unique features of the method described in this chapter include an integral approach to feature detection and recognition that is more efficient than considering them separately and improves robustness to noise. The approach includes scanning the image, so it avoids the combinatorial explosion of hypothesizing feature correspondences [52, 53, 54]. This approach simultaneously uses the data from all the points in a neighborhood (thus the robustness to noise and obscuration).

### 3.5.1 The Sensor Model

The sensor model characterizes the transformation group acting on the object, and the projection from $\mathbb{R}^3$ to $\mathbb{R}^3$. The model presented here is written for four points on an object undergoing the same rigid motion. This is a good starting point, with the articulation being incorporated subsequently.

The transformation group is the rigid transformation (rotation and translation). The relation between the measured feature location, $\{u, v, r\}$, and the model feature location, $\{x, y, z\}$ (expressed in rectangular coordinates) is

$$\begin{bmatrix} u_1 & u_2 & u_3 & u_4 \\ v_1 & v_2 & v_3 & v_4 \\ r_1 & r_2 & r_3 & r_4 \\ 1 & 1 & 1 & 1 \end{bmatrix} = \begin{bmatrix} & & & a \\ & \mathbf{R} & & b \\ & & & c \\ 0 & 0 & 0 & 1 \end{bmatrix} \begin{bmatrix} x_1 & x_2 & x_3 & x_4 \\ y_1 & y_2 & y_3 & y_4 \\ z_1 & z_2 & z_3 & z_4 \\ 1 & 1 & 1 & 1 \end{bmatrix}$$

such that $\mathbf{R} \in SO(3)$ (rotations), and $\{a, b, c\}$ is a vector denoting a rigid translation of $\mathbb{R}^3$. The trailing 1's are included to permit writing the translation as a matrix product. As written, both the front and back of the object would appear in the image. The fact that a ladar cannot see the through an object is why the projection is referred to as 2.5D in Section 3.3.1.

### 3.5.2 Invariants and the Object–Image Relations

Object–image relations express a geometric relation (constraint) between a 3D object and its image. This is a general approach and the particular invariants and number of points required for an invariant depends upon the transformation group associated with the sensor model.

Although constructing invariants is difficult, once an invariant has been found it is typically simple to compute. Object–image (O-I) relations are an application of invariance theory as applied to the world viewed by a sensor (ladar in this case). O-I relations provide a formal way of asking, "What are all possible images of this object?" and "What are all possible objects that could produce this image?" Clearly, this is a very powerful formalism and it is well suited to object recognition.

Recent research has yielded the fundamental geometric relation between "objects" and "images" (for RADAR, SAR, UHRR, EO, IR, and other sensors) [55, 56, 57, 58, 59, 52]. Although the object–image relations are very simple for ladar, they are useful for demonstrating the benefits of using covariants.

Fundamentally the O-I relations can be viewed as the result of elimination of the unknown parameters in the model describing the projection of the 3D world onto the sensor. Thus, the ladar O-I relation can be derived by eliminating the group parameters associated with object motion, and the parameters associated with the ladar look direction (for this case, the group actions are the same). The result is an equation relating the object to its ladar range image written in terms of their associated invariants. More specifically, the equation relates 3D rigid invariants (inter-scatterer 3D Euclidean distances, determinants, or inner products) to rigid invariants measured from the ladar image.

### 3.5.3 The ladar Object–Image Relation

The above model of a ladar is essentially an orthographic projection from 3D to 3D. An invariant of 3D objects (undergoing rigid transformations) requires a minimum of four points (in general position), $P_i = \{X_i, Y_i, Z_i, 1\}$. Let the point $P_i$ correspond to the $i$−th column of a $4 \times 4$ matrix. By translating and rotating appropriately, one can always transform the points into the standard position,

$$\begin{bmatrix} 0 & I_1 & I_2 & I_3 \\ 0 & 0 & I_4 & I_5 \\ 0 & 0 & 0 & I_6 \\ 1 & 1 & 1 & 1 \end{bmatrix}$$

where $\{I_1, I_2, I_3, I_4, I_5, I_6\}$ are the object model invariants. They are invariants under 3D rigid transformations. Linear algebra shows a unique transformation (up to sign) exists to make this change of basis. It is not obvious without further explanation, but the invariants are functions of the Euclidean distance, determinants, and inner products.

The range image can also be transformed to the standard position,

$$\begin{bmatrix} 0 & i_1 & i_2 & i_3 \\ 0 & 0 & i_4 & i_5 \\ 0 & 0 & 0 & i_6 \\ 1 & 1 & 1 & 1 \end{bmatrix}$$

where $\{i_1, i_2, i_3, i_4, i_5, i_6\}$ are the image invariants. They are also invariant under 3D rigid transformations.

The fundamental object–image relation can now be determined by solving the model projection equations with respect to the unknown rotation and translation. The resulting object–image relations are

$$i_j = I_j, \qquad \forall j \in \{1, 2, 3, 4, 5, 6\} \tag{3.6}$$

An algorithm that uses these object–image relations does not need to worry about any rigid transformations. In other words, use of the object–image relations guarantees that all the continuous group actions that were modeled have been factored out of the resulting equalities. What remains are the discrete permutations, i.e., the correspondence and ordering of sets of features between the object and image. However the formulation, as presented above, is not feasible for two reasons:

1. *Noise issues:* As presented, these equations treat the initial points as "special." A formulation is desired that treats all points equivalently, as this should inject some robustness to noise.
2. *Complexity issues:* Arnold [49] demonstrated that the complexity of a brute force search of the discrete permutations is not computationally feasible in general.

Therefore a modified formulation is desired that provides an invariant relation between the object and image, and also handles the noise and complexity issues.

### 3.5.4 Covariants

Covariants provide an alternative approach to determining O-I relations by considering equivariant functions. This section will formalize the approach, including how projection is naturally handled in the technique.

Let $G$ be a Lie group (a continuous transformation group) acting on an $n$-dimensional manifold $M$, $\phi : G \times M \to M : (g,p) \mapsto gp$. An important example of such an action is $G = \text{GL}_3(\mathbb{R})$, the set of invertible matrices, acting on the function space $M = \prod_{i=1}^{3} \mathbb{R}_i^3$. An element of this function space is simply an ordered triple, $\{p_i\}_{i=1}^{3}$, where each "point" $p_i$ is an ordered 3-tuple, $\{x_i, y_i, z_i\}^T$. Here the action is a componentwise matrix multiplication—hence the action is linear. A $G$-invariant function is a function $\Phi$ satisfying $\Phi(gp) = \Phi(p)\ \forall p \in M\ \forall g \in G$. The following figure characterizes an invariant function $\Phi$ under an action $\phi : G \times \mathbb{R}^n \mapsto \mathbb{R}^n$

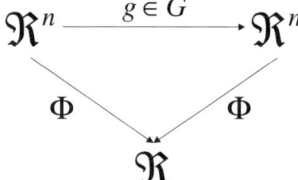

where the horizontal arrow denoted by $g$ is the induced map $\phi_g : \mathbb{R}^n \to \mathbb{R}^n : p \mapsto \phi(g, p)$.

With respect to how this linear action is used in applications, view $p$ as the object variables (i.e., 3D coordinates), and $q = \rho(gp)$ as the image variables (i.e., 2D coordinates), where $\rho$ is a projection from 3-space to 2-space, e.g.,

$$\rho = \begin{bmatrix} 1 & 0 & 0 \\ 0 & 1 & c \end{bmatrix}$$

and where $c$ is a fixed parameter. The variable $q$ is embedding into 3-space— thus enabling standard invariant theoretic techniques. Embedding $\rho \hookrightarrow \bar{\rho}$, where

$$\bar{\rho} = \begin{bmatrix} 1 & 0 & 0 \\ 0 & 1 & c \\ 0 & 0 & 1 \end{bmatrix}$$

and noting that this gives an automorphism of $G = \text{GL}_3(\mathbb{R})$ allows one to write $\bar{q} = (\bar{\rho}g)p = \bar{g}p$. Using the property characterizing an invariant function, $\Phi(\bar{g}p) = \Phi(p)\ \forall p \in M\ \ \forall \bar{g} \in G$, it follows $\Phi(\bar{q}) = \Phi(p)$. This equation provides the relationship between the "object" variables $p$ and the "image" variables $q$. Thus the O-I relation is $f(p,q) = \Phi(\bar{q}) - \Phi(p) = 0$. Computationally, this idea can be implemented by (1) finding the invariants of the given group action, and (2) "equating invariants" $\Phi(\bar{q}) = \Phi(p)$, and (3) eliminating any artificial components associated with the embedding (in the example, an artificial component to $\bar{q}$ was introduced, namely the third component). Note that the invariants found in step one have already eliminated the group parameters from the second step. It should be noted that there are a set of fundamental invariants $\Phi_i$ $i = 1, \ldots, k$ for each group action. Hence in practice there is a set of O-I relations $\Phi_i(\bar{q}) - \Phi_i(p) = 0$ for $i = 1, \ldots, k$. The technique of Lie group analysis makes determination of the invariants relatively simple. This directly avoids the conceptually simple but often computationally difficult task of eliminating the group parameters and camera parameters directly.

The latter approach naturally lends itself to consideration of covariant functions. A covariant involves two actions. The figure below characterizes the definition of a covariant function $\Phi$ under the two actions $\phi: G \times \mathbb{R}^n \to \mathbb{R}^n$ and $\psi: G \times \mathbb{R} \to \mathbb{R}$

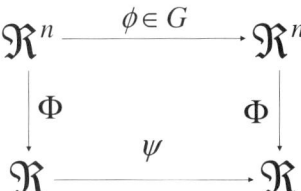

where the two horizontal arrows correspond to the two induced maps $\phi_g : \mathbb{R}^n \to \mathbb{R}^n : p \mapsto \phi(g, p)$ and $\psi_g : \mathbb{R} \to \mathbb{R} : q \mapsto \psi(g, q)$. Only nontrivial choices of $\Phi$, $\phi$, and $\psi$ are interesting. Choosing $\psi$ as the identity is equivalent to absolute invariants. As with invariants, the goal is to find a basis from which all other covariants can be written. Vector-valued covariants, $\Phi : \mathbb{R}^n \mapsto \mathbb{R}^m$, are a fairly simple extension that will be used below. The desire is to find a covariant with a small $m$.

The defining property of a covariant function $\Phi$ is $\Phi(gp) = g\Phi(p)$ $\forall p \in M$ $\forall g \in G$. The covariants are constructed into a set of O-I relations $\Phi_i(gp) = g\Phi_i(p)$ for $i = 1, \ldots, k$. Similar to the case with invariant O-I relations, any artificial components associated with the embedding must be eliminated. Note that the group parameters have not been eliminated. The benefit of using covariants is that it allows one to explicitly solve (estimate) for the group transformation to move back to the standard position. This advantage follows from the discussion on alignment versus voting techniques in Section 3.3.3. Similarly, noise analysis is easier to perform in the original parameter space.

Covariants arise with the rigid group, $E_n = SO_n \ltimes \mathbb{R}^n$, acting componentwise on the product space $\prod_{i=1}^m \mathbb{R}^n$. Consider the case $n = 2$. The two group actions are

$$\phi : (SO_2 \ltimes R^2) \times \prod_{i=1}^m \mathbb{R}_i^2 \to \mathbb{R}^2$$

$$: ((\mathbf{A}(\theta), b), \{p^i\}_{i=1}^m) \mapsto \{\mathbf{A}(\theta)p^i + b\}_i^m$$

and
$$\mu : (SO_2 \rtimes R^2) \times \mathbb{R}_i^2 \to \mathbb{R}^2$$
$$: ((\mathbf{A}(\theta), b), q) \mapsto \mathbf{A}(\theta)q + b,$$

where
$$\mathbf{A} = \begin{bmatrix} \cos(\theta) & -\sin(\theta) \\ \sin(\theta) & \cos(\theta) \end{bmatrix},$$
$$\mathbf{b} = \begin{bmatrix} t_x \\ t_y \end{bmatrix},$$

and
$$\mathbf{p}^i = \begin{bmatrix} x^i \\ y^i \end{bmatrix}.$$

For brevity, denote $\mathbf{p} = \{\mathbf{p}^i\}_{i=1}^m$.

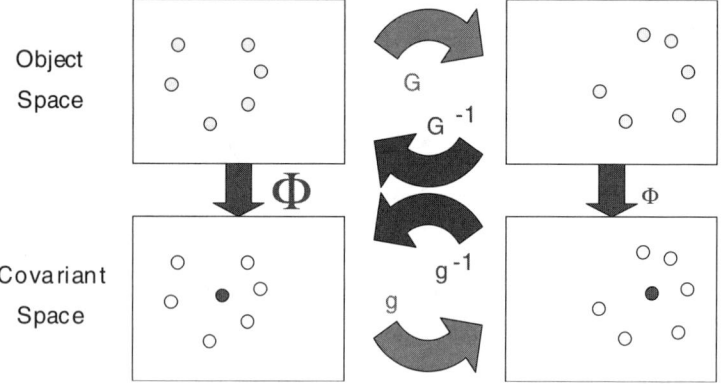

**Figure 3.6.** Transformations in the object space induce transformations in the covariant space. The top row, from left to right, represents points in object space being translated and rotated by $G$. The bottom row shows the covariant (the dark point in the middle; the object points are shown for reference) representation, specifically the centroid for this example. From left to right, the centroid is transformed by $g$, the action induced by $G$, the object transformation. The inverse $g^{-1}$ is readily calculated, and by applying $G^{-1}$ to the object it can be returned to its canonical coordinate system.

This system gives the covariant function "centroid"
$$\Phi = \frac{1}{m} \left\{ \begin{matrix} \sum_{i=1}^m x^i \\ \sum_{i=1}^m y^i \end{matrix} \right\}.$$

An example is shown in Fig. 3.6. This result easily generalizes to 3D.

Similarly it can be shown that the eigenvectors are covariant under rotation in 3D. These two covariants form the basis of the object recognition algorithm to be described subsequently.

## Advantages of Covariants

The advantage of using a covariant-based method is that for a large number of points the transformations can still be computed efficiently. Whereas invariants require searching $\binom{p}{r}$ combinations of features, where $p$ is the number of image points and $r$ is the number of features required to compute the invariants, the covariant-based technique scans the image (a neighborhood is defined by a sphere) thereby avoiding the combinatorial problem. Therefore, the covariant-based approach is computationally more efficient and more robust to noise. Also, as previously mentioned, the covariant-based approach is conducive to alignment versus voting techniques. Finally, noise analysis is easier to perform in the original parameter space.

## Covariants in the Algorithm

Covariants are simply an (improved) approach to achieving invariance. Specifically, calculating covariants is an intermediate step toward transforming the selected data into a canonical coordinate system. To summarize, covariants appear in two places in the algorithm that follows:

1. *Translation: A covariant of translation is the centroid.*
2. *Rotation: The eigenvectors are covariant with respect to rotation.*

### 3.5.5 Articulation Invariants?

Invariance can be used to reduce the articulation problem. When considering invariants, the imperative question is "what transformations should the function be invariant with respect to?" For instance, when the same object is viewed from different viewpoints, invariance with respect to the viewpoint transformation is useful. In ladar range images, the viewpoint invariants are the rigid invariants (in a rectangular coordinate system). This is a well-defined group of transformations and it applies to any (static) object in the ladar range images, i.e., the transformation group is independent of the object. Thus, the viewpoint invariants can be applied generically to all objects.

When it comes to articulation this is no longer the case. Each object has different articulation degrees of freedom, i.e., a different transformation group. An object's DoF (and therefore its transformation group) cannot be determined from a single image. Therefore, while it is mathematically possible to find articulation invariants for each individual object, generic articulation invariants that apply to all objects do not exist. Lacking articulation invariants, the goal is to turn as many of the articulation DoF into generic viewpoint DoF as possible. The remaining DoF are parameterized and used to define a manifold with respect to the canonical coordinate system.

### 3.5.6 Dividing the Object

To simplify the articulation problem, viewpoint invariants are applied to parts of the object (subobjects). To avoid explicit segmentation, these smaller parts are not necessarily the "functional" object parts. They are arbitrary sections of the object as partitioned by a sphere of a certain center and radius. For example, one sphere may contain part of the body of a backhoe, and another may contain a joint of a backhoe's arm. The sphere that contains a rigid body part has viewpoint DoF but no articulation. The sphere that contains a joint has both viewpoint and articulation DoF. However this joint has only one articulation parameter, namely the angle between the two segments of the arm. All other DoF of these arm segments have been turned into viewpoint DoF. In other words, using this approach, the articulation of each arm segment is independent of the backhoe's body. The joint within each sphere is viewed as a separate object that is seen from an unknown viewpoint. Consequently dividing the backhoe into subobjects reduces the number of articulation DoF from ten to usually at most two. Viewpoint invariance methods can now be used for each subobject such as the joint, obtaining invariants that depend only on the angle of the joint. These invariants are a smooth function of the angle.

A major advantage of the object division is the use of so-called "global" invariants of each subobject. A global invariant is a function that depends on the entire subobject rather than on isolated features such as points or lines, i.e., a global invariant is a function of all the voxels within the sphere. This achieves two purposes: (a) it avoids the problem of feature extraction, with the high sensitivity associated with feature-based methods; (b) it avoids calculating global invariants of the whole object, which would be sensitive to occlusion and missing parts.

### 3.5.7 Method Summary

1. Divide each modeled object into subobjects.
2. Find global viewpoint (rigid) invariants for each subobject as functions of its articulation parameters.
3. Use these invariant manifolds for storage, matching, and identification.

A brief description of these steps follows in the next section.

## 3.6 Implementation

### 3.6.1 Object Division

Each modeled object is divided into parts that are contained within spheres. Spheres are utilized since they are preserved under rotation. An important question is how to choose the spheres' centers and radii. Ideally all possible centers and radii would be used, but of course this is infeasible. Consequently, a finite set of radii starting from the biggest, containing the whole object, to the smallest, in which the data looks planar, are selected. This results in the description of the object on different scales. For each radius a set of centers is chosen. Sufficient numbers of centers are necessary to describe an object uniquely at a given scale. Although the object is 3D, its visible range image is described as a 2D array of ranges. Currently, a rectangular grid is draped over this array to describe where the sphere centers will be placed. Nominally, the grid spacing is some integer subsampling of the array at some fraction of the radius. The sphere centers are placed at the grid coordinates, $\{\theta, \phi, \rho\}$. Obviously the division is dependent on the image since each image will have a different grid and it is different from the grid on the image stored in the database. However, the invariant functions calculated on each sphere vary smoothly from one center to another. Consequently, it is easy to interpolate between grid points when matching is performed.

### 3.6.2 Finding Invariants

Once the object has been partitioned into (generally overlapping) spheres, the invariants of these subobjects are calculated. Note that invariants of a 3D object as a whole cannot be used since only partial views of the object are visible to the range sensor. Thus, at a minimum, a few different views of each object, such as front, back, and sides are required. The views are chosen such that any other view has the same invariants as one of these views. Thus, invariants are calculated for representative views, at various scales, and stored in a database. These are used as reference "models."

There are several ways to calculate such invariants. Ideally, the ones chosen are the least sensitive to changes in the boundaries of the subobjects, resulting from changing the radius or center of the sphere. Currently, the covariants explained in Section 3.5.4 are used to transform the sphere into a canonical, or standard, coordinate system. Specifically, the centroid of the subobject is the canonical origin, and the eigenvectors form the canonical axes. The new origin and axes are independent of the viewpoint; therefore the new coordinate system is viewpoint invariant. See Figure 3.7 for a simplistic example.

Transforming the subobject into this system, the grid point (sphere center) now has new coordinates that are invariant since they are given in the invariant coordinate system. Hence for each subobject (or sphere) three invariants are extracted, namely the 3D coordinates of the sphere's center in the invariant coordinate system. By using invariants to describe every grid point, the full description of the object is invariant. This description does not depend on point features and is insensitive to occlusion. A noteworthy complication is the affect of image discretization. This approach assumes planar patches connect the data points in order to facilitate the necessary integration and normalization to marginalize the affects of discretization [3].

**Figure 3.7.** A 2D analogy to the ellipsoid fit. The dashed circle determines the data to be fitted by the ellipse. The axes of the ellipse (eigenvectors) define a canonical coordinate system. Thus any point $X$ within the sphere is invariant in these coordinates.

The current approach and results shown here are based on this simple choice of invariants. This choice was made to provide maximum robustness to low numbers of pixels-on-target. As previously mentioned, all of the points are invariant once they have been expressed in the canonical coordinate system. The results are excellent; however, further discrimination and robustness would be achievable by using more information about the data. Obvious alternative approaches include using the spin images, spherical harmonics, or moments of the "sphere of data." Ultimately, the chosen method should include a shape metric, such as the Procrustes metric presented in Section 3.4.2.

### 3.6.3 Indexing

The invariants found above are functions of the articulation parameters. Denoting the invariants of each grid point by a vector $\mathbf{x}$ and the articulation parameters by a vector $\mathbf{u}$, the invariants are $\mathbf{x}(\mathbf{u})$. Indexing amounts to inverting these functions, i.e., given the invariant coordinates $\mathbf{x}$ in the image, find the articulation parameters $\mathbf{u}(\mathbf{x})$. To do that, the above relations are represented as a surface in a hyperspace, namely $f(\mathbf{x}, \mathbf{u}) = 0$. To build this surface, the articulation parameters are varied, and for each vector $\mathbf{u}$ all the corresponding vectors $\mathbf{x}$ are found. In this hyperspace, the voxels lying on the surface $f$ are marked by 1 and all other voxels are left as 0. This is a digital representation of the hypersurface. This is done off-line for every model in the database. The functions for all models are thus represented in the hyperspace. Thus indexing has been obtained such that given the invariant coordinates $\mathbf{x}$, the corresponding models can be found with articulations $\mathbf{u}$. This can be done by intersecting all the hypersurfaces with the hyperplane $\mathbf{x} = $ constant. For most points $\mathbf{x}_i$ there will be relatively few corresponding models, since most models do not go through all points in the hyperspace even with articulation. Thus the indexing space is rather sparse.

### 3.6.4 Matching

Given an image, the algorithm should match it to the closest model in the database. An initial scale is picked, which fixes the spheres' radii. Then the

Chapter 3 Three-Dimensional Laser Radar Recognition Approaches    109

invariant coordinates $x_i$ are computed at each grid point as described above. Note that the grid used for the matching step can be much more sparse than the grid used to build the database. The next step is to find a model in the database with an articulation $u$ that has the same invariant spatial coordinates $x_i$. The algorithm starts with the invariant coordinates $x_1$ of one point of the given range image.

For the given point $x_1$, the surfaces $u(x_1)$ for all models in the neighborhood of this point are extracted. The whole hypersurface need not be extracted, only the portion in the neighborhood of $x_1$ is necessary. Several techniques make it possible to reconstruct the surface in the neighborhood of $x_1$ although the original surface was constructed using a slightly different grid [3]. Next, the intersection of all the model surfaces with the hyperplane $x_1 = $ constant is computed. This provides a list of all models with all articulation parameters $u_j$ that match the given point $x_1$. The process is repeated for another image point $x_2$. This new point will have a different set of matching models. Continuing with other points $x_i$, all the models (and articulations) are collected in a voting table. Each additional point $x_i$ will contribute votes to certain models. The models with the most votes will be the best candidates for possible further verification.

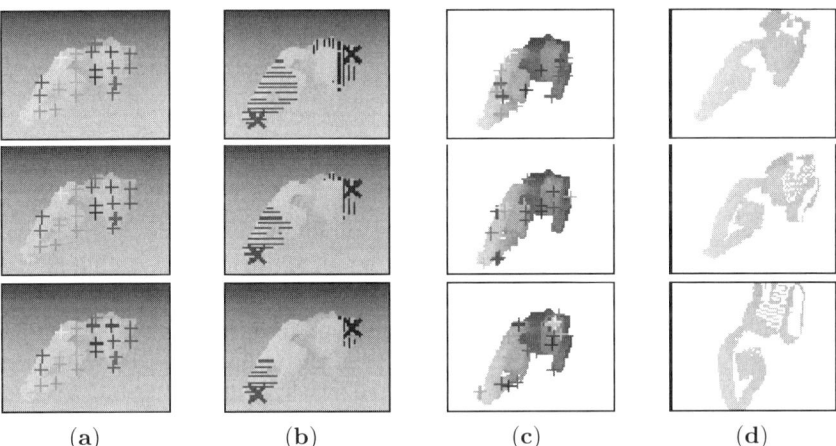

(a)    (b)    (c)    (d)

**Figure 3.8.** Matching of a backhoe. In each row matching is done using a progressively smaller sphere radius. Column (**a**) shows crosses at the spheres' centers. Column (**b**) shows representative subobjects (hatched areas). Column (**c**) shows crosses on a matching model. The invariant coordinates match in the neighborhood of each cross. Column (**d**) is a projection of the model to match the images' pose.

## 3.7 Conclusions

The current algorithm, as presented, efficiently handles nine of the ten primary technical challenges. These include translation, rotation, surface projection, surface resampling, varying numbers of pixels-on-target, point correspondence, obscuration, articulation, and fidelity. The final technical challenge, unknown objects, is partially achieved.

The extensive verification step would reject unknown objects, but to do this reliably requires a metric. Ideally, the verification will contain more information than the location of the grid point in the canonical coordinate system. Substantially different surfaces could produce the same grid point, so shape metrics are required to remove this possibility and further refine the indexing procedure.

The major contribution of this work is an approach that simultaneously addresses segmentation, recognition, and articulation in an efficient manner. The efficiencies are achieved by decomposing the image into subobjects, applying invariants, using a multiresolution decomposition, and hypothesis voting. The final algorithm explicitly avoids segmenting objects from the background, reduces the articulation parameters to a small number that are found within small subobjects, and encapsulates the benefits of a scanning-based approach (i.e., it is not combinatoric). This approach has demonstrated robustness in experiments.

## 3.8 Future Research

Future research includes developing and testing an alignment-based technique for comparison with the current voting technique. Future goals include developing an improved data comparison technique that has the desirable properties of metrics. However, this will in turn require readdressing the questions of point correspondence, surface resampling, and varying numbers of pixels-on-target. Ultimately, the goal is to minimize the overall computational cost for the same performance. The tradeoff is computational complexity versus saliency of the invariant features.

Quantifying and qualifying the affects of noise on the system have not been studied extensively, and this is necessary in order to understand the robustness of the system. Initial results have been promising. At the lowest level, this would include estimating the confidence region of the range and angular measurements from the ladar (this is dependent upon the range and the relative surface orientation). Finally, a large-scale assessment of the algorithm on both background and modeled objects remains to be completed.

## 3.9 Acknowledgements

This work was supported in part by the U.S. Air Force Office of Scientific Research under laboratory tasks 93SN03COR and 00MN01COR. The authors also acknowledge and appreciate the input of Dr. Isaac Weiss and Dr. Manjit Ray for their description of the algorithm and Figure 3.8. Dr. Peter Stiller has been instrumental in forming the viewpoint expressed in this chapter.

## References

[1] Jelalian, A.V.: Laser Radar Systems. Artech House (1992)
[2] Ross, T.D., Bradley, J.J., Hudson, L.J., O'Conner, M.P.: SAR ATR — so what's the problem? — an MSTAR perspective. In Zelnio, E., ed.: Proceedings SPIE International Conf. Algorithms for Synthetic Aperture Radar Imagery VI, Bellingham, WA, SPIE—The International Society for Optical Engineering (1999)
[3] Ray, M.: Model-based 3D Object Recognition using Invariants. PhD thesis, University of Maryland, College Park (2000)
[4] Small, C.G.: The Statistical Theory of Shape. Springer-Verlag New York (1996)
[5] Papadimitriou, C.H., Stieglitz, K.: Combinatorial Optimization: Algorithms and Complexity. Prentice-Hall (1982)
[6] Ettinger, G.J., Klanderman, G.A., Wells, W.M., Grimson, W.E.L.: A probabilistic optimization approach to SAR feature matching. In Zelnio, E., Douglass, R., eds.: Proceedings SPIE International Conf. Volume 2757 of Algorithms for Synthetic Aperture Radar Imagery III., Bellingham, WA, SPIE—The International Society for Optical Engineering (1996) 318–329
[7] Hung-Chih, C., Moses, R.L., Potter, L.C.: Model-based classification of radar images. IEEE Transactions on Information Theory **46** (2000) 1842–1854
[8] van Wamelen, P.B., Li, Z., Iyengar, S.S.: A fast expected time algorithm for the 2D point pattern matching problem. see http://www.math.lsu.edu/~wamelen/publications.html (2002)
[9] Maciel, J., Costeira, J.P.: A global solution to sparse correspondence problems. IEEE Transactions on PAMI **25** (2003) 187–199
[10] Jones III, G., Bhanu, B.: Recognition of articulated and occluded objects. IEEE Transactions on PAMI **21** (1999) 603–613
[11] Hetzel, G., Leibe, B., Levi, P., Schiele, B.: 3D object recognition from range images using local feature histograms. In: Proc. IEEE CVPR. Volume 2., Seattle, WA, IEEE Computer Society (2001) 394–399
[12] Campbell, R.J., Flynn, P.J.: A survey of free-form object representation and recognition techniques. Computer Vision and Image Understanding **81** (2001) 166–210

[13] Arman, F., Aggarwal, J.K.: Model-based object recognition in dense-range images: A review. ACM Computing Surveys **25** (1993) 5–43
[14] Bhanu, B., Ho, C.C.: CAD-based 3D object representation for robot vision. Computer **20** (1987) 19–35
[15] Piegl, L.A., Tiller, W.: The Nurbs Book (Monograph in Visual Communications). 2nd edn. Springer-Verlag (1997)
[16] Fan, T.J., Medioni, G., Nevatia, R.: Recognizing 3D objects using surface descriptions. IEEE Transactions on PAMI **11** (1989) 1140–1157
[17] Solina, F., Bajcsy, R.: Recovery of parametric models from range images: The case for superquadrics with global deformations. IEEE Transactions on PAMI **12** (1990) 131–147
[18] Binford, T.O., Levitt, T.S.: Quasi-invariants: Theory and exploitation. In Firschein, O., ed.: DARPA Image Understanding Workshop Proceedings, Washington DC, Morgan Kauffman (1993) 819–830
[19] Jain, A.K., Duin, R.P.W., Mao, J.: Statistical pattern recognition: A review. IEEE Transactions on PAMI **22** (2000) 4–37
[20] Moses, Y., Ullman, S.: Limitations of non-model-based recognition schemes. A.I. Memo 1301, Massachusetts Institute of Technology (1991)
[21] Jacobs, D.W., Alter, T.: Uncertainty progation in model-based recognition. A.I. Memo 1476, Massachusetts Institute of Technology (1994)
[22] Basri, R., Weinshall, D.: Distance metric between 3D models and 2D images for recognition and classification. A.I. Memo 1373, Massachusetts Institute of Technology (1992)
[23] Grimson, W.E.L., Huttenlocher, D.P., Jacobs, D.W.: Affine matching with bounded sensor error: A study of geometic hashing and alignment. A.I. Memo 1250, Massachusetts Institute of Technology (1991)
[24] Stockman, G.: Object recognition and localization via pose clustering. Computer Vision Graphics Image Processing **40** (1987) 361–387
[25] Huttenlocher, D.P., Ullman, S.: Object recognition using alignment. In: Proc. 1st Int. Conf. Computer Vision, Cambridge, MA, IEEE Computer Society (1987) 102–111
[26] Rigoutsos, I., Hummel, R.: A bayesian approach to model matching with geometric hashing. Computer Vision and Image Understanding **62** (1995) 11–26
[27] Lamdan, Y., Schwartz, J.T., Wolfson, H.J.: Affine invariant model-based object recognition. IEEE Trans. on Robotics and Automation **6** (1990) 578–589
[28] Sonka, M., Hlavac, V., Boyle, R.: Image Processing: Analysis and Machine Vision. 2nd edn. Brooks Cole (1998)
[29] Binford, T., Levitt, T., Mann, W.: Bayesian inference in model-based vision. In Kanal, L., Levitt, T., Lemmer, J., eds.: Uncertainty in AI, 3, Elsevier (1989) 920–925
[30] Hummel, R.A., Landy, M.S.: A statistical viewpoint on the theory of evidence. IEEE Transactions on PAMI **10** (1988) 235–247

[31] Haykin, S.: Neural Networks: A Comprehensive Foundation. 2nd edn. Prentice-Hall (1999)
[32] Duda, R.O., Hart, P.E., Stork, D.G.: Pattern Classification. 2nd edn. Wiley-Interscience (2001)
[33] Turk, M.A., Pentland, A.P.: Face recognition using eigenfaces. In: Proc. IEEE CVPR, New York, IEEE, Computer Society (1991) 586–591
[34] Pope, A.R., Lowe, D.G.: Probabilistic models of appearance for 3D object recognition. Int. Journal of Computer Vision **40** (2000) 149–167
[35] Dryden, I.L., Mardia, K.V.: Statistical Shape Analysis. John Wiley & Sons (1998)
[36] Johnson, A.: Spin Images: A Representation for 3D Surface Matching. PhD thesis, Robotics Institute, Carnegie Mellon University, Pittsburgh, PA (1997)
[37] Funkhouser, T., Min, P., Kazhdan, M., Chen, J., Halderman, A., Dobkin, D., Jacobs, D.: A search engine for 3D models. ACM Transactions on Graphics **22** (2003) 83–105
[38] Lo, C.H., Don, H.S.: 3D moment forms: Their construction and application to object identification and positioning. IEEE Transactions on PAMI **11** (1989) 1053–1064
[39] Flusser, J., Boldyš, J., Zitová, B.: Moment forms invariant to rotation and blur in arbitrary number of dimensions. IEEE Transactions on PAMI **25** (2003) 234–246
[40] Lele, S.R., Richtsmeier, J.T.: An Invariant Approach to Statistical Analysis of Shape. Chapman and Hall/CRC (2001)
[41] Kendall, D.G.: Shape manifolds, procrustean metrics, and complex projective spaces. Bulletin of the London Mathematical Society **16** (1984) 81–121
[42] Bookstein, F.L.: Size and shape spaces for landmark data in two dimensions (with discussion). Statistical Science **1** (1986) 181–242
[43] Arnold, G., Sturtz, K., Velten, V.: Similarity metrics for ATR. In: Proc. of Defense Applications of Signal Processing (DASP-01), Adelaide, AU, AFOSR (2001)
[44] Mumford, D.: Mathematical theories of shape: Do they model perception? In Vemuri, B.C., ed.: Proceedings SPIE, Geometric Methods in Computer Vision. Volume 1570., Bellingham, WA, SPIE—The International Society for Optical Engineering (1991) 2–10
[45] Csiszár, I.: Why least squares and maximum entropy? an axiomatic approach to inference for linear inverse problems. The Annals of Statistics **19** (1991) 2032–2066
[46] Santini, S., Jain, R.: Similarity measures. IEEE Transactions on PAMI **21** (1999) 871–883
[47] Wallis, G., Rolls, E.T.: Invariant face and object recognition in the visual system. Progress in Neurobiology **51** (1997) 167–194
[48] Weisstein, E.W.: The CRC Concise Encyclopedia of Mathematics. 2nd edn. CRC Press (2002) see http://www.mathworld.com.

[49] Arnold, G., Sturtz, K.: Complexity analysis of ATR algorithms based on invariants. In: Proceedings Computer Vision Beyond the Visible Spectrum (CVBVS), Hilton Head, SC, IEEE Computer Society (2000)
[50] Zhang, X., Zhang, J., Walter, G.G., Krim, H.: Shape space object recognition. In Sadjadi, F., ed.: Proceedings SPIE International Conf. Volume 4379 of Automatic Target Recognition XI., Bellingham, WA, SPIE–The International Society for Optical Engineering (2001)
[51] Levine, L., Arnold, G., Sturtz, K.: A label-invariant approach to procrustes analysis. In Zelnio, E., ed.: Proceedings SPIE International Conf. Volume 4727 of Algorithms for Synthetic Aperture Radar Imagery IX., Bellingham, WA, SPIE–The International Society for Optical Engineering (2002) 322–328
[52] Weiss, I.: Model-based recognition of 3D objects from one view. In: Proc. DARPA Image Understanding Workshop, Monterey, CA, Morgan Kauffman (1998) 641–652
[53] Weiss, I.: Noise-resistant invariants of curves. IEEE Transactions on PAMI **15** (1993) 943–948
[54] Rivlin, E., Weiss, I.: Local invariants for recognition. IEEE Transactions on PAMI **17** (1995) 226–238
[55] Stiller, P.F.: General approaches to recognizing geometric configurations from a single view. In: Proceedings SPIE International Conf., Vision Geometry VI. Volume 3168., Bellingham, WA, SPIE–The International Society for Optical Engineering (1997) 262–273
[56] Stiller, P.F., Asmuth, C.A., Wan, C.S.: Single view recognition: The perspective case. In: Proceedings SPIE International Conf., Vision Geometry V. Volume 2826., Bellingham, WA, SPIE–The International Society for Optical Engineering (1996) 226–235
[57] Stiller, P.F., Asmuth, C.A., Wan, C.S.: Invariants, indexing, and single view recognition. In: Proc. ARPA Image Understanding Workshop, Monterrey, CA (1994) 1432–1428
[58] Stuff, M.A.: Three dimensional invariants of moving targets. In Zelnio, E., ed.: Proceedings SPIE International Conf. Algorithms for Synthetic Aperture Radar Imagery VII, Bellingham, WA, SPIE–The International Society for Optical Engineering (2000)
[59] Stuff, M.A.: Three-dimensional analysis of moving target radar signals: Methods and implications for ATR and feature aided tracking. In Zelnio, E., ed.: Proceedings SPIE International Conf. Algorithms for Synthetic Aperture Radar Imagery VI, Bellingham, WA, SPIE–The International Society for Optical Engineering (1999)

# Chapter 4

# Target Classification Using Adaptive Feature Extraction and Subspace Projection for Hyperspectral Imagery

Heesung Kwon, Sandor Z. Der, and Nasser M. Nasrabadi

U.S. Army Research Laboratory, ATTN: AMSRL-SE-SE, 2800 Powder Mill Road, Adelphi, MD 20783  {hkwon,sder,nnasraba}@arl.army.mil

**Summary.** Hyperspectral imaging sensors have been widely studied for automatic target recognition (ATR), mainly because a wealth of spectral information can be obtained through a large number of narrow contiguous spectral channels (often over a hundred). Targets are man-made objects (e.g., vehicles) whose constituent materials and internal structures are usually substantially different from natural objects (i.e., backgrounds). The basic premise of hyperspectral target classification is that the spectral signatures of target materials are measurably different than background materials, and most approaches further assume that each relevant material, characterized by its own distinctive spectral reflectance or emission, can be identified among a group of materials based on spectral analysis of the hyperspectral data.

We propose a two-class classification algorithm for hyperspectral images in which each pixel spectrum is labeled as either target or background. The algorithm is based on a mixed spectral model in which the reflectance spectrum of each pixel is assumed to be a linear mixture of constituent spectra from different material types (target and background materials). In order to address the spectral variability and diversity of the background spectra, we estimate a background subspace. The background spectral information spreads over various terrain types and is represented by the background subspace with substantially reduced dimensionality. Each pixel spectrum is then projected onto the orthogonal background subspace to remove the background spectral portion from the corresponding pixel spectrum.

The abundance of the remaining target portion within the pixel spectrum is estimated by matching a data-driven target spectral template with the background-removed spectrum. We use independent component analysis (ICA) to generate a target spectral template. ICA is used because it is well suited to capture the structure of the small targets in the hyperspectral images. For comparison purposes a mean spectral template is also generated by simply averaging the target sample spectra. Classification performance for both of the above-mentioned target extraction techniques are compared using a set of HYDICE hyperspectral images.

## 4.1 Introduction

Progress in automatic target recognition (ATR) techniques is crucial to the rapid development of digital battlefield technologies. Targets are generally man-made objects whose constituent materials and internal structures are substantially different from natural objects (i.e., backgrounds). The differences in materials and appearance between the targets and backgrounds lead to distinctive statistical and structural characteristics in images.

A large number of target classification/detection techniques have been developed based on broadband passive forward-looking infrared (FLIR) sensors that sense electromagnetic radiation in the 3–5 or 8–12 $\mu$m bands [1]. A commonly used multiple-band CFAR (constant false alarm rate) detection algorithm for a known signal pattern was introduced in [2]. An adaptive feature fusion technique has been recently developed, in which several local properties of the targets and backgrounds were jointly exploited within a local dual window and then fused to integrate different local features [3]. Statistical approaches have also been studied, in which the local statistical characteristics of the target and background regions were exploited by estimating, e.g., a probability density function [4], a co-occurrence matrix [5], or an eigen transformation matrix [6]. However, most previously attempted IR-based target detection techniques fail to produce satisfactory performance in the presence of high thermal clutter backgrounds or camouflaged targets.

Hyperspectral imaging sensors have been used in many remote sensing applications, such as reconnaissance, surveillance, environmental monitoring, and agriculture, because they capture more spectral information than a broadband IR sensor. Spectral information is obtained through a large number (often over a hundred) of narrow contiguous spectral channels, each channel capturing electro-magnetic reflectance or emission within the corresponding spectral range. The basic premise of hyperspectral target classification is that the spectral signatures of target materials are measurably different than background materials, and most approaches further assume that each relevant material, characterized by its own distinctive spectral reflectance or emission signature, can be identified among a group of materials. Anomaly detectors assume only the first, while spectral matching algorithms assume the latter as well.

We describe a two-class classification algorithm in which each pixel is labeled either target or background. The hyperspectral images we use were generated with the HYDICE (HYperspectral Digital Imagery Collection Experiment) sensor whose spectral range spans from 0.4 to 2.5 $\mu$m. In this spectral range the signal energy is dominated by reflected solar radiation rather than photon emission. The operation of the HYDICE sensor is normally limited to the daytime and the sensor cannot distinguish between hot and cold targets. Targets in our HYDICE images are combat platforms whose spectral characteristics depend largely on paint put onto a surface of the target.

Based on the linear spectral mixing models discussed in [7], the algorithm assumes that a reflectance spectrum of each pixel is a linear mixture of constituent spectra from disparate material types present in the pixel area. If a pixel spectrum includes a target spectrum, the pixel is labeled a target pixel. The pixel is classified as the background, if the target spectral contribution within the pixel spectrum is zero. Consequently, decomposition of the pixel spectrum into the constituent spectra is required in order to determine whether the target spectrum is present within the corresponding pixel spectrum. In order to successfully unmix the pixel spectrum, the spectral characteristics of the constituent materials have to be known a priori or be obtained from the data. In the proposed algorithm we obtain spectral information of the material types directly from the data.

The reflectance spectra in a spatial neighborhood, even from pixels within the same class of materials, are not identical mainly because of material variations and different illumination conditions. In addition, the background spectra generally consist of the combined reflectance spectra from a number of different natural objects, such as trees, grass, water, soil, etc. In order to address the spectral variability and diversity of the background spectra, a background subspace model is used to integrate key spectral signatures that spread over the background areas, into a subspace with substantially reduced dimensionality. Principal component analysis (PCA) is used to reduce the dimensionality of the original background space [8, 9, 10]; a relatively small number of significant eigenvectors are used to generate the background subspace.

In the proposed algorithm a target spectral template is created based on sample spectra. The target spectral template is then matched with the target spectral portion of the input pixel spectrum. Two methods are used to generate the target spectral template. A target spectral template generated by averaging the target sample spectra will not closely represent the true spectral characteristics of the target. Instead, we use independent component analysis (ICA) [11, 12, 13] to generate the target spectral template. ICA is applied to a portion of the HYDICE training images containing a small number of targets. The associated basis vector of the independent component showing the highest target prominence, is considered to represent the spectral feature of the target.

The target spectral portion within each pixel spectrum is obtained by removing the background spectral portion from the pixel spectrum. The removal of the background is performed by projecting the pixel spectrum onto the orthogonal background subspace. The orthogonal subspace projection for the subpixel target classification was first introduced by Harsanyi and Chang [14]. The amount of target spectrum within each pixel spectrum is then estimated by measuring the correlation between the target spectral portion and the target spectral template obtained from the ICA method.

This chapter is organized as follows. The HYDICE imaging system is briefly introduced in Section 4.2. A linear spectral mixing model for subpixel target classification is described in Section 4.3. In Section 4.4 we present a brief introduction to ICA and ICA-based feature extraction for the hyperspectral data. In Section 4.5 an adaptive target classification algorithm based on orthogonal subspace projection is presented. Localized subspace estimation for the background is also introduced in Section 4.5. Two-class target classification results for the HYDICE images using the proposed algorithm are presented in Section 4.6. Conclusions are summarized in Section 4.7.

## 4.2 Hyperspectral Imaging System

Hyperspectral sensors operate in a variety of spectral bands, such as the visible, short-wave IR (SWIR: 1–2.5 $\mu$m), mid-wave IR (MWIR:3–5 $\mu$m), and long-wave IR (LWIR: 8–12 $\mu$m). The spectral range of the HYDICE sensors spans the visible and short-wave IR wavelengths where reflected solar radiation rather than object emission is the predominant source of energy. In the HYDICE imaging system, a group of spatially registered images, called a hyperspectral cube, is taken by an imaging spectrometer at a spectral range of 0.4 to 2.5 $\mu$m with a step size of 10 nm. The imaging spectrometer splits the reflected light (e.g., reflected sunlight and skylight) into narrow contiguous spectral channels, each channel generating an associated band image, as shown in Figure 4.1. Two-hundred-ten images are generated over the whole spectral range. For every pixel in the cube a spectral curve is then formed, which represents the spectral characteristics of the corresponding pixel, as shown in Figure 4.2.

## 4.3 Spectral Mixing Model

Airborne or tower-based hyperspectral imaging sensors normally view large areas on the ground, so the spatial resolution of the sensors is coarse in many applications. The area covered by each pixel in the hyperspectral images includes several constituent materials rather than a single material type. Consequently, a pixel spectrum is a mixture of the spectral signatures of the different constituent materials present. The detection (or classification) problem that deals with a mixed-pixel (subpixel) model has been intensively researched [7, 14, 15, 16]. We propose a target classification algorithm based on the spectral mixing model that estimates the abundances of the constituent spectra via spectral unmixing and matching to find whether the spectrum of the material of interest is present.

Our algorithm focuses on a two-class classification problem in which the pixel spectrum is considered a linear combination of individual spectra of two different material types — the target and background. In order to effectively address the spectral variability of the background spectra, we estimate a low-dimensional background subspace. The subspace is spanned by linearly

# Chapter 4 Target Classification Using Adaptive Feature Extraction

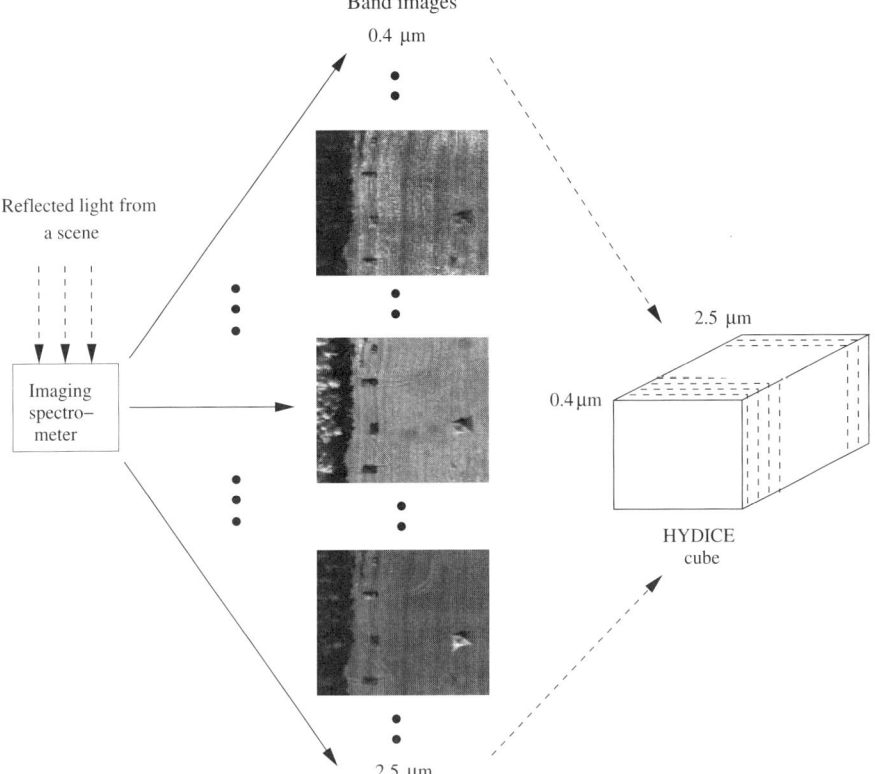

**Figure 4.1.** Hyperspectral imaging system

independent vectors $\mathbf{b}_1, \mathbf{b}_2, \ldots, \mathbf{b}_K$, which are the first $K$ eigenvectors of a background covariance matrix; $K$ is equal to the intrinsic dimensionality of the original background space.

The target spectrum varies, due mainly to the material variations and different illumination conditions. However, a single target spectral feature vector $\mathbf{s}_t$, an $M \times 1$ column vector, may still be capable of representing the target spectral characteristics in the data; $M$ is the number of the spectral bands. Based on [17], the pixel spectrum $\mathbf{x}$, an $M \times 1$ column vector, is given by

$$\mathbf{x} = \begin{cases} \mathbf{s}_t a + \mathbf{Bc} + \mathbf{n}, & \text{if } 0 < a < 1, \\ \mathbf{Bc} + \mathbf{n}, & \text{if } a = 0, \end{cases} \quad (4.1)$$

where $a$ is a scalar for target abundance, $\mathbf{B} = [\mathbf{b}_1 \mathbf{b}_2 \cdots \mathbf{b}_K]$ is an $M \times K$ matrix, $\mathbf{c} = (c_1, c_2, \ldots, c_K)^T$, $0 \leq c_1, c_2, \ldots, c_K < 1$, is a $K$-dimensional column vector whose components are the coefficients that account for the abundances of the corresponding endmember spectra $\mathbf{b}_1, \mathbf{b}_2, \ldots, \mathbf{b}_K$, and $\mathbf{n}$ is an $M$-dimensional column vector, representing Gaussian random noise. If $\mathbf{x}$ is

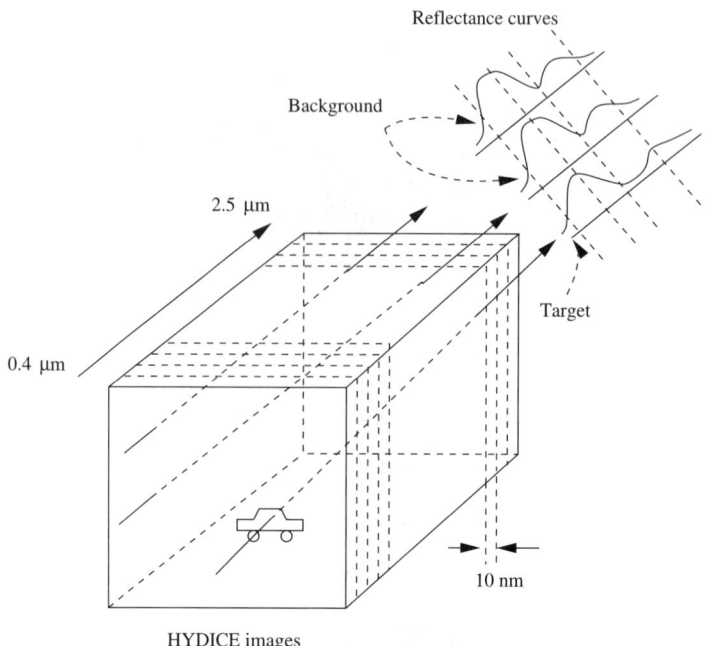

**Figure 4.2.** Creation of spectral reflectance curves from a hyperspectral cube.

the pixel spectrum from the target regions, the target abundance $a$ is a nonzero value. On the other hand, $a$ is zero if **x** is from the background region; in this case, **x** is a mixture of the background spectra only, as described in Equation (4.1). An accurate decision for the presence of the target spectrum within the pixel spectrum, therefore, can lead to successful classification of the individual pixel spectra.

The target feature vector $\mathbf{s}_t$ can be obtained from a spectral library or spectral samples in the given hyperspectral data. In the proposed algorithm the target feature is created from the sample spectra via an ICA-based feature extraction technique introduced in Section 4.4.

## 4.4 ICA-Based Feature Extraction

ICA, a relatively new concept in data representation, exploits high-order statistical dependencies among the data [11, 12, 13]. In this section, we introduce the ICA mixing and unmixing models and an ICA-based feature extraction technique. The feature extraction technique is based on the well-known ICA unmixing process [12], designed to find a feature vector that can closely represent the target spectral characteristics of the given data.

### 4.4.1 Independent Component Analysis (ICA)

Data representation in signal processing generally entails efficient removal of correlations present in the data. Principal component analysis (PCA) [8] is a well-known data transform technique that decorrelates the input data by exploiting pairwise second-order dependencies (e.g., covariance) and produces linearly independent variables, called principal components (or eigenvectors). Because the eigenvectors are only linearly independent from one another, non-linear (high-order) statistical dependencies of the input data — as can be easily observed from sparse structures in an image, such as edges and small areas with relatively high contrast — cannot be removed [18]. ICA is a generalization of PCA where the input data is expressed as a linear mixture of statistically independent components which are nonlinearly decorrelated [11]. The independent components are, therefore, highly nongaussian (e.g., the Laplace distribution) and well suited to represent the sparse structures of the input data. We use ICA to better capture targets scattered in HYDICE images, each target occupying a small number of pixels. Detailed information on ICA and its applications for signal and image processing can be found in [19, 11, 12, 13, 20, 21, 18].

Suppose we are given the data $X$, consisting of $M$ input images; the number of pixels in each image is $N$. Then $X$ can be represented by an $M \times N$ matrix $[\mathbf{x}_1 \mathbf{x}_2 \cdots \mathbf{x}_N]$, each row, a $N$-dimensional vector, representing the corresponding input image. Each column $\mathbf{x}_i = (x_1, x_2, \ldots, x_M)^T$ is an $M$-dimensional vector whose components are the pixel values of the $M$ images at the corresponding pixel location $i$. We denote by $S = [\mathbf{s}_1, \mathbf{s}_2, \ldots, \mathbf{s}_M]^T$ an $M \times N$ matrix, each row $\mathbf{s}_i^T$, an $N$-dimensional vector, representing the corresponding independent component image. The ICA mixing model is then defined by

$$X = AS, \qquad (4.2)$$

where $A = [\mathbf{a}_1, \mathbf{a}_2, \ldots, \mathbf{a}_M]^T$ represents the $M \times M$ mixing matrix, each row $\mathbf{a}_i^T$, an $M$-dimensional basis vector, representing mixing coefficients that account for the abundances of the independent components for the $i$th input image. Equation (4.2) indicates that any given input image can be expressed as a linear combination of the independent component images whose abundances are specified in the corresponding row of $A$. Figure 4.3 shows a simplified schematic diagram of the ICA mixing and unmixing processes. In order to calculate $A$, we first estimate the unmixing matrix $W$ of $X$ using the basic assumption of ICA — the statistical independence of the components. Let the unmixing matrix $W = [\mathbf{w}_1, \mathbf{w}_2, \ldots, \mathbf{w}_M]^T$ be the inverse of the mixing matrix $A$. Equation (4.2) can then be expressed as

$$WX = S. \qquad (4.3)$$

Figure 4.4 illustrates how matrix multiplications in Equation (4.3) are performed to unmix $X$. The unmixing process is basically a linear projection

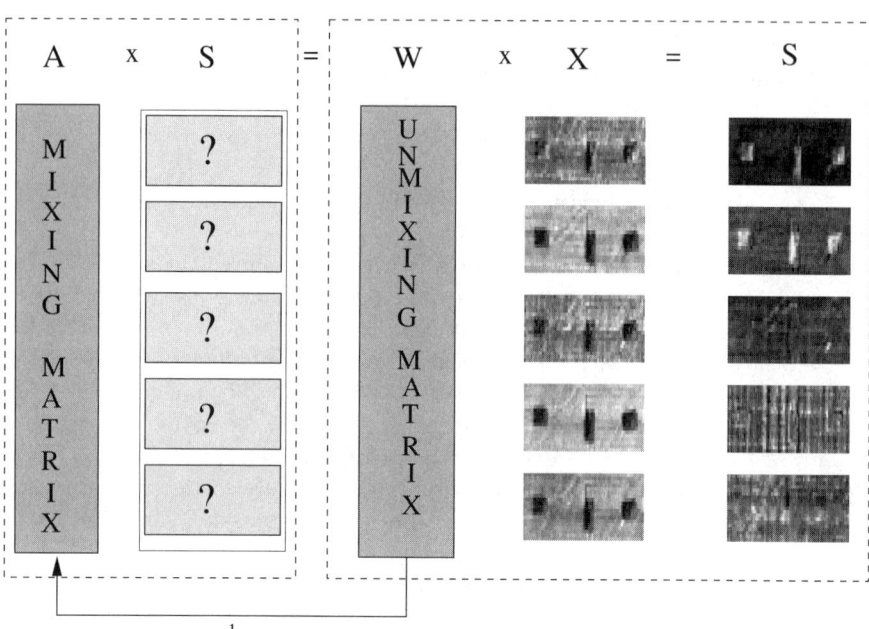

**Figure 4.3.** Simplified schematic diagram for the ICA mixing and unmixing processes

of $X$ onto $\mathbf{w}_i^T$, $i = 1, 2, \ldots, M$, which are the basis vectors (axes) in a new coordinate system represented by Equation (4.3). $\mathbf{w}_i^T$ are actually the orthogonal basis vectors, and the input images are realigned along $\mathbf{w}_i^T$ to create the corresponding ICA component images. The independent components should be nongaussian, otherwise ICA cannot be achieved [11]. Accordingly, the directions of $\mathbf{w}_i^T$ are set such that nongaussianity of the $i$th independent component $\mathbf{s}_i = \mathbf{w}_i^T X$ is maximized.

### 4.4.2 ICA-Based Target Feature Extraction

Sample reflectance spectra collected from the pixels in the target regions (the targets in the scene are assumed to be the same kind) can be used directly to estimate the target spectral signature, e.g., by averaging them. However, in practice the material variations and different illumination conditions result in a broad range of spectral variability, as shown in Figure 4.5. Therefore, the individual spectra from the target regions, even from the same region, could be different. In this context the arithmetic mean of the sample reflectance spectra is not an adequate representation of the target spectral signature. We use an ICA-based feature extraction technique that focuses especially on extracting

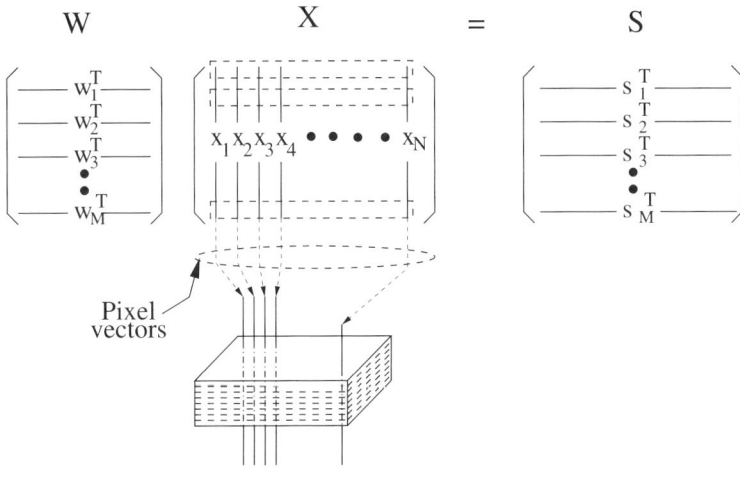

**Figure 4.4.** Matrix multiplications for the ICA unmixing process.

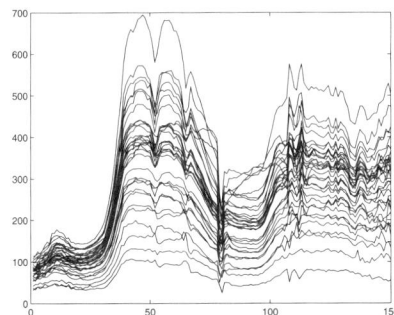

**Figure 4.5.** Example of spectral variability of the target reflectance spectra.

the independent components that form the target. In this technique a small hyperspectral cube is used to generate the target spectral feature vector.

Suppose we are given a relatively small hyperspectral cube $X = [\mathbf{x}_1 \mathbf{x}_2 \cdots \mathbf{x}_N]$, an $M \times N$ matrix, including the targets and the neighboring backgrounds. Normally, before unmixing $X$ two preprocessing steps need to be taken — dimensionality reduction and whitening. Both the steps are based on the eigenvalue–eigenvector factorization described in [11]. The number of independent sources present in the data is closely associated with the intrinsic dimensionality $L$ of the input data. $L$ can be successfully determined by finding an optimum subspace of the input data, which is equivalent to a problem of finding a number of eigenvectors to be retained. Important studies have been performed to determine $L$ based on probabilistic PCA models [9, 22]. We use the maximum likelihood PCA model introduced by Bishop [9] to calculate $L$. $L$ is normally far smaller than $M$; the input vectors in $X$ are the column vectors $\mathbf{x}_i = (x_1, x_2, \ldots, x_M)^T$, $i = 1, 2, \ldots, N$. $L$ has to be known a priori to decide how many independent components need to be created in the unmixing process.

The dimensionality reduction of $X$ is normally accomplished by projecting $X$ onto the first $L$ eigenvectors of the covariance matrix of $X$. Whitening is a process in which the input data are uncorrelated and their variances become unity. Both the dimensionality reduction and whitening are performed by

$$X_s = U^{-1/2} E^T X, \qquad (4.4)$$

where $X_s = [\mathbf{x}_1^s \mathbf{x}_2^s \cdots \mathbf{x}_N^s]$, an $L \times N$ matrix, is the transformed input data in the $L$-dimensional space, $E$ is an $L \times M$ matrix, representing the first $L$ eigenvectors, and $U$ is a diagonal matrix whose main diagonal elements are the eigenvalues of $E$. The corresponding independent components $S_s = [\mathbf{s}_1^s, \mathbf{s}_2^s, \ldots, \mathbf{s}_L^s]$ can be obtained by

$$W_s X_s = S_s, \qquad (4.5)$$

where $W_s = [\mathbf{w}_1^s \mathbf{w}_2^s \cdots \mathbf{w}_L^s]$ is the unmixing matrix in the $L$-dimensional space. Figure 4.6 illustrates the dimensionality reduction of the sample hyperspectral cube and the associated ICA unmixing process.

Each independent component image represents a set of corresponding projection values $\mathbf{s}_i = \mathbf{w}_i^T X$. The distributions of $\mathbf{s}_i$ are close to the Laplace distribution which has more pixels in tails and less pixels around the mean than the Gaussian distribution. Therefore, edges and small areas with high contrast are better represented by the independent component images. The structural details of the small targets in the sample hyperspectral cube are detected during the unmixing process, generating one or more ICA component images with greater target prominence (See the ICA component inside the rectangular box in Figure 4.6). The target prominence is measured by calculating the difference between the two values — the averages of the background pixels and target pixels in the ICA component image. As shown in Figure 4.6, the component images with higher target prominence include highly suppressed background where the mean and variation in pixel values are generally low. Once the component image with the highest target prominence $\mathbf{s}^{s*}$ is found, the corresponding $\mathbf{w}^{s*T}$, the unmixing vector onto which $X_s$ is projected to create $\mathbf{s}^{s*}$, is identified. The associated dewhitened vector of $\mathbf{w}^{s*}$ is then obtained by

$$\mathbf{g}^* = E U^{1/2} \mathbf{w}^{s*}. \qquad (4.6)$$

$\mathbf{g}_i$, $i = 1, 2, \ldots, L$, are the column vectors of the mixing matrix $A$ ($\mathbf{g}_i$ are shown in Figure 4.6). Each component of the $M$-dimensional vector $\mathbf{g}^*$ accounts for the abundance of $\mathbf{s}^{s*}$ in the associated band image. Therefore, $\mathbf{g}^*$ naturally represents the spectral characteristics of the target relative to those of the background in the given data. If the background spectral characteristics in

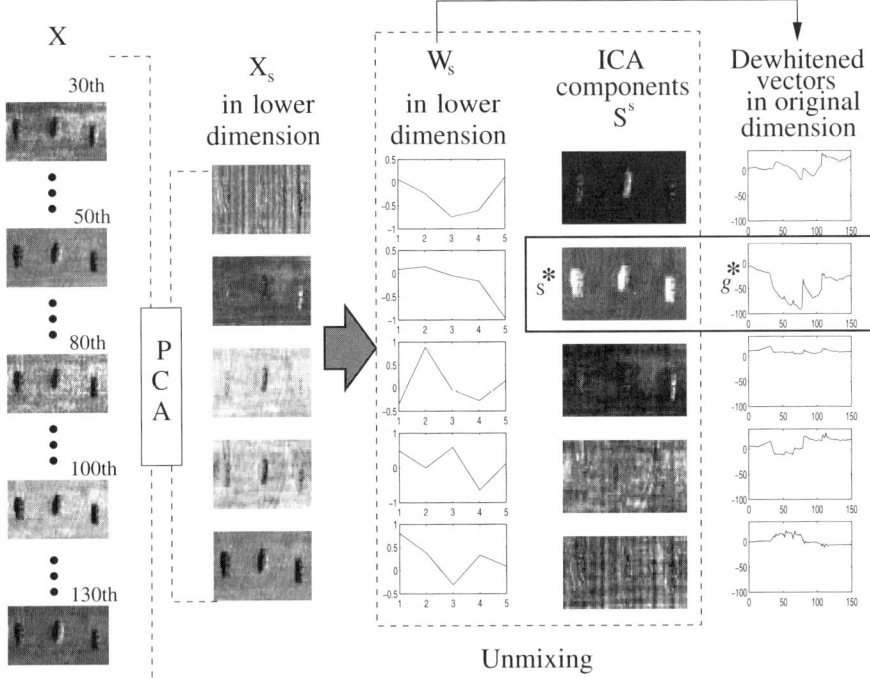

**Figure 4.6.** Schematic diagram of ICA-based feature extraction via the ICA unmixing process.

test images are not substantially different from those of the training images, we can apply $\mathbf{g}^*$ as a spectral template to the test images. If the targets are frequently situated among different background types, e.g., grass and soil, we generate multiple spectral templates, each template being adapted to the corresponding background type.

## 4.5 Subspace-Based Adaptive Target Classification

In this section, we estimate a low-dimensional subspace to address the background spectral variability. The subspace is spanned by a small number of linearly independent vectors (eigenvectors) directly estimated from the background spectra. Each pixel spectrum is projected onto the orthogonal background subspace to remove the spectral contribution due to the background materials. The target abundance of the pixel spectrum is then estimated by adaptive template matching.

### 4.5.1 Background Subspace Model

Suppose $\mathbf{x} = (x_1, x_2, \ldots, x_M)^T$ is an $M$-dimensional vector, representing a background reflectance spectrum. We denote by $C_\mathbf{x}$ the covariance matrix of the vector population of $\mathbf{x}$. $M$ is equal to the number of spectral bands. We estimate the low-dimensional subspace $\langle \mathbf{B} \rangle$ spanned by first $K$ eigenvectors $\mathbf{B} = [\mathbf{b}_1, \mathbf{b}_2, \ldots, \mathbf{b}_K]$ of $C_\mathbf{x}$; $K$ is equal to the dimension of the subspace ($K \ll M$), and is set such that $\mathbf{B}$ accounts for most of the energy of the original background space. The construction of $\langle \mathbf{B} \rangle$ ensures that the most of the spectral information from the input space is integrated into the compact subspace with greatly reduced dimensionality.

### 4.5.2 Orthogonal Subspace Projection and Adaptive Spectral Matching

Orthogonal subspace projection is a geometrical approach originally devised to determine the least squares approximations to inconsistent linear algebraic equations (or linear systems) [23]. It has also been used for hyperspectral target classification/detection applications [14, 16]. Because the pixel spectrum is assumed to be a linear mixture of the target and the background constituent spectra, and $\langle \mathbf{B} \rangle$ is spanned by the key eigenvectors of the background covariance matrix, we can estimate the background portion $\mathbf{Bc}$ within the pixel spectrum $\mathbf{x}$ by projecting $\mathbf{x}$ onto the background subspace $\langle \mathbf{B} \rangle$, as shown Figure 4.7. The spectral contribution by the target and noise $\mathbf{w}$ is then estimated by simply subtracting the background contribution $\mathbf{P}_B \mathbf{x}$ from $\mathbf{x}$:

$$\mathbf{x} - \mathbf{P}_B \mathbf{x} = (I - \mathbf{P}_B)\mathbf{x}, \tag{4.7a}$$
$$= \mathbf{P}_B^\perp \mathbf{x}, \tag{4.7b}$$

where $I$ is the identity matrix and $\mathbf{P}_B = \mathbf{B}(\mathbf{B}^T \mathbf{B})^{-1} \mathbf{B}^T$ is a background projection matrix, and $\mathbf{P}_B^\perp$ is an orthogonal projection matrix. For every pixel, adaptive template matching is performed to estimate the correlation between the target spectral contribution $\mathbf{P}_B^\perp \mathbf{x}$ within the pixel spectrum and the target spectral template $\mathbf{g}^{*T}$, which was obtained from Section 4.4.2:

$$F(\mathbf{x}) = \mathbf{g}^{*T} \mathbf{P}_B^\perp \mathbf{x}. \tag{4.8}$$

The pixels with higher values of $F(\mathbf{x})$ than a predefined threshold are then classified as the target.

### 4.5.3 Localized Background Subspace

The background subspace is estimated based on the sample spectra collected from various terrain types in the hyperspectral training images. Successful subspace modeling ensures robust and enhanced classification performance.

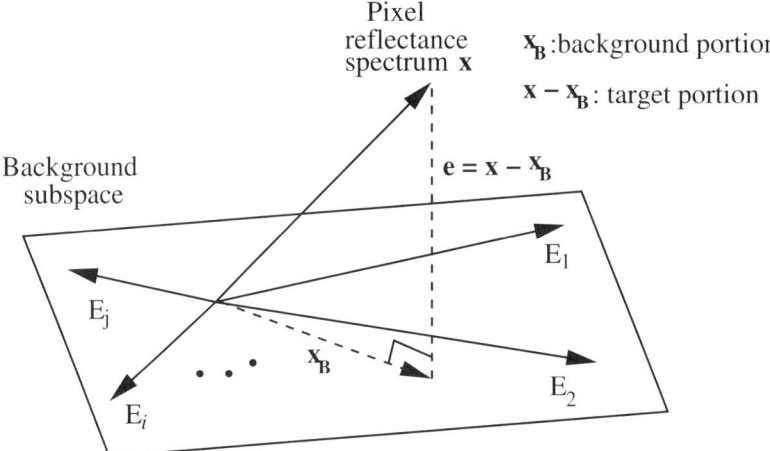

**Figure 4.7.** Orthogonal background projection.

Collecting the background spectra from the training images is normally performed by a human operator. Figure 4.8 shows the various background regions from which the sample spectra were collected. However, human intervention in a target classification process often results in inconsistent performance. It can be eliminated by replacing it with an automated procedure, in which the background subspace is estimated using the sample spectra directly collected from the hyperspectral sensor. In this section we introduce an unsupervised segmentation technique developed by the authors [24] to estimate the background subspace adapted to the local characteristics of the hyperspectral test images. In the technique the reflectance spectra in the test images are first segmented into two classes based on spectral dissimilarity.

The spectral dissimilarity is measured within a local window. A relatively large fixed-size local window is placed around each pixel. Note that each pixel corresponds to a pixel vector (spectrum) in the spectral domain. The spectral dissimilarity $d_i$ associated with pixel location $i$ is defined as

$$d_i = \frac{\sum_{j \in \mathcal{B}} \|\mathbf{s}_j - \mathbf{s}_i\|}{N_i}, \tag{4.9}$$

where $\|\cdot\|$ represents the Euclidean norm, $\mathcal{B}$ represents a set of pixels selected randomly from within the local window, as shown in Figure 4.9. The vectors $\mathbf{s}_i$ and $\mathbf{s}_j$ represent the corresponding pixels at locations $i$ and $j$, respectively, and $N_i$ represents the number of the selected pixels (the randomly selected pixels and neighboring pixels) required to estimate $d_i$. After the spectral dissimilarity of every pixel is obtained, a spectral dissimilarity image $D$ is formed, as shown in Figure 4.10. Each element of the image represents the amount of the average spectral difference between the center pixel and its randomly selected neighbors. The spectral dissimilarity image provides a set of spectral-

**Figure 4.8.** Sample background spectra from the hyperspectral images to estimate the sample-based background subspace; the samples were collected from the rectangular regions.

feature values $d_i$ suitable for clustering. The pixels in the targets regions tend to have greater values than those of the background pixels, because of the spectral difference between the two material types. The pixels associated with smaller $d_i$ are then identified as the background whose spectra are used to estimate the background subspace. Segmentation is performed by applying a simple thresholding method to the feature values in $D$ whose statistics are closely associated with the contents of the HYDICE images — mainly the complexity of the backgrounds. The threshold is calculated by averaging the mean and the maximum value of $D$.

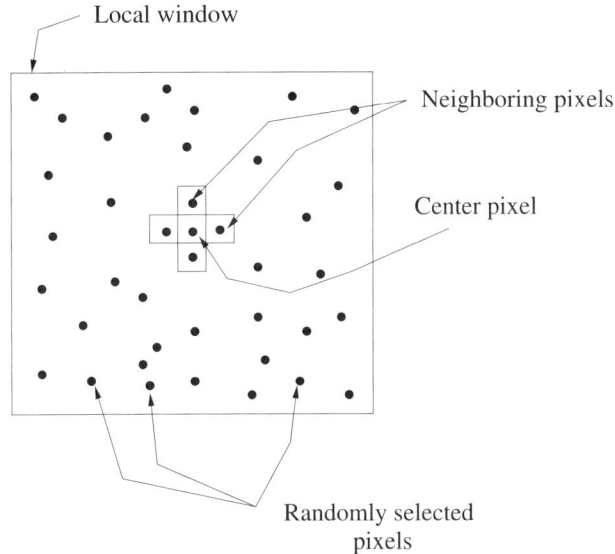

**Figure 4.9.** Randomly selected pixels and neighboring pixels within the local window for calculation of spectral dissimilarity.

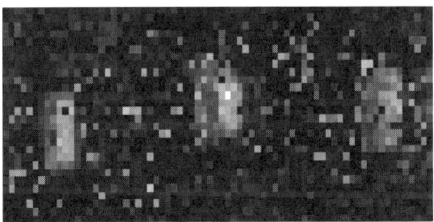

**Figure 4.10.** Spectral dissimilarity image $D$ of the small hyperspectral cube.

## 4.6 Experimental Results

In this section we apply the proposed ICA-based adaptive matching technique to the HYDICE images to detect targets of interest. The HYDICE imaging sensor generates 210 band images across the whole spectral range (0.4–2.5 $\mu$m), each band covering a narrow spectral range of 10 nm. We, however, use only 150 band images by discarding water absorption and low signal-to-noise ratio (SNR) bands; the band images used are the 23rd–101st, 109th–136th, and 152nd–194th. The low SNR bands were identified by human observation of sample spectra. The test cubes were provided with the associated ground truth maps. The coordinates of the centers of the targets in the ground truth map were compared with those of the detected targets to check if the classification was accurate.

### 4.6.1 Target Feature Extraction

Figures 4.11–4.13 show typical reflectance curves from the regions of the disparate material types, such as targets, trees, and grass. Even though a distinguishable spectral pattern can be found among a group of reflectance curves of the same material type, a wide range of spectral variability, prevailing in all the material types, hinders correct and accurate classification of the targets from the surrounding backgrounds. In particular, the reflectance curves from the target region, as shown in Figure 4.11, display substantial intraclass variability, showing how seriously material variations and different illumination conditions interfere with the hyperspectral target classification process.

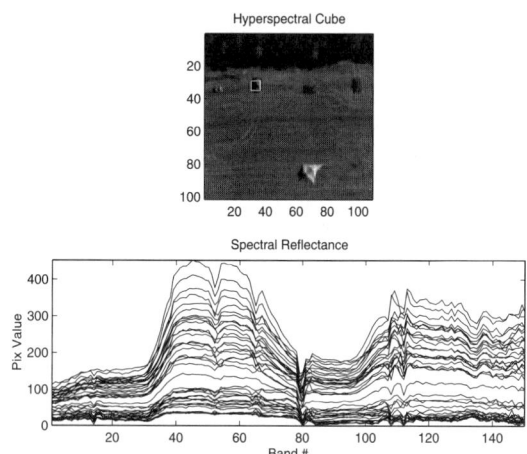

**Figure 4.11.** Spectral variability in the target region.

In order to estimate the target abundance within the individual pixel spectra, we generate a spectral feature vector, used as a correlation template in the classification process, to represent the target spectral characteristics in the given data. Two methods are used to create the spectral feature vector — mean curve generation and ICA-based feature extraction. We compare the target classification performance associated with the two feature generation methods. Figure 4.14 shows the sample target spectral reflectance curves and the corresponding mean spectral curve $\mathbf{m}_t^T$. The ICA-based spectral feature extraction technique, based on [20], uses a set of training images $X = [\mathbf{x}_1 \mathbf{x}_2 \cdots \mathbf{x}_N]$ which are usually a small-sized hyperspectral cube with the single background type, as shown in Figure 4.6. The intrinsic dimensionality $L$ of the input vectors $\mathbf{x}_i$ (the pixel spectrum vector) was obtained based on the maximum likelihood PCA model [9]; $L$ was 5 instead of 150, meaning that the column vector $\mathbf{x}_i^s$, $i = 1, 2 \ldots, N$, are in the 5-dimensional space, and there exists not more than five independent sources in the training im-

Chapter 4 Target Classification Using Adaptive Feature Extraction    131

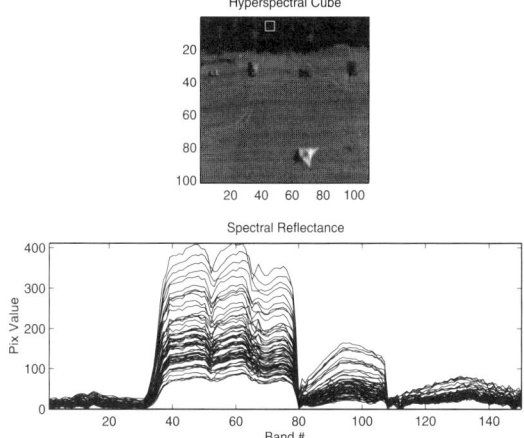

**Figure 4.12.** Spectral variability in the tree region.

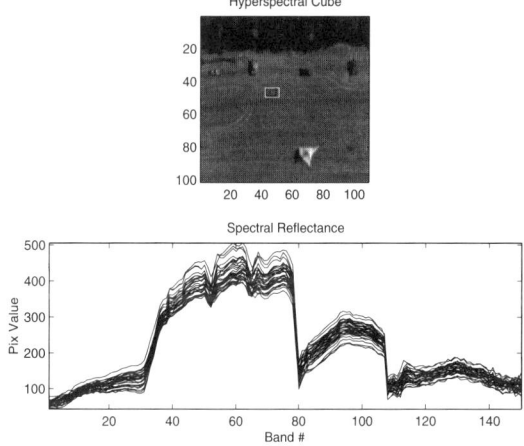

**Figure 4.13.** Spectral variability in the grass region.

ages. Figure 4.6 shows the training images $X$ and a schematic diagram of the ICA unmixing process, in which the training images are decomposed into the five ICA component images. In the unmixing process the first five principal component images $X_s$ were actually used to calculate the corresponding ICA component images. The direction of each of the five unmixing vectors $\mathbf{w}_i^{sT}$, $i = 1, 2, \ldots, 5$, was set such that nongaussianity of the corresponding ICA component $\mathbf{s}_i^s$ was maximized. The 150-dimensional vectors $\mathbf{g}_i$ were then obtained by first dewhitening $\mathbf{w}_i^s$ and then projecting them back to the original measurement space. The ICA component image with the highest target prominence $\mathbf{s}^{s*}$ and its associated basis vector $\mathbf{g}^{*T}$ are shown in Figure 4.15.

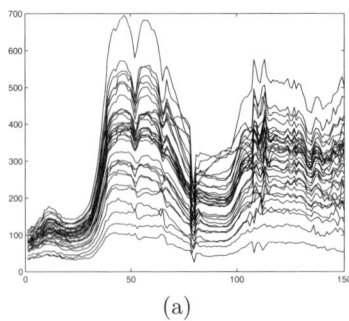
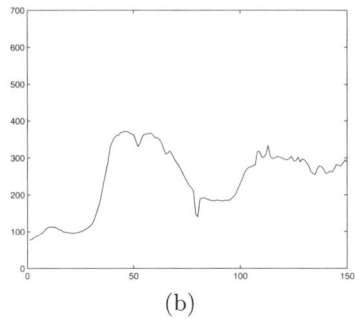

**Figure 4.14.** (a) Sample target reflectance curves and (b) the corresponding mean curve $\mathbf{m}_t^T$.

$\mathbf{g}^{*T}$ is used as a target spectral template in the target classification process

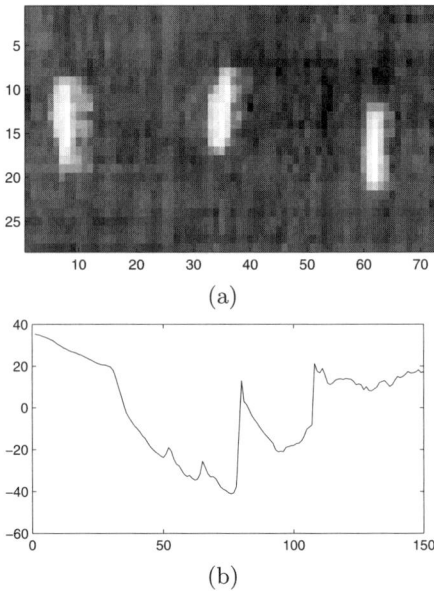

**Figure 4.15.** (a) ICA component image with the highest target prominence and (b) its corresponding basis vector $\mathbf{g}*^T$.

to measure the significance of the target contributions within the individual background-removed pixel spectra. The spectral shape of $\mathbf{g}^{*T}$ is quite different from that of $\mathbf{m}_t^T$. This is because the direction of $\mathbf{g}^{*T}$ is set such that it suppresses the background spectra, while emphasizing the target spectra. $\mathbf{m}_t^T$, in contrast, is generated only to detect the target spectra, regardless of the spectral characteristics of the backgrounds of the given hyperspectral data.

Chapter 4 Target Classification Using Adaptive Feature Extraction  133

### 4.6.2 Background Subspace Estimation

Figure 4.16 shows two hyperspectral band images, each from either of the two hyperspectral cubes, Cube I and II, to be used as test images for target classification. Cube I includes five targets (four vehicles and one man-made

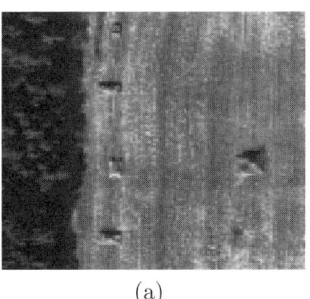
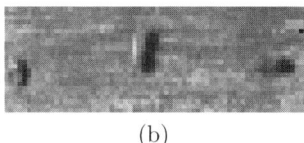

(a)  (b)

**Figure 4.16.** Sample band images from two hyperspectral test cubes, (a) Cube I and (b) Cube II.

object) and two different types of background (trees and grass). Cube II is smaller than Cube I in size and includes three targets (all of them are vehicles) and the single background type of grass.

We first estimate the background subspace to remove the background portion from every pixel spectrum in the test cubes. Two different techniques were used to estimate the background subspace: localized subspace estimation and sample-based subspace estimation. In order to estimate the localized subspace $\langle \mathbf{B}_l \rangle$, we first apply the unsupervised segmentation, described in Section 4.5.3, to the corresponding hyperspectral test cube. For each test cube, $\langle \mathbf{B}_l \rangle$ is spanned by the first eight eigenvectors of the covariance matrix of the pixel spectra in the regions classified as the background.

Figure 4.17 shows the spectral dissimilarity matrices and unsupervised segmentation results for the two test cubes. The size of the local sliding window was half the size of the test image. We used 80 randomly selected pixels and four neighboring pixels of the corresponding input pixel in the local window to estimate the spectral dissimilarity. A simple threshold-based technique was used to segment the dissimilarity matrices. The segmentation of Cube II was successful since most of the background areas were segmented out of the image. Cube I includes more diverse backgrounds than Cube II. In the segmentation of Cube I the area covered by grass was successfully classified as background, while most of the tree regions were not. This is mainly because of the irregular spectral reflectivity of the tree regions. Accordingly, for Cube I most of the background spectra used to estimate the localized subspace were collected from the grass areas.

The sample-based background subspace $\langle \mathbf{B}_s \rangle$ was estimated for both the test cubes based on the sample spectra collected from various types of the

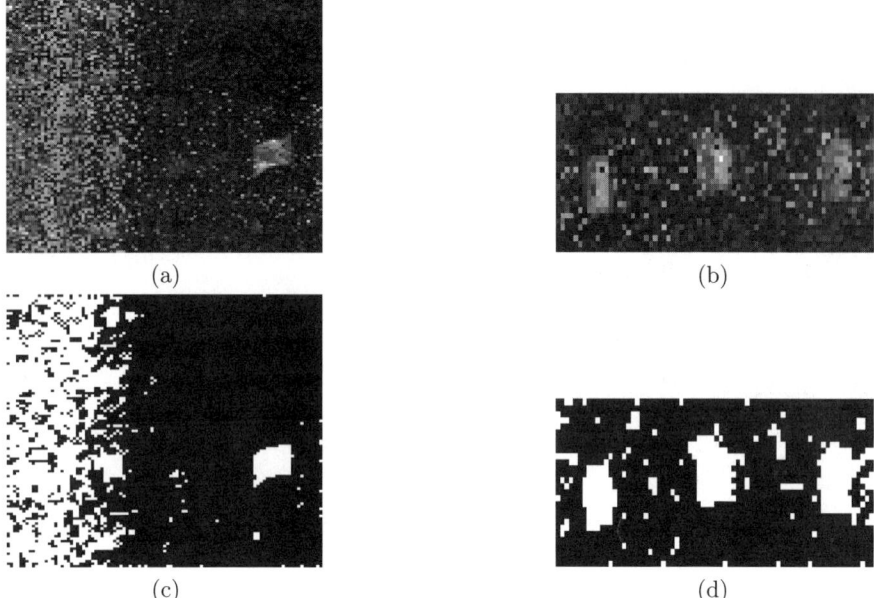

**Figure 4.17.** Spectral dissimilarity images for (a) Cube I and (b) Cube II and the corresponding segmentation results for (c) Cube I and (d) Cube II.

backgrounds in the hyperspectral training images (inside the rectangular boxes), as shown in Figure 4.8; the same number of eigenvectors used to estimate $\langle \mathbf{B}_l \rangle$ was used to estimate $\langle \mathbf{B}_s \rangle$. It should be noted that modeling of $\langle \mathbf{B}_s \rangle$ involves a human operation, incurring inconsistent classification performance.

Once $\langle \mathbf{B}_l \rangle$ and $\langle \mathbf{B}_s \rangle$ are estimated, the associated orthogonal projection matrixes $\mathbf{P}_{B_l}^{\perp}$ and $\mathbf{P}_{B_s}^{\perp}$, respectively, are readily calculated. Figure 4.18 shows three band images from Cube I and their corresponding background-removed images, in which the background portion of every pixel spectrum is discarded by projecting the pixel spectrum $\mathbf{x}$ onto $\mathbf{P}_{B_l}^{\perp}$ and $\mathbf{P}_{B_s}^{\perp}$. The background removal was quite successful for both the subspace estimation methods. The localized subspace estimation is preferable for the given test cubes because it provides comparable performance to that of the sample-based method and allows automated target classification. However, as HYDICE images include increasingly complex backgrounds (e.g., urban areas), the dissimilarity measure for the localized subspace estimation fails to work properly because of a high level of the spectral dissimilarity inside the background. In general, the choice between the sample-based and the localized subspace estimations needs to be made in the context of background complexity. Therefore, it is necessary to further develop an appropriate measure of the background complexity for hyperspectral imagery as future work.

Chapter 4 Target Classification Using Adaptive Feature Extraction 135

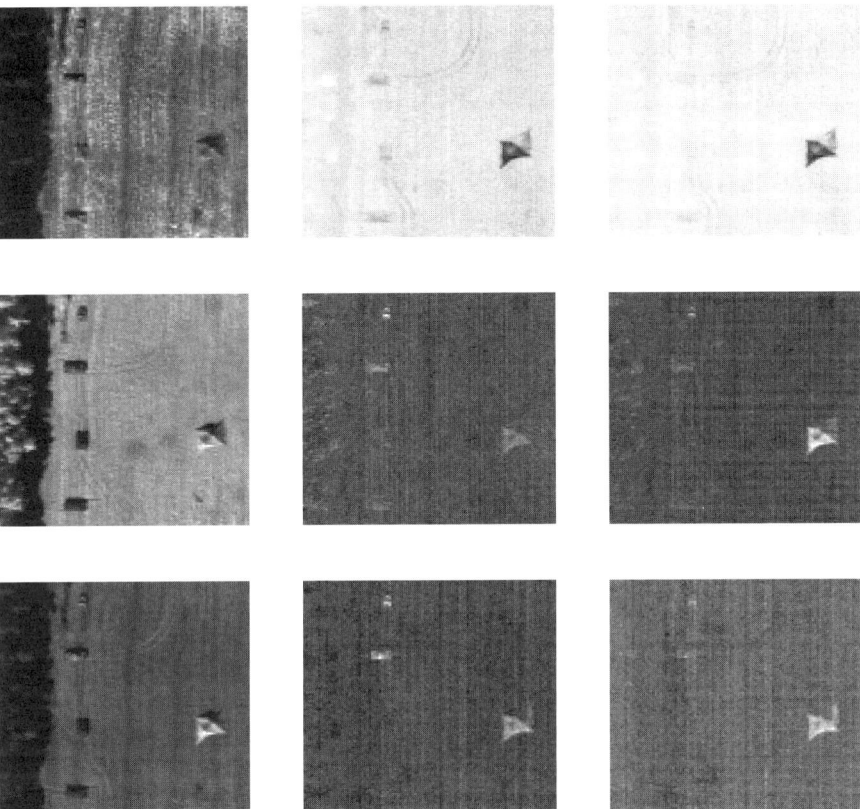

**Figure 4.18.** Example of the background-removed images based on the localized and sample-based subspace estimation. The images in the first column are the three band images from Cube I. The images in the second and third column are the corresponding background-removed band images based on $\mathbf{P}_{B_l}^{\perp}\mathbf{x}$ and $\mathbf{P}_{B_s}^{\perp}\mathbf{x}$, respectively.

### 4.6.3 Adaptive Target Classification

After removing the background portion from every pixel in the hyperspectral test cube, we estimate the target abundance within the pixel spectrum by calculating $\mathbf{m}_t^T\mathbf{P}_B^{\perp}\mathbf{x}$ and $\mathbf{g}^{*T}\mathbf{P}_B^{\perp}\mathbf{x}$; $\mathbf{m}_t^T$ and $\mathbf{g}^{*T}$ serve as the target spectral templates. Quantitative performance of the proposed algorithm, such as the receiver operating characteristic (ROC) curves, was not evaluated mainly because of a lack of HYDICE images. Figure 4.19 shows the target abundance images and the corresponding classification results for Cube I and Cube II using the mean-based spectral template $\mathbf{m}_t^T$, which is shown in Figure 4.14. The sample-based background subspace $\langle\mathbf{B}_s\rangle$ was used in the mean-based target classification. Because the mean curve cannot inherently address a wide range of spectral variability of the target, only partial targets were detected for both

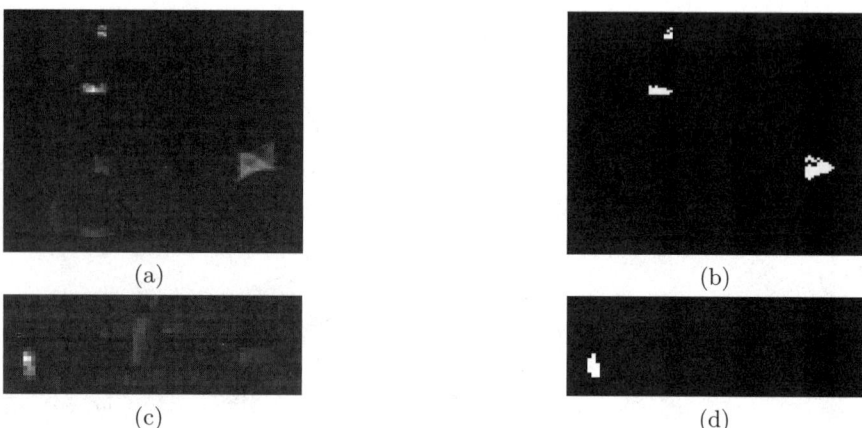

**Figure 4.19.** Target classification based on the mean-based spectral template and the sample-based subspace estimation. (a) Target abundance image and (b) classification results for Cube I. (c) Target abundance image and (d) classification results for Cube II.

the test cubes. Figures. 4.20 and 4.21 show the target abundance images and the corresponding classification results using the ICA-based spectral template $\mathbf{g}^{*T}$ for both the sample-based and the localized background subspace models, respectively. A complete set of the targets was successfully detected for both the subspace models, demonstrating the superiority of the ICA-based classification over the mean-based technique.

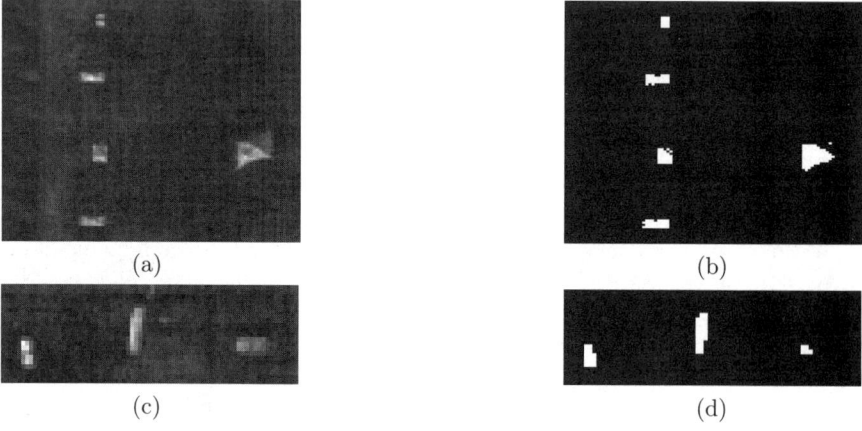

**Figure 4.20.** Target classification results based on the ICA-based spectral template and the sample-based subspace estimation. (a) Target abundance image and (b) classification results for Cube I. (c) Target abundance image and (d) classification results for Cube II.

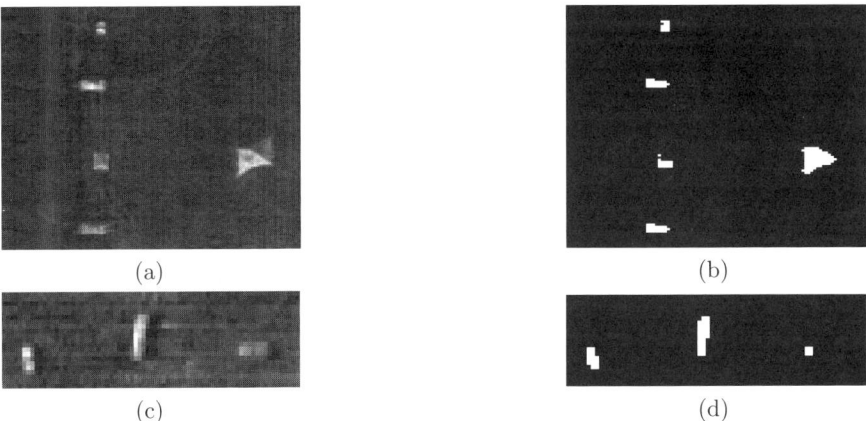

**Figure 4.21.** Target classification results based on the ICA-based spectral template and the localized subspace estimation. (a) Target abundance image and (b) classification results for Cube I. (c) Target abundance image and (d) classification results for Cube II.

The ICA-based target classification without background removal from the individual spectra was also applied to the test cubes by calculating $\mathbf{g}^{*T}\mathbf{x}$. The experiments were performed to find how significantly the background removal process contributed to the target classification process. The projection $\mathbf{g}^{*T}\mathbf{x}$ was not able to suppress the background portion within the pixel spectra, drastically increasing the false classification in the background regions, as shown in Figure 4.22.

## 4.7 Conclusions

We have presented a two-class target classification algorithm, in which an individual pixel spectrum is linearly decomposed into constituent spectra and then labeled either target or background. The background and target spectral characteristics are represented by the low-dimensional background subspace and the target feature vector generated by the ICA-based feature extraction, respectively.

The background subspace model has been used to address a wide range of spectral variability by integrating key spectral features of various terrain types into the low-dimensional subspace. In order to fully automate the target classification process, the localized background subspace model has been also developed, in which the eigenvectors spanning the subspace are calculated based on the reflectance spectra directly from the background regions of the test images. Unsupervised segmentation based on the spectral dissimilarity is applied to the test images to identify the potential background regions. In our

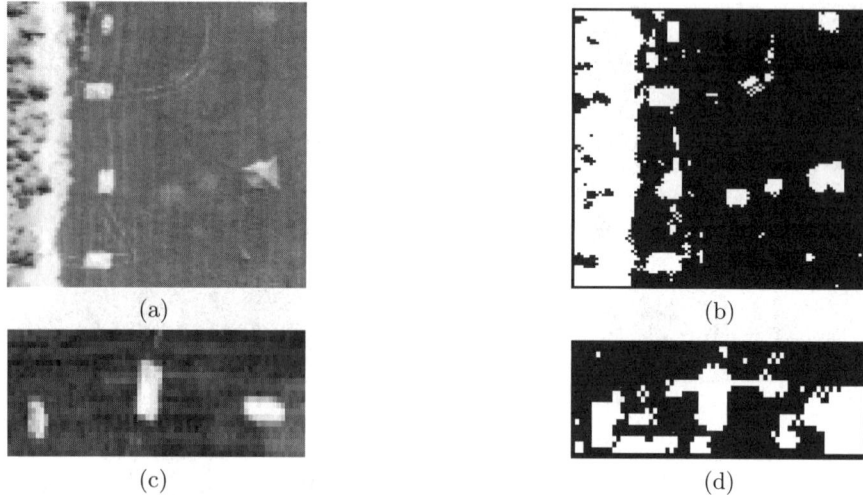

**Figure 4.22.** Target classification results without background removal. (a) Target abundance image and (b) classification results for Cube I. (c) Target abundance image and (d) classification results for Cube II.

experiments, the localized and sample-based subspace models in the proposed classification technique provides comparable classification performance.

The background portion is removed from each pixel spectrum by projecting it onto the orthogonal background subspace. The background removal process greatly improves target classification performance, as supported by the simulation results. The abundance of the remaining target spectral portion is estimated by measuring the correlation between the background-removed individual pixel and the target template to identify the intrinsic spectral nature of the corresponding pixel surface. The target features obtained by both the ICA-based and mean-based feature extraction techniques are used as the target spectral templates. The nonlinear data structures of the targets are exploited by the ICA-based feature extraction. Consequently, ICA-based feature extraction results in better classification performance than mean-based extraction.

Due to the lack of HYDICE images available for experiment, only the qualitative performance was evaluated. In order to further investigate the usefulness of the proposed algorithm, a quantitative performance measure such as the ROC curve needs to be calculated using a larger data set that includes blind test images.

The types of paint used for the targets in our data set is unknown (it may be the three-color camouflage paint). Analysis of the spectral characteristics of different types of paint on the surface of military targets is desirable and needs to be conducted as future work. It will help develop robust target detection/classification methods that can be used in various battlefield environments, e.g., wooded (and/or grassy) areas and barren desert.

## 4.8 Acknowledgment

This research was sponsored by the U.S. Army Research Laboratory (ARL) and was accomplished under the ARL/ASEE postdoctoral fellowship program, contract DAAL01-96-C-0038. The views and conclusions contained in this document are those of the authors and should not be interpreted as representing the official policies, either expressed or implied, of ARL or the U.S. government. The U.S. government is authorized to reproduce and distribute reprints for government purposes, notwithstanding any copyright notation herein.

## References

[1] Bhanu, B.: Automatic target recognition: State-of-the-art survey. IEEE Transactions on Aerospace and Elect. Syst. **22** (1986) 364–379
[2] Reed, I.S., Yu, X.: Adaptive multiple-band CFAR detection of an optical pattern with unknown spectral distribution. IEEE Transactions Acoustics, Speech and Signal Process. **38** (1990) 1760–1770
[3] Kwon, H., Der, S.Z., Nasrabadi, N.M.: Adative multisensor target detection using feature-based fusion. Optical Engineering **41** (2002) 69–80
[4] Schachter, B.J.: A survey and evaluation of FLIR target detection/segmentation algorithm. In: Proc. of DARPA Image Understanding Workshop. (1982) 49–57
[5] Aviram, G., Rotman, S.R.: Evaluating human detection performance of targets and false alarms, using a statistical texture image metric. Optical Engineering **39** (2000) 2285–2295
[6] Chan, L., Nasrabadi, N.M., Torrieri, D.: Eigenspace transformation for automatic clutter rejection. Optical Engineering **40** (2001) 564–573
[7] Manolakis, D., Shaw, G., Keshava, N.: Comparative analysis of hyperspectral adaptive matched filter detector. In: Proc. SPIE. Volume 4049. (2000) 2–17
[8] Joliffe, I.T.: Principal Component Analysis. Springer-Verlag, New York (1986)
[9] Bishop, C.M.: Baysian PCA. In: Neural Information Processing System. Volume 11. (1998) 382–388
[10] Rajan, J.J., Rayner, P.: Model order selection for the singular value selection and the discrete Karhunen-Loeve transform using a bayesian approach. IEE Proc. Image Signal Process **144** (1997) 116–123
[11] Hyvärinen, A., Oja, E.: Independent component analysis: Algorithms and applications. Neural Networks **13** (2000) 411–430

[12] Hyvärinen, A.: Survey on independent component analysis. Neural Computing Surveys **2** (1999) 94–128
[13] Hyvärinen, A.: Sparse code shrinkage: Denoising of nongaussian data by maximun likelyhood estimation. Neural Computation **11** (1999) 1739–1768
[14] Harsanyi, J.C., Chang, C.I.: Hyperspectral image classification and dimensionality reduction: An orthogonal subspace projection approach. IEEE Transactions Geosci. Remote Sensing **32** (1994) 779–785
[15] Ashton, E.L.: Detection of subpixel anomalies in multispectral infrared imagery using an adaptive bayesian classifier. IEEE Transactions Image Process. **36** (1998) 506–517
[16] Chang, C.I., Zhao, X.L., Althouse, M., Pan, J.J.: Least squares subspace projection approach to mixed pixel classification for hyperspectral images. IEEE Transactions Geosci. Remote Sensing **36** (1998) 898–912
[17] Manolakis, D., Shaw, G.: Detection algorithms for hyperspectral imaging applications. IEEE Signal Processing Magazine **19** (2002) 29–43
[18] Bartlett, M.S.: a dissertation: Face Image Analysis by Unsupervised Learning and Redundancy Reduction. University of California, San Diego (1998)
[19] Comon, P.: Independent component analysis: A new concept? Signal Processing **36** (1997) 287–314
[20] Hyvärinen, A.: Fast and robust fixed-point algorithms for independent component analysis. IEEE Transactions Neural Networks. **10** (1999) 626–634
[21] Lee, T.W., Girolami, M., Sejnowski, T.J.: Independent component analysis using an extended infomax algorithm for mixed subgaussian and supergaussian sources. Neural Computation **11** (1999) 417–441
[22] Minka, T.P.: Automatic choice of dimensionality for PCA. M.I.T Media Laboratory Perceptual Computing Section Technical Report (2000)
[23] Strang, G.: Linear Algebra and Its Applications. Harcourt Brace & Company (1986)
[24] Kwon, H., Der, S.Z., Nasrabadi, N.M.: An adaptive unsupervised segmentation algorithm based on iterative spectral dissimilarity measure for hyperspectral imagery. In: Proc. SPIE. Volume 4310. (2001) 144–152

# Chapter 5

# Moving Object Detection and Compression in IR Sequences *

Namrata Vaswani, Amit K Agrawal, Qinfen Zheng, and Rama Chellappa

Center for Automation Research, University of Maryland, College Park, MD, {namrata,aagrawal,qinfen,rama}@cfar.umd.edu

**Summary.** We consider the problem of remote surveillance using infrared (IR) sensors. The aim is to use IR image sequences to detect moving objects (humans or vehicles), and to transmit a few "best-view images" of every new object that is detected. Since the available bandwidth is usually low, if the object chip is big, it needs to be compressed before being transmitted. Due to low computational power of computing devices attached to the sensor, the algorithms should be computationally simple. We present two approaches for object detection — one which specifically solves the more difficult long-range object detection problem, and the other for objects at short range. For objects at short range, we also present techniques for selecting a single best-view object chip and computationally simple techniques for compressing it to very low bit rates due to the channel bandwidth constraint. A fast image chip compression scheme implemented in the wavelet domain by combining a non-iterative zerotree coding method with 2D-DPCM for both low-and high-frequency subbands is presented. Comparisons with some existing schemes are also included. The object detection and compression algorithms have been implemented in C/C++ and their performance has been evaluated using the Hitachi's SH4 platform with software simulation.

## 5.1 Introduction

Remote monitoring of activities of stationary or moving vehicles and humans is a critical component in surveillance applications. The sensors used in practice are typically of low quality and the available bandwidth for transmission is quite limited. We consider the problem of remotely monitoring a battlefield with IR sensors and present two approaches for object detection. The first approach is useful when the objects are at large distances, are very small, appear

---

* Prepared through collaborative participation in the Advanced Sensors Consortium sponsored by the U.S. Army Research Laboratory under the Collaborative Technology Alliance Program, Cooperative Agreement DAAD19-01-2-0008. The U.S. Government is authorized to reproduce and distribute reprints for Government purposes, notwithstanding any copyright notation thereon.

to move slowly, and their signatures have low contrast over the background. In such cases, traditional methods based on the analysis of the difference between successive frames and/or image and background or image intensity change will not work. We present a novel algorithm (referred to as the MTI algorithm) based on variance analysis which is useful at long ranges, when the object is small and slowly moving. The algorithm uses temporal variance to detect potential moving spots, spatial variance analysis to suppress false alarms caused by sensor vibration, and object geometrical constraints to filter out false alarms caused by tree branches and sensor noise.

In other scenarios, when the objects are not so far away, the problem of moving object detection is formulated as one of segmenting an image function using a measure of its local singularity as proposed in [1]. When the detected objects are very small (see Figures. 5.1, 5.2), all views are equally good or bad and hence the problem of best-view selection becomes irrelevant. Also, since the object chip is already very small, compression is not necessary and hence any chipped image of the object can be transmitted. Thus, in such cases all the computational power available on the sensor can be used to solve the object detection problem. When the object chips are larger so that over time the object pose changes, we choose one best-view of the object, compress it using computationally simple techniques and then transmit the compressed best view chip. The challenge here is to solve the best view selection problem and to develop techniques which can compress the image to very low bit rates and are yet computationally simple since they have to be implemented in realtime and on hardware with limited computing power. Note that in this work, we do not study techniques for channel coding since we assume a reliable channel is available for transmission.

Performance evaluation of algorithms is critical for determining the memory and computation needs of the algorithm. We present these measures for the algorithms using the Hitachi's SH-4 microprocessor.

The rest of the chapter is organized as follows. Section 5.2 describes the MTI object detection algorithm for long-range, small-size object detection along with results followed by the technique for detecting short-range and larger objects. Section 5.3 describes techniques for best-view selection for large objects. Section 5.4 discusses compression techniques for real-time, low-power, and very low bit rate compression of object chips. Simulations results on SH4–7751 microprocessor for performance characterization are presented in Section 5.5 and conclusions in Section 5.6.

## 5.2 Object Detection

We present two object detection algorithms, the first one deals with the more difficult problem of long-range object detection when objects are very small and the second one for detecting objects at short range.

## 5.2.1 Long-Range Object Detection (MTI)

Techniques based on optical flow and image intensity analysis may not work well for long-range object detection. In typical scenarios, a moving object size can be as small as $2 \times 3$ pixels and the motion between two adjacent frames can be less than 0.1 pixels. Several challenging issues need to be addressed. The first challenge is to compensate for sensor motion using electronic stabilization methods. Over the years, two dominant approaches have been developed — flow-based and feature based. The feature-based methods extract point, edge or line features and solve for an affine transformation between successive frames, while the flow-based methods compute optical flow and then estimate an affine model. These methods do not perform well in IR images as feature extraction in IR images is not reliable, and flow estimates for IR images suffer from severe bias due to noise. Also, these methods are too complex for real-time computation. Even if stabilization can be solved using one of these approaches, the problem of separating moving objects from the stabilized background is challenging due to low signal-to-clutter ratio in surveillance applications. Other factors that complicate the problem are changes in the background as the sensor is panning, false motion detection due to atmospherics, and other confusing motion (motion of tree leaves, etc).

We assume a stationary sensor. The proposed algorithm uses a variance analysis based approach for moving object detection that can effectively integrate the object motion information over both temporal and spatial domains. Each input frame is first checked for possible errors and then used to update the temporal variance. When the variance for a pixel increases above a given threshold, the pixel is labelled as corresponding to a potential moving object. All candidate-moving pixels are then grouped to form disjoint moving objects. The detected potential moving objects (regions) are verified for their size, shape, and orientation to filter out false detection. After false detection verification, the remaining change regions are reported as moving objects.

The temporal variance at a pixel $(i, j)$ at frame $k$ is computed as

$$\sigma^2_{i,j,k} = S_{i,j,k} - \mu^2_{i,j,k}, \tag{5.1}$$

where

$$S_{i,j,k} = \frac{1}{L} f^2_{i,j,k} + \frac{L-1}{L} S_{i,j,k-1}, \tag{5.2}$$

$$\mu_{i,j,k} = \frac{1}{L} f_{i,j,k} + \frac{L-1}{L} \mu_{i,j,k-1}, \tag{5.3}$$

$$S_{i,j,1} = f^2_{i,j,1}, \tag{5.4}$$

$$\mu_{i,j,1} = f_{i,j,1}. \tag{5.5}$$

Here, $f_{i,j,k}$ is the image value at frame $k$ and location $(i, j)$ and $L$ is the temporal window parameter. A pixel is detected as belonging to potential moving object if the following conditions are satisfied:

$$\sigma_{i,j,k}^2 \geq \sigma_{i,j,k-1}^2, \tag{5.6}$$

$$\sigma_{i,j,k}^2 \geq T_0(i,j) + T_h, \tag{5.7}$$

or

$$\sigma_{i,j,k}^2 \geq \sigma_{i,j,k-1}^2, \tag{5.8}$$

$$\sigma_{i,j,k}^2 \geq T_0(i,j) + T_l, \tag{5.9}$$

$$\max_{-1 \leq \delta_x \leq 1, -1 \leq \delta_y \leq 1} \sigma_{i+\delta_x, j+\delta_y, k}^2 \geq T_0(i,j) + T_h, \tag{5.10}$$

where

$$T_0(i,j) = \max_{-2 \leq \delta_x \leq 2, -2 \leq \delta_y \leq 2} |f_{i+\delta_x, j+\delta_y, 0} - f_{i,j,0}|, \tag{5.11}$$

$$\sigma_t = var_{i,j \in N \times N}(T_0(i,j)), \tag{5.12}$$

$$T_h = \gamma_h \times \sigma_t, \tag{5.13}$$

$$T_l = \gamma_l \times \sigma_t. \tag{5.14}$$

In our experiments, we set $\gamma_h = 3.4$, $\gamma_l = 0.8 * \gamma_h$, and $L = 16$.

### 5.2.2 MTI Algorithm Results

Figure 5.1 shows the object detection result for frame 200 and 300 on an IR sequence (IRseq10) for the case of people walking at a long distance. Even though the width of the objects (people) is 2 to 3 pixels, the algorithm correctly detects and tracks them. In this sequence, the sensor was stationary and its direction was fixed. Figure 5.2 shows the detection of a moving vehicle for frames 160 and 250. In this case the sensor pans and stops and is quickly able to find the object again. Hence, the algorithm is capable of recovering after sensor panning motion.

As a comparison with background subtraction techniques, Figure 5.3 shows the result using background subtraction based on nonparametric background modelling [2] on IRseq10. We can see that there are a lot of false detections whose sizes are the same or even greater than the object size. So, the objects cannot be detected reliably. Also background subtraction techniques need some number of frames for background modelling before they can start the detection process. Figure 5.4 shows the ROC curve for the MTI algorithm for different IR sequences for $\gamma_h$ varying from 3.0 to 5.5. The detection percentage is calculated as the ratio of object detections in all frames and the total objects actually present in all frames. The false alarms percentage is calculated as the ratio of the false alarms in all frames and the total objects in all frames. Since we do not have ground truth, it was generated manually as follows. For all the sequences considered, the objects are far away and are moving approximately parallel to the image plane with uniform speed. The objects locations were hand-picked every thirty frames and are interpolated

# Chapter 5 Moving Object Detection and Compression in IR Sequences

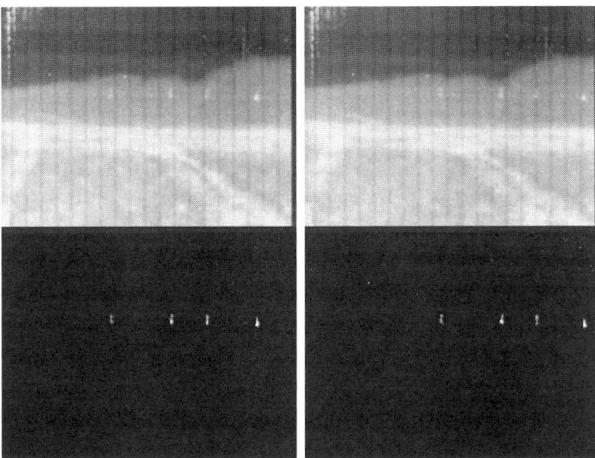

**Figure 5.1.** Object detection: Frames 200 and 300 of IRseq10 along with detected objects.

**Figure 5.2.** Frames 160 and 250 of a moving vehicle IR sequence along with detected objects (sensor recovering after panning motion).

in between assuming constant object velocity. Table 5.1 shows the effect of parameter $L$ on detection rates and false alarms for different sequences.

The MTI algorithm has been implemented on a Dual Pentium 550-MHz personal computer. It can perform moving object detection for $974 \times 436$ image sequences at a sustained rate of 12 frames per second (including reading data from the hard drive and sending image to a video display). The algorithm has been tested on fourteen $974 \times 436 \times 3600$ sequences with objects at various distances, moving speeds, moving directions, and motion patterns, and with different numbers of moving objects. From the ROC curves (Figure 5.4) and

**Figure 5.3.** Detection result using background subtraction on IRseq10.

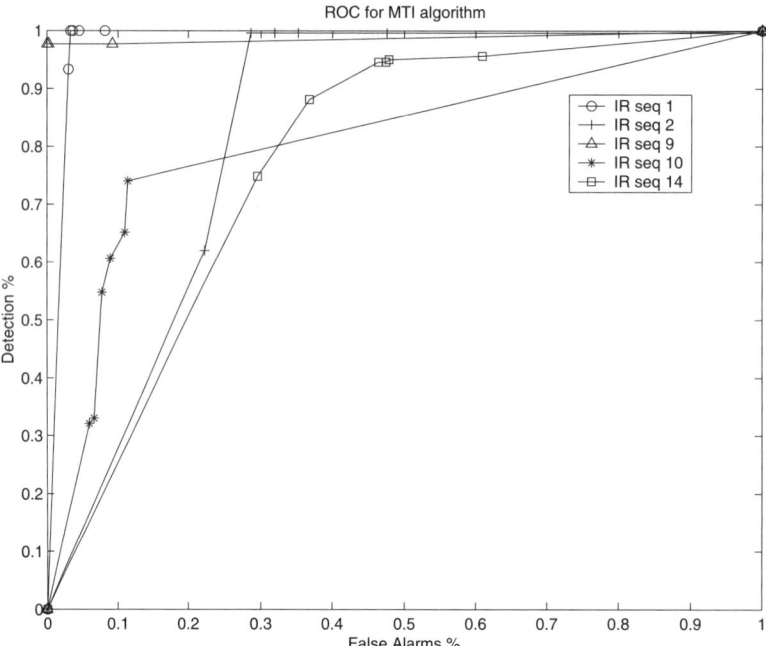

**Figure 5.4.** ROC curve for the MTI algorithm for different IR sequences.

**Table 5.1.** Effect of parameter L on detections and false alarms for various IR sequences.

| Sequence | L  | Detections(%) | False Alarms(%) |
|----------|----|---------------|-----------------|
| IRSeq1   | 8  | 98.33         | 13.8            |
| IRSeq1   | 16 | 99.83         | 28.3            |
| IRSeq9   | 8  | 97.70         | 2.24            |
| IRSeq9   | 16 | 97.70         | 0.05            |
| IRSeq2   | 8  | 99.80         | 32.09           |
| IRSeq2   | 16 | 99.60         | 57.33           |

Table 5.1, we see that the MTI algorithm gives good results (high detection rate and very low false alarm rate) on almost all of the test sequences with no running parameters to be adjusted.

### 5.2.3 Short-Range Object Detection

The problem of object detection is formulated as one of segmenting an image function using a measure of its local singularity as proposed in [1]. The method combines the problem of stabilization, object detection, and tracking into a single process when interframe motions are restricted to lateral translations or tilts and scale changes, and has the advantage of exploiting these sensor motion constraints for performing simultaneous activity detection and stabilization. The algorithm makes use of the Holder exponent of a hybrid capacity (derivative of Gaussian along the X- and Y- axis). Using this measure, Lipschitz signatures which reflect the singularity of the image function along each spatial axis are defined. The Lipschitz signatures are used for detection and tracking of objects. The proposed measure is obtained by applying the operators $G_{x,\sigma}$ and $G_{y,\sigma}$ to the images which are the derivatives of the Gaussian applied along the $x$- and $y$- axes, respectively. The Lipschitz signatures can then be defined as the projection of the these measure along the $x$- and $y$- axes. The main assumption of the algorithm is that the "active regions" of the image exhibit some higher level of singularity in the Lipschitz signatures. In other words, the singularities can be detected and tracked over time. The algorithm is robust to image scale variations and can handle multiple moving objects. It also involves projection along spatial axis and hence can be done in real time. Spatio-temporal information on the objects in the scene can also be inferred.

## 5.3 Best View Selection

As mentioned earlier, the best-view selection is not required when the detected object chip is small (few pixels), but becomes important in case the object chip is large and the object pose changes from time to time.

Three different techniques were tried for best-view selection. We work on the assumption that the side view is the "best view" since it has most of the identifiable features (See Figure 5.5). Eigenspace classification has successfully been used for pose detection [3] and face recognition applications [4, 5]. In the first approach, we formulate the best-view selection problem as a pose matching problem in eigenspace. Another approach to best-view selection would be to wait while the object size keeps increasing (it is approaching the sensor) and transmit the largest-sized image or transmit the last frame before the object takes a turn. This can done by estimating the focus of expansion (FOE) which can be used to calculate the velocity direction. Since both techniques mentioned above are computationally intensive, they are not suitable for real-time hardware implementation. The eigenspace technique also suffers from the drawback that it is not generalizable, i.e., to use it for a different type of vehicle would require a new training phase. It works well only when similar objects are available in its training database. Hence, the third approach, a size-based detection method, was finally implemented. In what follows, we describe these techniques in detail.

### 5.3.1 Eigenspace Classification

This technique is useful for best-view selection when there are multiple sensors capturing an object from different orientations and a database of multiple views of the object is available. In our experiments we have used multiple views of tanks taken from different directions. A view closest to the side view is classified as the best view.

**Figure 5.5.** Tank2 (side view).

**Algorithm**

Construct an eigenspace using images of tanks in various orientations. Place the camera at 45-degree spacings around the tanks to obtain eight possible views of each tank. Obtain the mean image of each of the eight orientations and save its coordinates in eigenspace. For a query image, classify it in eigenspace and calculate the distance from each of the orientations using a distance metric.

Since the required data for doing this was not available, we constructed an eigen-space using all the tank images available and obtained class means corresponding to front, side and back views. On classification of a query image, it got mapped to the correct class most of the times.

**Eigenspace Construction**

Instead of directly obtaining the eigenvectors of an $N^2 \times N^2$ covariance matrix, the first $M$ ($M$ is the number of training vectors) eigenvectors can be obtained by first calculating the eigenvectors of an $M \times M$ covariance matrix of the transpose of the image data and then obtaining the image eigenvectors by taking linear combinations of the training images weighted by the eigenvector component as described in [4, 5, 6].

**Scale and Intensity Invariance**

Since eigenspace classification is sensitive to scale variations, all images were scaled down to a fixed size before classifying and since the image chip contained the tank only, scaling down the image to a fixed size implied obtaining the tank image of a fixed size. Both for eigenspace construction and classification, the images were normalized by their total energy.

**Distance Metrics**

The distance metric can either be the simple Euclidean distance (ED) or the distance along each component normalized by the corresponding eigenvalue (ND). The latter gives better results since it gives more weight to those directions where the noise variance is lower.

A better solution is to obtain variance along each component in each class and calculate the distance from a particular class using those eigenvectors which have low variance in that class but overall high variance in eigenspace. For scaling the distance from class k, the variance in class k is used rather than the global eigenvalue for scaling. In this way, the intraclass variance can be suppressed while the interclass variance can be emphasized. We call this measure the class normalized distance (CND). All the three distance metrics for some sample images are tabulated in Table 5.2. It can be seen from the results that the CND is the best metric for classification for the reason stated above.

**Table 5.2.** Eigenspace Classification Results

| Class | Class Normalized Distance(CND) | Euclidean Distance(ED) | Normalized Distance(ND) |
| --- | --- | --- | --- |
| tank2 | 4 | 10 | 24 |
| tank6 | 15 | 20 | 35 |
| tank9 | 15 | 19 | 28 |
| btank12 | 30 | 44 | 56 |
| sftank5 | 17 | 21 | 31 |

### 5.3.2 Focus of Expansion Estimation

The focus of expansion is the point in the image sequence of a moving body from which all the motion vectors appear to diverge. The FOE may not always lie inside the image boundary. Mathematically, the FOE($X_f$,$Y_f$) is defined as

$$X_f = \frac{T_x}{T_z}, \tag{5.15}$$

$$Y_f = \frac{T_y}{T_z}, \tag{5.16}$$

where $T_x, T_y, T_z$ are the translational motion vectors. Therefore, assuming the ground is the X-Z plane, the direction of motion (velocity angle) of the tank is given by

$$\theta = \tan^{-1}\frac{V_x}{V_z} = \tan^{-1}\frac{T_x}{T_z} = \tan^{-1} X_f. \tag{5.17}$$

A modification of the partial search technique for FOE estimation developed by Srinivasan in [7] was used. The equations to be solved are

$$u(x,y) = -(x - x_f)h(x,y) + xy\omega_x - (1 + x^2)\omega_y + y\omega_z, \tag{5.18}$$

$$v(x,y) = -(y - y_f)h(x,y) + (1 + y^2)\omega_x - xy\omega_y - x\omega_x, \tag{5.19}$$

where $u(x,y)$ and $v(x,y)$ are the optical flow estimates at $(x,y)$, $x_f, y_f$ are the $x$ and $y$ coordinates of the FOE in the image(pixel) coordinates, $h(x,y)$ is the inverse of the depth at point $(x,y)$ and $\omega_x, \omega_y, \omega_z$ are the $x$-, $y$-, and $z$- direction rotations. The FOE estimation algorithm requires calculation of the optical flow which is done using the overlapped basis functions technique developed by Srinivasan and Chellappa [8]. The FOE in the world coordinates is given by

$$X_f = \frac{x_f + x_{\text{offset}} - (N-1)/2}{f} \tag{5.20}$$

which is used for direction of motion calculation in (5.17).

In the modified partial search FOE estimation technique, we use only the optical flow estimates of the part of the image containing the moving object since the flow estimates of the background are unreliable. Since the size of the image for which optical flow is available is smaller, the FOE is also searched over a smaller region. This speeds up the FOE calculation. Also, the FOE estimate of the previous frame can be used to select an initial offset for the FOE in the current frame to speed up the FOE calculation over successive frames. The direction of motion is estimated at regular intervals and we keep waiting if the tank is approaching towards the camera. The frame at which the tank takes a turn away from the camera or the tank's size increases above a certain threshold could be chosen as the "best view" in this approach.

### 5.3.3 Size-Based Best-View Selection

Since the application requires best view selection and compression to be done in realtime on hardware with low computing power, we need very simple techniques for best-view selection. Thus, the final algorithm that was implemented simply waits till the size of the image chips exceeds a predefined threshold. If the size starts decreasing, it simply chooses the maximum size frame in the last 90 frames and sends it. Actually a single frame buffer is used to store the maximum sized chip in the last few frames, and as soon as the size starts decreasing or increases beyond the threshold the stored frame is compressed and transmitted. The algorithm rejects chips which are very close to the image boundary (the tank may not be complete). Spurious frames (produced as a result of wrong object detection) are also rejected based on thresholding the height-to-width ratio.

### 5.3.4 Best-View Selection Results

In this section, we present the results of best-view selection using the three approaches discussed above.

**Eigenspace Classification**

An eigenspace of front, side, and back views of various tanks is constructed and the class means for each class are precalculated. In Table 5.2, results for distances from the perfect side view ("tank2") class are shown.

Figure 5.6 shows the mean image of the tank2 class. The distance of a query tank2 image (side view, see Figure 5.5) is a minimum. Distance of tank6 (side-back view) shown in Figure 5.8 is higher than tank2 but lower than tank12 (back view) shown in Fig. 5.7. Hence tank2 is the "best-view" in this case. As can be seen from Table 5.2, the CND (class normalized distance) has the maximum variation (4 for perfect side view and 30 for back view), and thus it is the best metric for classification among the metrics used.

The eigenspace classification is not a very good method for best-view selection because it is sensitive to the lighting conditions, the type of IR sensor

**Figure 5.6.** Mean image of the itank2 class.

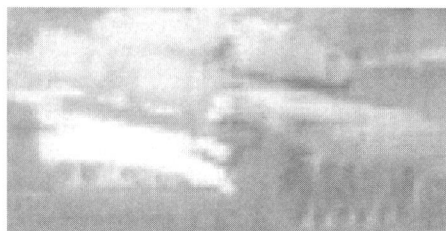

**Figure 5.7.** Query image 3: tank12 (back view).

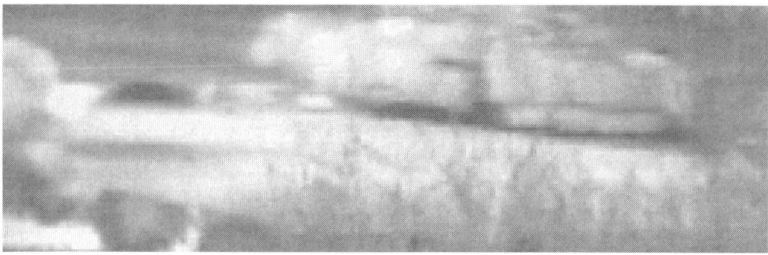

**Figure 5.8.** Query image 2: tank6 (back-side view).

used, and to the scale of the image. If the actual sensor is different from the sensor used in the database, classification could fail. Moreover a very large database of tank images in various poses is required for a robust eigenspace construction which may not be feasible.

**Focus of Expansion Estimation**

The FOE estimation algorithm was run on IR video sequences of the tanks. Since most of the tanks are moving almost horizontally in front of the camera in the test sequences, the FOE values are very large (tending to infinity). The FOE in pixel coordinates for a sample sequence is shown in Table 5.3.

Chapter 5 Moving Object Detection and Compression in IR Sequences  153

**Table 5.3.** FOE estimates using the optical flow shown in Figure 5.10.

| Frame No. | $FOE_x$ | $FOE_y$ |
|---|---|---|
| 121 | −244 | 145 |
| 123 | −369 | 152 |
| 125 | −201 | 140 |
| 127 | −253 | 136 |
| 129 | −553 | 110 |
| 131 | −583 | 110 |

5.9.

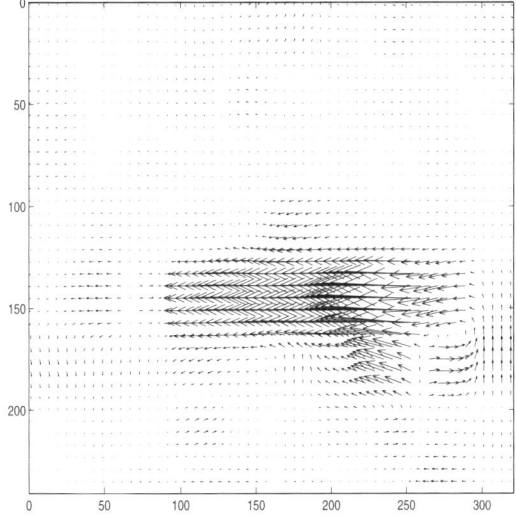

**Figure 5.9.** Optical flow estimate of a typical frame (125) for FOE Estimation.

The optical flow estimate for the full frame 125 is shown in Figure The region of significant motion that is segmented out and used for FOE estimation finally is shown in Figure 5.10. The FOE estimation technique is not very suitable for our application because both optical flow calculation and FOE estimation are computationally intensive. Also for the IR images, the optical flow estimates are not very accurate and as a result the FOE estimates are also not accurate enough.

**Size-Based Best-View Selection**

Figure 5.11 shows a best view selected by this approach.

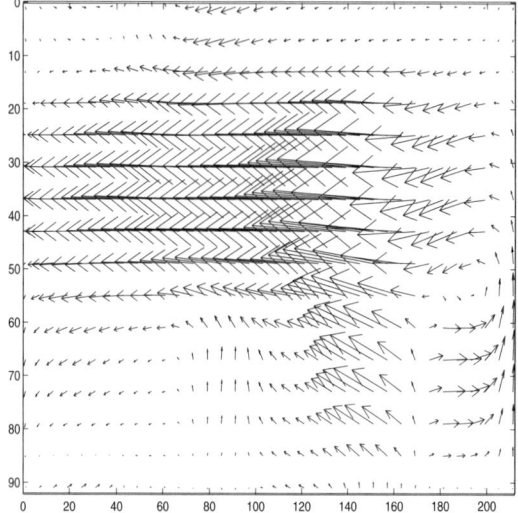

**Figure 5.10.** Segmented optical flow estimate for frame 125 which is finally used in FOE estimation (magnified view).

**Figure 5.11.** Size-based best-view selection.

## 5.4 Compression

For long-range objects, since the detected image chip is already very small, compression is not really necessary. Also since the object chip contrast can be very low compared to the background, it is preferable to just transmit the binary image of the object (instead of transmitting a grayscale one) which itself provides an 8 : 1 compression ratio. If more computational power is available, the binary image could be compressed by run-length coding else just the raw bits can be transmitted.

For objects at short range (which are large), the image chip chosen by the "best-view selection" algorithm has to be compressed before transmission.

Chapter 5 Moving Object Detection and Compression in IR Sequences    155

Since the available channel bandwidth is low, we have tried to develop compression schemes which can provide very high compression ratios while at the same time maintaining a reasonable image quality. Since the algorithm is to be implemented in real-time using limited computing hardware, the computational complexity should be low. We have developed realtime algorithms for image compression in the wavelet domain. We first provide a background of existing image compression schemes and discuss the theoretical background for wavelet transforms and the compression techniques. Following that, we discuss the compression schemes implemented and compare against other techniques. We compare our scheme, combined zerotree and DPCM coding, against three existing schemes all of which have low computational complexity, viz. scalar quantization (SQ), zerotree coding, and DPCM coding. The more efficient compression schemes like JPEG and LZW are not compared here because they have a much higher computational cost associated with their implementation. We then provide a performance analysis of the coding schemes. Finally, experimental results are provided followed by computational cost analysis. We use PSNR (peak signal-to-noise ratio) which is a standard metric for image compression schemes to compare decompressed image quality.

### 5.4.1 Previous Work

A wavelet zerotree coding scheme for compression is presented in [9], but since it uses vector quantization (VQ), it cannot be used for our application due to high computational cost. Reference [10] presents an embedded predictive wavelet image coder. But it uses arithmetic coding which is not suitable for implementation on a embedded processor such as Hitachi's SH4. So we have developed algorithms using the Haar wavelet transform followed by noniterative zerotree coding [9] and 2D-DPCM for all subbands. The computational complexity of these algorithms is only marginally higher than simple scalar quantization (SQ) of the entire image.

### 5.4.2 Wavelet Transform Properties

The wavelet transform is an atomic decomposition that represents a signal in terms of shifted and dilated versions of a prototype bandpass wavelet function and shifted versions of a low-pass scaling function. In discrete time, the bandpass wavelet function is a high-pass filter at different scales and the scaling function is a low-pass filter. The image is low-pass and high-pass filtered first along rows and then along columns to generate LL, LH, HL, and HH images each of which is subsampled by two. This process is repeated on the subsampled LL image. The wavelet transform has the following properties which make it suitable for compression.

- *Multiresolution:* The image is decomposed into wavelets at $N$ scales, and only the top few coarsest scales need to be transmitted to obtain a reasonable image quality. Depending on the available channel bandwidth, more

finer scale coefficients can be transmitted to improve the reconstructed image quality.
- *Entropy Reduction:* The wavelet transform of a real image generates a large number of small coefficients (which can be set to zero) and a small number of large coefficients which can be encoded. This property is based on the fact that a real world image will not have information in all frequencies at all points in space. At most points except edges, the higher-frequency information is almost zero.
- *Clustering and Persistence:* The wavelet transform attempts to decorrelate the image, but the decorrelation is not complete (since the filters are constant, not data dependent). There is a residual dependency between adjacent coefficients at the same scale (clustering) and between coefficients in adjacent scales but the same spatial location (persistence). Our coding schemes attempt to remove these correlations in the image.

### 5.4.3 A-DPCM for Scaling Coefficient(LL) Encoding

The scaling coefficients (LL subband) contain the maximum information and thus more bits are allocated for its encoding. But it is also the most highly correlated subband and this fact can be exploited to maximize compression. An adaptive-DPCM scheme is used for encoding the LL subband. The current pixel is predicted based on a linear combination of three causal nearest neighbors. The predicted value of the pixel, $\hat{X}$ is obtained as

$$\hat{X} = l(\bar{Q}) = \bar{w}.\bar{Q} = \sum w_k Q_k. \tag{5.21}$$

The predictor coefficients $\bar{w}$ are calculated to minimize the mean squared prediction error as

$$\bar{w} = E(\bar{Q}\bar{Q}^T)^{-1} E(X.\bar{Q}). \tag{5.22}$$

where X is the pixel to be predicted, $Q_i$ are the quantities based on which the pixel would be predicted (in this case the nearest neighbors), and $\bar{w}$ are the predictor coefficients. Instead of quantizing the pixel value, the error between the actual and the predicted value $(X - \hat{X})$ is quantized, which requires fewer bits since the error would be much smaller than the original pixel value if the prediction is good. Calculation of LMSE predictor coefficients can be done offline on a set of similar images.

### 5.4.4 Zerotree Coding

In multiresolution wavelet decomposition, each coefficient $X_i$, except those in the LL subband and the three highest subbands, is exactly related to $2 \times 2$ coefficients of the immediately higher subband. These four *children* coefficients correspond to the same orientation and spatial location as the parent coefficient $X_i$. Each of the four children coefficients is in turn related to $2 \times 2$

coefficients in the next higher subband, and so on. These coefficients are collectively called the descendants of the parent $X_i$. All coefficients with magnitude less than threshold $T$ are called *insignificant coefficients* and their collection is known as a zerotree. In order to obtain a real-time implementation [9], the search for insignificant coefficients is started in the lowest-frequency subbands except baseband and continued in higher-frequency subbands. When a coefficient is decided as insignificant and set to zero, all its descendants are also set to zero. Thus, one needs to transmit only the escape code for the zerotree root vector besides encoding the nonzero coefficients. The zerotree root positions at each scale can be encoded efficiently using the Run-length Coding(RLC). The non-zero coefficients can be scalar quantized and transmitted. This type of simple threshold based zerotree coding, RLC and SQ are computationally simple algorithms for hardware implementation.

However, it is possible that there are significant descendants even though their parent is insignificant. These mispredictions are inevitable in a non-iterative search method, but the conditional probability of this happening is very small, as discussed in [9]. The misprediction error can be reduced by predicting the value of a "zero" coefficient based on its nearest nonzero neighbors (causal and noncausal) while decoding.

### 5.4.5 DPCM on Wavelet Coefficients

The zerotree coding exploits the persistence property of wavelet coefficients. But there is also a residual correlation in the high-frequency subbands, especially the LH and the HL bands with horizontal and vertical neighbor, respectively. Hence applying an A-DPCM scheme like that discussed for the LL subband can give additional compression. Also while obtaining the prediction value for the current pixel we can exploit both clustering and persistence properties, i.e., obtain a prediction for the current pixel based on its vertical (for HL) or horizontal neighbor(for LH) and its parent coefficient. Again as in the case of the LL subband, the predictor coefficients can be calculated offline for a sequence of similar images using (5.22). In this case the predictors are the parent coefficient at the same spatial location and the horizontal or vertical neighbor. This scheme is motivated by a similar scheme discussed in [10] for visual images.

### 5.4.6 Compression Schemes

Four different schemes for encoding the wavelet coefficients were compared. In all cases the LL subband was encoded using the A-DPCM scheme discussed in Section 5.4.3.

## Scalar Quantization

This scheme involves scalar quantization (SQ) of the wavelet coefficients and DPCM encoding of the LL coefficients. Variable bits are allocated to the subbands based on their variances as discussed in [9].

## Zerotree Coding

Zerotree coding is applied as discussed in Section 5.4.4. This not only gives a significantly reduced bits per pixel (BPP) value than the SQ (as expected), but also gives reduced MSE value compared to SQ. The reason is that the quantization error is higher than the thresholding error for high-frequency subbands which are coarsely quantized.

## 2D Predictive DPCM on Wavelet Subbands

Only DPCM coding is applied as discussed in Section 5.4.5 with no zerotree coding. The performance of this scheme is bad because the "noisy data" close to zero, cannot be predicted correctly and hence the prediction errors obtained are sometimes larger than the original pixel value. Hence the MSE is significantly higher.

## Combined Zerotree and DPCM Coding

We propose to combine zerotree coding and the DPCM encoding (ZT/DPCM) of wavelet coefficients to achieve maximal compression. First a simple zerotree coding is applied to the subbands. This is followed by DPCM coding of the "nonzeroed" coefficients. The value of a "zeroed" neighbor is predicted as follows. If we predict $C_{x,y}$ based on $C_{x-1,y}$ which is "zeroed" and the zeroing threshold is $T$, we estimate $C_{x-1,y}$ as follows

$$S = C_{x-2,y} + C_{x-1,y-1},$$

$$\hat{C}_{x-1,y} = \begin{cases} 0 & \text{if } S = 0, \\ -T & \text{if } S < 0, \\ +T & \text{if } S > 0. \end{cases}$$

This is based on the assumption that since the next coefficient is nonzero, the previous one would be close to the threshold. DPCM combined with zerotree coding works much better because the noisy coefficients have been set to "zero" and we do not try to predict their value. The prediction model is applicable only to those subbands for which enough ($> 2$) bits have been allocated and the prediction error energy obtained while calculating the predictor coefficients is less than 25% of the subband energy. For other subbands, SQ is used.

### 5.4.7 Performance Analysis

The aim of any compression scheme is to minimize the mean squared error (maximize the PSNR) and the entropy per pixel (entropy rate, ER). In SQ, each pixel is coded independently and the correlation in the image is not exploited. So the entropy rate is higher. Entropy rate will be minimized if each pixel is coded based on all past pixels on which it depends, i.e. (for a 1D signal)

$$h(X_n) > h(X_n|X_{n-1}) > h(X_n|X_{n-1},...,1). \tag{5.23}$$

If we assume a one-step Markov model,

$$h(X_n|X_{n-1}\cdots 1) = h(X_n|X_{n-1}) \tag{5.24}$$

For 2D data (assuming a Markov random field model), this translates to $X_{n,n}$ depending only on $X_{n-1,n}$ and $X_{n,n-1}$.

The quantization MSE will be minimized for a given bit rate if the mean square value of the quantity to be quantized is minimum. Hence instead of quantizing $X_{n,n}$, in 2D predictive DPCM, we predict a value $(\hat{X}_{n,n})$ based on past values and quantize the difference $(X_{n,n} - \hat{X}_{n,n})$. $\hat{X}_{n,n}$ is calculated as discussed in (5.21) to minimize $E[X_{n,n} - \hat{X}_{n,n}]^2$ and hence the quantization MSE over all linear estimators. Also for a given quantization step size (fixed MSE), reduced data variance means reduced entropy.

In zerotree coding, the PSNR is higher than SQ because the zeroing error is lower than the quantization error for high-frequency subbands which are coarsely quantized. Zeroing also reduces entropy since the number of symbols to be compressed is reduced. The 2D MRF model with second-order dependencies (correlations) fits well for the LL subband, but does not fit well for the wavelet subbands and the prediction fails completely for very small values (only noise). This is the reason why DPCM on wavelet subbands gives the worst PSNR values. Combined zerotree and DPCM (ZT/DPCM) gives best results in terms of PSNR and entropy rate. The noisy coefficients are zeroed and hence not predicted and thus the quantization error remains low. Because of LMSE prediction, the entropy is minimum and zerotree coding further reduces the entropy rate by reducing the number of symbols to be coded.

### 5.4.8 Image Compression Results

Various types of low-pass and high-pass filters satisfying the prefect reconstruction property can be used. In our implementation, the Haar transform is used because of its simplicity and ease of hardware implementation. Using a longer length filter will not be useful because the low-pass filter will tend to average over a very large area and thus lose the localization property of wavelet transforms. The Haar wavelet is built using a two-tap low-pass filter $[1, 1]$ and a two-tap high-pass filter $[1, -1]$.

Figure 5.5 shows the original tank2 image and Figure 5.12 shows the compressed tank2 images using combined zerotree/DPCM coding and zerotree coding. Table 5.4 shows the compression results for two sample IR images and the Lena image.

**Table 5.4.** The bpp, PSNR[$10\log_{10} 255^2/MSE$] and entropy for three sample images using zerotree (ZT), zerotree and DPCM (ZT/DPCM), scalar quantization (SQ) and only DPCM coding schemes.

| Image  | Coder    | Total BPP | PSNR  | Entropy (Non-zero) | RLC BPP |
|--------|----------|-----------|-------|---------------------|---------|
| tank2  | ZT/DPCM  | 0.5628    | 31.73 | 0.0920              | 0.2757  |
|        | ZT       | 0.5628    | 31.61 | 0.2112              | 0.2757  |
| tank12 | ZT/DPCM  | 0.5232    | 31.75 | 0.0880              | 0.2649  |
|        | ZT       | 0.5232    | 31.65 | 0.2045              | 0.2649  |
| lena   | ZT/DPCM  | 0.5066    | 29.40 | 0.0851              | 0.2286  |
|        | ZT       | 0.5066    | 29.30 | 0.1542              | 0.2286  |
|        | SQ       | 0.7947    | 13.07 | 0.3156              |         |
|        | DPCM     | 0.7947    | 25.34 | 0.1508              |         |

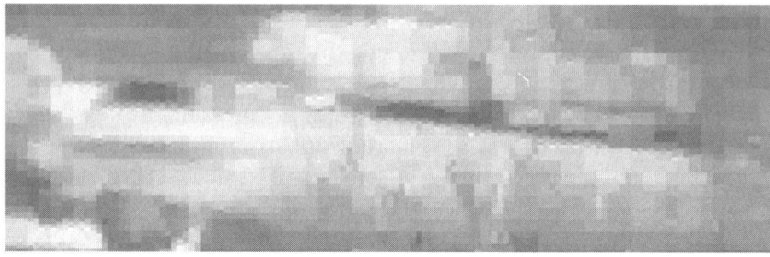

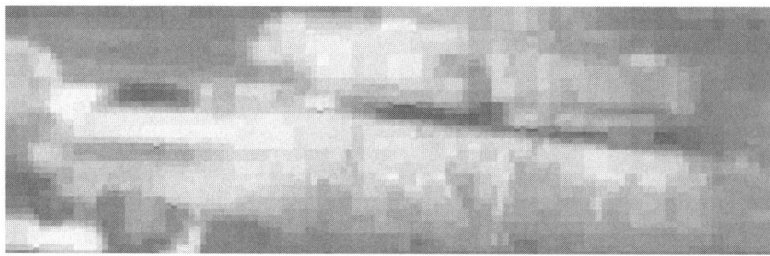

**Figure 5.12.** Compressed tank2 by (a) combined zerotree and DPCM coding (b) zerotree coding.

Chapter 5 Moving Object Detection and Compression in IR Sequences    161

The results have been obtained by allocating a total of 0.5 BPP to various subbands proportional to the logarithm of their variances. Since for this low value of BPP, the lowermost subbands get negative bits allocated to them (which are set to zero), the actual BPP obtained is higher than 0.5.

In Table 5.4 we have compared the total BPP, PSNR, BPP, for RLC coding and entropy rate for three different images (two from the IR sequence and the Lena image). The entropy rate is the minimum bits/pixel that can be theoretically achieved for the image. Due to hardware constraints we have not implemented any form of entropy coding (Arithmetic/Huffman).

As can be seen from the values of RLC BPP, almost half the bits are used up in encoding the zerotree information. More efficient binary encoding schemes can be employed to reduce this value and this could considerably improve the BPP. Also, in most cases the combined zerotree and DPCM scheme gives the best results both in terms of PSNR values and entropy. In some cases like the Lena image, the PSNR is higher for simple zerotree coding, but the entropy of the combined scheme is less. Ideally, one would assume that applying a DPCM encoding would cause a significant reduction in PSNR. This is definitely true for the LL subband, but the reduction for higher-frequency subbands is not so much because of lesser correlation. Another reason is the uncertainty in predicting the value of a pixel based on a neighboring "zeroed" pixel. We are experimenting with better methods to improve the prediction model for combined zerotree and DPCM encoding.

The BPP without zerotree coding is consistently higher for all the images and hence the zerotree coding is advantageous even though half the BPP is used up in RLC coding of the zerotree. Also surprisingly, the PSNR for SQ is lower than for zerotree coding even when the BPP is higher. The reason for this is that the quantization error in SQ is higher than the zeroing error for the high-frequency subbands which are coarsely quantized. From Table 5.4, we observe that DPCM encoding without zeroing is the worst scheme. This is because a lot of the coefficients below the zeroing threshold in the high-frequency subband are actually "noise." Thus, in DPCM we are trying to predict the value of these "noise" pixels or use them to predict other pixels and hence the predictions are very bad, thus leading to a higher PSNR. Hence the DPCM model fails in the absence of zerotree coding, while it provides a reasonably good model for the image when combined with zerotree coding.

### 5.4.9 Computational Complexity: Hardware Issues

The entire coding scheme is computationally very simple. The Haar wavelet transform involves a single addition operation per pixel. Zerotree coding requires one comparison to a threshold (per pixel) and a multiplication by two (a shift operation) to calculate the descendant position. The DPCM operation involves three real multiplications and two additions to calculate the predicted vale and one subtraction to obtain the error. Run-length coding is again a counting operation requiring one addition per pixel. Thus the additional cost over scalar quantizing the entire image is (which is the minimum one has to do to compress an image) is three multiplications and a few additions per pixel. For an $N^2 * N^2$ image with a three-level wavelet decomposition, the additional cost for our scheme is given by

$$AC_{\text{Haar}} = (N^2 + N^2/4 + N^2/16)C_A, \qquad (5.25)$$

$$AC_{\text{Zeroing}} = (N^2 + N^2/4 + N^2/16)(C_C + C_S), \qquad (5.26)$$

$$AC_{RLC} = N^2 C_A, \qquad (5.27)$$

$$AC_{DPCM} = N^2(3C_M + 3C_A), \qquad (5.28)$$

where $AC$ is additional cost. $C_A$ is the cost for one addition, $C_M$ is the cost for one multiplication, $C_C$ is the cost for one comparison, and $C_S$ is cost for one shift (multiply by two) operation . Since comparison and shift are single operations, $C_C = C_S = 1$. Hence, total additional cost is

$$AC = (N^2 + N^2/4 + N^2/16)(C_A + 2) + N^2(3C_M + 4C_A). \qquad (5.29)$$

## 5.5 Performance Evaluation of Algorithms

We have developed a C/C++ implementation of the object detection and compression/decompression algorithms. The C/C++ implementation of image compression part of the system currently uses zerotree coding with SQ and RLC. Performance evaluation of the C/C++ code was done using Hitachi's SH4–7751 microprocessor. SH4–7751 is a high-performance superscalar RISC microprocessor designed for embedded applications. Some of the features of SH4 include 167-mHZ clock frequency, upto 360 MIPS capability, and on-chip cache for instruction and data. More information on SH4 can be found at http://semiconductors.hitachi.com. We evaluated the code performance using the Hitachi Embedded Workshop (HEW) which is an SH4 simulator provided by Hitachi. The results were obtained using the profile utility of HEW. Note that these results are obtained by directly cross-compiling the C code using Hitachi's cross-compiler. Hence, certain features of SH4 architecture (such as floating point unit) which can enhance real-time performance are not used. In practice, the performance can be improved by designing the assembly language code for computationally expensive procedures so as to take advantage of such features of the object microprocessor. The performance results of the algorithms in terms of code size, run-time memory requirement and instruction cycles on SH4 are as follows.

## 5.5.1 MTI Algorithm

The code size required for the MTI algorithm was 9.8 KB. For an input frame size of 120 × 160 maximum run-time memory required was 1.086 MB. Cycles required per frame were equal to 8.5 M which corresponds to 19.54 frames per second (30 fps frame rate).

## 5.5.2 Image Chipping Algorithm

### Motion Detection and Best-View Selection

For motion detection using the technique described in Section 5.2.3 and best-view selection, the code size required was 16 KB. The run-time memory required for frame size of 240 × 320 was 2.4 MB and the processing frame rate (for 120 × 160 frame size) was 4–5 frames per second.

### Compression

The compression part of algorithm required a code size of 30.5 KB. For a typical image chip of size 60×120, run-time memory required was 0.75 MB and cycles on SH4 needed to compress the images was 31.5 M which corresponds to run-time of 0.188 sec. The run-time is calculated as cycles/clock frequency.

### Decompression

Decompression required a code size of 21.5 KB and for a typical image chip of size 60 × 120, run-time memory required was 0.75 MB. Note that this memory requirement will increase with chip size. Table 5.5 gives the instruction cycles required for different PSNR values of reconstructed sample tank chip.

**Table 5.5.** Cycles and run-time required for decompression of a sample target chip of size 60 × 120 for different PSNR.

| PSNR(db) | Cycles(million) | Runtime on SH4(sec) |
|---|---|---|
| 30.21 | 20.73 | 0.120 |
| 34.36 | 21.00 | 0.125 |

## 5.6 Conclusions

A novel variance-analysis-based algorithm has been presented for moving object detection. The algorithm is especially suitable for detection of long-range, small, slow moving objects. An initial test on several IR sequences has revealed high detection rates with low false alarms for vehicles at distances of several kilometers. The algorithm is robust and requires no tuning of parameters by the operator. For objects at short distances, technique based on detecting and tracking image singularities have been discussed. Also for objects at short range, methods for best-view selection and compression were presented. Three different approaches to the best-view selection problem were compared. The approach based on classification in eigenspace is suitable when multiple views of the same object are available but has limitations when significant scale and illumination variations are present. The FOE estimation approach would be a useful method for direction of motion calculation but is computationally expensive. The size-based technique is the fastest and gives reasonable results and is used in our implementation. A new scheme combining noniterative zerotree coding with 2D DPCM for LL and for the high-frequency subbands was presented. This method gives better results than simple scalar quantization and simple zerotree coding both in terms of BPP and PSNR at a marginally increased computational cost. The algorithms were implemented in C and their performance results on SH4 processor were presented.

The image compression results can be further improved by using some form of entropy coding (since the entropy rate of our scheme is significantly lower) and by replacing the run length coding method with more efficient binary coding techniques. Also the zeroing thresholds can be calculated for the required PSNR values.

## References

[1] Shekarforoush, H., Chellappa, R.: A multi-fractal formalism for stabilization, object detection and tracking in FLIR sequences. In: International Conference on Image Processing. Volume 3. (2000) 78–81
[2] Elgammal, A., Harwood, D., Davis, L.: Nonparametric background model for background subtraction. In: Proc. 6th European Conf. Computer Vision. Volume 2. (2000) 751–767
[3] Murase, S., Nayar, S.: Visual learning and recognition of 3-d objects from appearance. International J. Computer Vision **14** (1995) 5–24
[4] Turk, M., Pentland, A.: Eigenfaces for recognition. J. Cognitive Neuroscience **3** (1991) 71–86
[5] Turk, M., Pentland, A.: Face recognition using eigenfaces. In: Proc. IEEE Conf. on Computer Vision and Pattern Recognition. (1991) 586–591
[6] Sirovich, L., Kirby, M.: Low dimensional procedures for the characterization of human faces. J. Optical Society of America **4** (1987) 519–524

[7] Srinivasan, S.: Extracting structure from optical flow using fast error search technique. International J. Computer Vision **37** (2000) 203–230
[8] Srinivasan, S., Chellappa, R.: Noise-resilient optical flow estimation using overlapped basis functions. J. Optical Society of America **16** (1999) 493–509
[9] Paek, S., Kim, L.: A real-time wavelet vector quantization algorithm and its vlsi architecture. IEEE Transactions on CSVT **10** (2000) 475–489
[10] Buccigrossi, R., Simoncelli, E.: Image compression via joint statistical characterization in the wavelet domain. IEEE Transactions on Image Processing **8** (1999) 1688–1700

# Chapter 6
# Face Recognition in the Thermal Infrared [*]

Lawrence B. Wolff, Diego A. Socolinsky, and Christopher K. Eveland

Equinox Corporation, 9 West 57th Street, New York, New York 10019
{wolff,diego,eveland}@equinoxsensors.com

**Summary.** Recent research has demonstrated distinct advantages of using thermal infrared imaging for improving face recognition performance. While conventional video cameras sense reflected light, thermal infrared cameras primarily measure emitted radiation from objects such as faces. Visible and thermal infrared image data collections of frontal faces have been on-going at NIST for over two years, producing the most comprehensive face database known to involve thermal infrared imagery. Rigorous experimentation with this database has revealed consistently superior recognition performance of algorithms when applied to thermal infrared, particularly under variable illumination conditions. Physical phenomenology responsible for this observation is analyzed. An end-to-end face recognition system incorporating simultaneous coregistered thermal infrared and visible has been developed and tested indoors with good performance.

## 6.1 Introduction

Accelerated developments in camera technology over the last decade have given computer vision researchers a whole new diversity of imaging options, particularly in the infrared spectrum. Conventional video cameras use photosensitive silicon that is typically able to measure energy at electromagnetic wavelengths from 0.4 $\mu$m to just over 1.0 $\mu$m. Multiple technologies are currently available, with dwindling cost and increasing performance, which are capable of image measurement in different regions of the infrared spectrum, as shown in Figure 6.1. Figure 6.2 shows the different appearances of a human face in the visible, shortwave infrared (SWIR) midwave infrared (MWIR), and longwave infrared (LWIR) spectra. Although in the infrared, the near-infrared (NIR) and SWIR spectra are still reflective and differences in appearance between the visible, NIR and SWIR are due to reflective material properties. Both NIR and SWIR have been found to have advantages over imaging in the visible for face detection [1] and detecting disguise [2].

---

[*] This research was supported by the DARPA Human Identification at a Distance (HID) program under contract #DARPA/AFOSR F49620-01-C-0008.

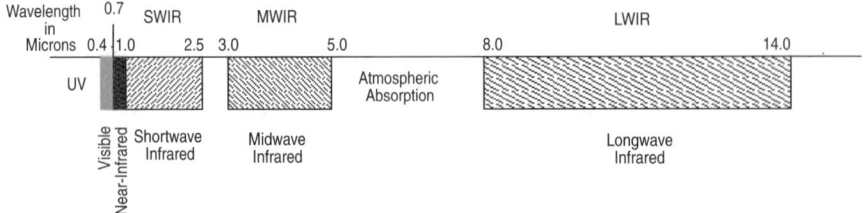

**Figure 6.1.** Nomenclature for various parts of the electromagnetic spectrum.

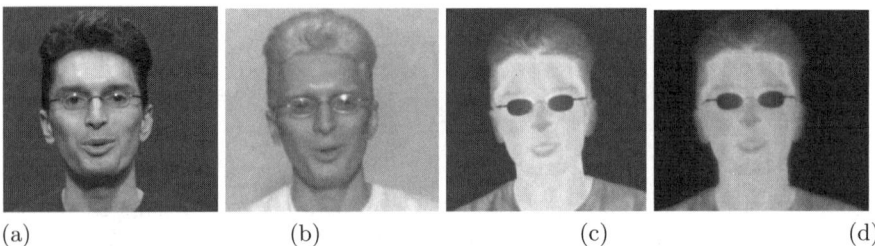

**Figure 6.2.** A face simultaneously imaged in the (a) visible spectrum, 0.4–0.7 μm, (b) shortwave infrared, 0.9–1.7 μm, (c) midwave infrared, 3.0–5.0 μm, and (d) longwave infrared, 8.0-14.0 μm.

At wavelengths of 3 μm and longer imaged radiation from objects becomes significantly emissive due to temperature, and is hence generally termed the thermal infrared. The thermal infrared spectrum is divided into two primary spectra, the MWIR and LWIR. Between these spectra lies a strong atmospheric absorption band between approximately 5 and 8 μm wavelength, where imaging becomes extremely difficult due to nearly complete opaqueness of air. The range beyond 14 μm is termed the very longwave infrared (VLWIR) and although in recent years it has recieved increased attention, it remains beyond the scope of this chapter. The amount of emitted radiation depends on both the temperature and the *emissivity* of the material. Emissivity in the thermal infrared is conversely analogous to the notion of reflective albedo used in the computer vision literature [3, 4]. For instance, a Lambertian reflector can appear white or grey depending on its efficiency for reflecting light energy. The more efficient it is in reflecting energy (more reflectance albedo) the less efficient it is in thermally emitting energy respective to its temperature (less emissitivity). Objects with perfect emissivity of 1.0 are completely black. Many materials that are poor absorbers transmit most light energy while reflecting only a small portion. This applies to a variety of different types of glass and plastics in the visible spectrum.

As detailed in the following section, the spectral distribution of energy emitted by an object is simply the product of the Planck distribution for a given temperature, with the emissivity of the object as function of wavelength [5]. In the vicinity of human body temperature (37° C), the Planck distribution has a maximum in the LWIR around 9 $\mu$m, and is approximately one-sixth of this maximum in the MWIR. As we will show through empirical measurement, the emissivity of human skin in the MWIR is at least 0.91, and at least 0.97 in the LWIR. Therefore, face recognition in the thermal infrared favors the LWIR, since LWIR emission is much higher than that in the MWIR. Thermal infrared imaging for face recognition first used MWIR platinum silicide detectors in the early 1990s [6]. At that time, cooled LWIR technology was very expensive. By the late 1990s, uncooled microbolometer imaging technology in the LWIR became more accessible and affordable, enabling wider experimental applications in this regime. At that time, cooled MWIR technology was about ten times more sensitive than uncooled microbolometer LWIR technology, and even though faces are more emissive in the LWIR, in the late 1990s MWIR could still discern more image detail of the human face. At present, uncooled microbolometer LWIR technology coming off the assembly lines is rapidly approaching one-half of the sensitivity of cooled MWIR. For face recognition in the thermal infrared, this is a turning point as for the first time the most appropriate thermal infrared imaging technology (i.e. LWIR) for studying human faces is also the most affordable.

For over two years, data collections of both visible and thermal infrared imagery of faces have been taken and continue to take place at regular intervals of 6 months at the National Institute of Science and Technology (NIST). This effort is supported by the DARPA HID program [7]. Section 6.2 describes the comprehensive database resulting from these collections, consisting of over 100,000 images of over 300 individuals so far. This database has provided the empirical foundation with which to rigorously compare the performance of various face recognition algorithms between visible and thermal infrared imagery. Some of these results are summarized in Section 6.6. Also described in Section 6.6 is a recently completed full-working prototype of the equinox access control environment (ACE) face recognition system, which uses fused coregistered visible and LWIR imagery from a novel sensor system.

The main advantage of thermal infrared imaging for boosting face recognition performance is its apparent invariance to changing illumination. Section 6.4 attempts to characterize how well thermal infrared images of the human face are invariant to illumination changes. Section 6.5 delves deeper into explaining the physical phenomenology responsible for this invariance by computing the emissivity of human skin in the MWIR and the LWIR. This culminates in a preliminary thermal model for human skin. Finally, Section 6.7 briefly overviews some of the remaining challenges that thermal infrared imaging does not immediately remedy for face recognition.

## 6.2 The Equinox Visible/Infrared Face Database

Figure 6.3 shows the experimental set-up for imagery being collected on a regular basis at NIST, simultaneosuly in the visible, SWIR, MWIR, and LWIR spectra. The objectives for these ongoing data collections are as follows:

1. To be able to directly and rigorously compare the performance of face recognition algorithms between visible imagery and imagery in the various modalities of the infrared spectrum.
2. To produce face imagery simultaneously in these modalities under variable illumination conditions.
3. To produce face imagery with significant intrapersonal variation for each imaging condition and modality.

Towards the first objective a configuration of four different sensors of respective modalities and interface software has been set-up for simultaneous acquisition of visible, SWIR, MWIR, and LWIR imagery. All infrared cameras are of the Indigo Merlin Series with $320*240$ resolution. The visible camera is a Pulnix 6710 with $640*480$ resolution. A special optical design insures precise pixel coregistration of visible and LWIR imaging modalities [8, 9]. Although the SWIR and MWIR imaging modalities are boresighted[2], impairing precise coregistration at close distances, the physical separation between these cameras has been minimized beyond what is shown in Figure 6.3 so that views are nearly identical. Optically coregistered sensors in the NIR were in use [2, 1] for face detection. The complexity of coregistering visible and LWIR wavelengths, however, is much greater due to the larger disparity between them.

Towards the second and third objectives, collection of image data was repeated for three different illumination conditions: (i) Frontal, (ii) frontal-Left, (iii) frontal-Right, with lamps shown in Figure 6.3 using standard 3200 K color temperature photographic bulbs. Figure 6.4 shows the emission curve for these bulbs in the wavelengths of interest. Forty image frame sequences of visible, SWIR, MWIR, and LWIR were digitized simultaneously at 10 frames/second (i.e., 4-seconds duration), while a human subject was reciting the vowels "a," "e," "i," "o," "u." This creates a continuous image sequence with changes in expression throughout providing significant intrapersonal variation over the course of multiple frames. At the same time there is little facial movement between consecutive image frames 1/10 second apart, allowing for analysis of image variations due to temporal sensor noise. Figure 6.2 is an example of one such multimodal frame within this 40-frame sequence. After the acquisition of each 40-frame contiguous image sequence, for each illumination three more static images are taken of individuals told to make extreme expressions of "smile," "frown," and "surprise."

---

[2] The term *boresighted* typically refers to cameras that have been placed alongside each other and aimed in the same direction. Due to the separation between the cameras, it is impossible to obtain the same view of a 3D object from both sensors. This is often exploited in stereo vision to compute depth maps.

Chapter 6 Face Recognition in the Thermal Infrared    171

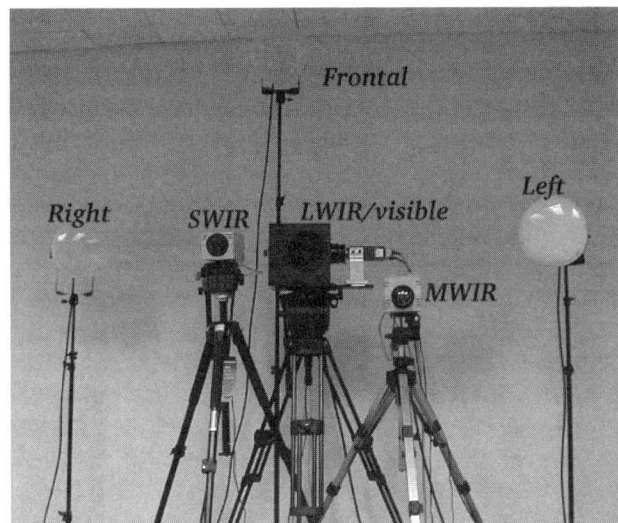

**Figure 6.3.** Camera and illumination equipment set-up used for simultaneous data collection of visible, SWIR, MWIR, and LWIR imagery.

Prior to data collection, the radiometric calibration procedure described in Section 6.3 was performed for the Indigo Merlin series MWIR and LWIR cameras using a Model 350 Mikron blackbody source. Software was developed to convert raw MWIR and LWIR image grayscale values directly into respective thermal emission values from ground-truth blackbody images. Raw image gray values for the MWIR and LWIR cameras are 12-bit integers from which floating point thermal emission values were computed and then rounded back to 12-bit values with appropriate dynamic range.

At present, the Equinox visible/infrared database consists of over 300 individuals imaged over five separate data collections at NIST. At least 60 individuals have participated in two or more of these data collections so that intrapersonal variations over 6 months or more can be analyzed. To summarize, for each individual a 40-frame sequence plus three static images were taken for three different illuminations and four spectral image modalities. Not including duplicate individuals, at least 340 individuals $*$ 43 images/modality $*$ 3 illuminations/individual $*$ 4 modalities $=$ 175,440 images are contained in the Equinox database. Almost all of this database was collected indoors, with outdoor imagery beginning to be collected during the last data collection at NIST in April 2002. A portion of this database is available on the Internet at http://www.equinoxsensors.com/hid.

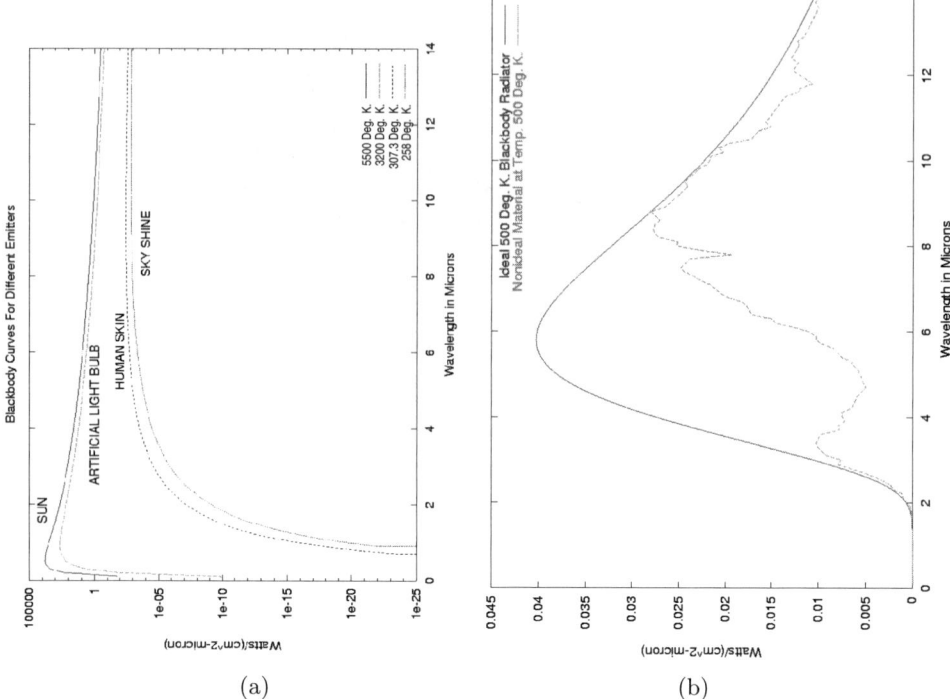

**Figure 6.4.** (a) Blackbody Planck curves comparing thermal IR emission from common natural and artificial illumination sources to thermal IR emission from human skin. (b) Comparison of an ideal blackbody Planck curve with a nonideal emitter at the same temperature.

## 6.3 Calibration Of Thermal IR Sensors

All objects above absolute zero temperature emit electromagnetic radiation. In the early 1900s Planck was the first to characterize the spectral distribution of this radiation for a *blackbody*, which is an object that completely absorbs electromagnetic radiation at all wavelengths [5]. According to Planck's law, the spectral distribution of emission from a blackbody at temperature $T$, is given by

$$W(\lambda, T) = \frac{2\pi h c^2}{\lambda^5 (e^{\frac{hc}{\lambda k T}} - 1.0)} \quad [\text{Watts/cm}^2]\mu\text{m}^{-1}, \quad (6.1)$$

$$Q(\lambda, T) = \frac{2\pi c}{\lambda^4 (e^{\frac{hc}{\lambda k T}} - 1.0)} \quad [\text{Photons/cm}^2 - \text{sec}]\mu\text{m}^{-1}, \quad (6.2)$$

expressed in two different units of energy flux which are commonly used. In the above formulas, $h$ is Planck's constant, $k$ is Boltzmann's constant, $c$ is the speed of light, and $\lambda$ is wavelength. Figure 6.4(a) shows a comparison of blackbody spectral distributions corresponding to the various temperatures for the Sun, artificial lightbulb illumination at 3200 K color temperature [10], human skin, and average temperature for the atmosphere.

In reality, only very few objects are near perfect energy absorbers, particularly at all wavelengths. The proportional amount of energy emission with respect to a perfect absorber is called the *emissivity* $\epsilon(T, \lambda, \psi)$, which takes values in the range $[0, 1]$. In addition to temperature $T$ and wavelength $\lambda$, this can also be a function of emission angle $\psi$. Kirchoff's law states that the emissivity at a point on an object is equal to the *absorption* $\alpha(T, \lambda, \psi)$, namely:

$$\epsilon(T, \lambda, \psi) = \alpha(T, \lambda, \psi).$$

This is a fundamental law that effectively asserts the conservation of energy. Blackbody objects are therefore the most efficient radiators, and for a given temperature $T$ emit the most energy possible at any given wavelength. Expressed in the same units as equations ( 6.1) and ( 6.2) above, the spectral distribution of emission from an object with emissivity $\epsilon(T, \lambda, \psi)$, is given by:

$$\epsilon(T, \lambda, \psi) \times W(\lambda, T), \qquad \epsilon(T, \lambda, \psi) \times Q(\lambda, T).$$

For illustrative purposes, Figure 6.4(b) compares the spectral distribution of the emission of an ideal blackbody at 500 K (227°C) with that of a nonideal emitter (e.g., could be a piece of bare metal) also at the same temperature. In this case the nonideal emitter has low emissivity at wavelengths in the MWIR spectral region (3-5 $\mu$m) and generally high emissivity in the LWIR spectral region (8-14 $\mu$m).

Under most practical conditions, 2D imaging array thermal IR sensors (i.e., what are termed *staring arrays*) measure simultaneously over broadband wavelength spectra, as opposed to making measurements at narrow, almost monochromatic, wavelengths (e.g., an IR spectrophotometer which measures only one point in a scene). With a staring array sensor it is possible to measure *average emissivity* over a broadband spectrum (e.g., 3-5 $\mu$m, 8-14 $\mu$m), which in Figure 6.4(b) is simply the ratio of the area under the nonideal curve to the area under the Planck curve over the respective wavelength spectrum.

Some of these principles can be observed in Figure 6.5 (in Section 6.4). Plastic materials transparent in the visible spectrum that compose glasses are opaque in the LWIR and appear dark. Emissivity of this material is small in the visible spectrum while being significantly above 0.80 in the MWIR and LWIR spectral regions. The dark appearance of glasses in the LWIR and the MWIR relative to thermal emission from human facial skin is mostly due to the glasses being close to room temperature, about 15°C cooler than body temperature. We performed simple experiments whereby these same pair of glasses were heated close to body temperature. Sure enough, the glasses appeared thermally much brighter, but did not show as much thermal emission as facial skin at the same temperature. Also, from Figure 6.5 the influence of reflection of external illumination from glasses is far more prominent than that from facial skin. All of this initially suggests that facial skin has very high emissivity, significantly higher than that of the material comprising glasses. A quantitative estimate of the average emissivity of facial skin in the MWIR and LWIR is developed in Section 6.5, supporting this assertion.

Just like visible video cameras, thermal IR cameras measure energy of electromagnetic radiation, the main difference being that because thermal IR cameras sense at such long wavelengths, they measure radiation that has been typically thermally emitted. Of course, visible cameras see radiation emitted from very hot sources (e.g., the sun or artificial lightbulbs which are thousands of degrees Kelvin) but the primary scene elements of interest in the visible are objects from which such light is reflected. Sometimes there is the misconception that thermal IR cameras directly measure temperature, which would be true if all objects were blackbodies. Temperature can be determined indirectly from a thermal IR camera by measurement of energy of emitted radiation, using precise knowledge of emissitivity of the object, which is dependent upon a number of parameters.

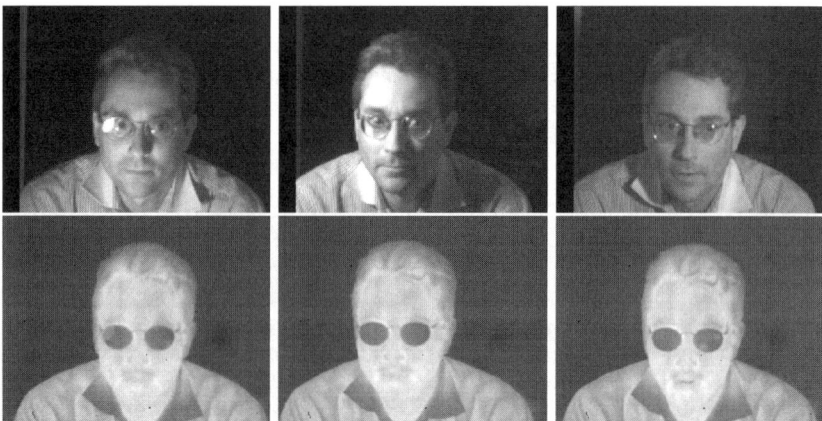

**Figure 6.5.** A qualitative demonstration of the illumination invariance for LWIR imagery of a face under different illuminations. Top row: Visible imagery of a face under three illumination conditions respectively front, left, and right. Bottom row: Co-registered thermal IR imagery simultaneously acquired for each of the three images in top row respectively.

Thermal IR cameras can be radiometrically calibrated using a blackbody ground-truth source. Radiometric calibration achieves a direct relationship between the gray value response at a pixel and the absolute amount of thermal emission from the corresponding scene element. This relationship is called *responsivity*. Depending on the type of thermal IR camera being used, thermal emission flux is measured in terms of watts/cm$^2$ or photons/(cm$^2$ − second) [11]. The gray value response of pixels for a MWIR camera with an indium antimonide (InSb) focal plane array is linear with respect to photons/(cm$^2$−second). The gray value response of pixels for an LWIR camera using a microbolometer focal plane array is linear with respect to watts/cm$^2$. Two-point radiometric calibration uses a blackbody plate filling the field of view of the thermal IR camera and capturing images for the blackbody at two different temperatures. Given that human body temperature is 37°C, two good temperatures to use for calibrating the imaging of humans in a room temperature scene would be 20°C and 40°C (293 K and 313 K), as these are relatively evenly spread about the temperature of skin. A relatively large difference between the calibration temperatures will insure numerical stability of the linear regression, while a choice of temperatures nearby the temperature of interest minimizes possible effects from a secondary nonlinear response of the focal plane array.

Since absolute thermal emission is known by computing the area under the Planck curve for the corresponding temperature and wavelength spectrum, a responsivity line is generated at each pixel by two (greyvalue, thermal emission) coordinate values. The slope of this responsivity line is called the "gain" and the vertical translation of the line is "offset." The gain and offset for each pixel on a thermal IR focal plane array can be significantly variable across the array. Radiometric calibration standardizes thermal emission measurement by generating a responsivity line for each pixel.

Figure 6.6 shows responsivity lines respective to different integration times, for a single pixel near the center of a MWIR InSb focal plane array that was used to collect face imagery. Eight different temperatures of a blackbody were used to generate multiple data points demonstrating the highly linear response. It is clearly important to record all thermal IR camera parameters for a given radiometric calibration. Note that the responsivity lines for different integration times intersect at the same point, related to various DC bias control settings on the camera. Beyond camera parameters, if an MWIR or LWIR camera is originally radiometrically calibrated in an indoors environment, taking it outdoors where there is a significant ambient temperature difference, the gain and offset of linear responsivity of focal plane array pixels will change as the optical lens temperature in front of the focal plane array changes. Radiometric calibration standardizes all thermal IR data collections, whether they are taken under different environmental factors or with different thermal IR cameras or at different times.

## 6.4 Measuring Illumination Invariance

Variation in illumination is one of the biggest factors that confounds face recognition algorithms in the visible spectrum [12, 13]. It has been recognized in the past [2, 14, 6, 15] that changes in illumination appear to play less of a role in the thermal infrared, but how does one quantify this invariance in terms that are meaningful to face recognition? One way is to quantitatively compare the effect that variation in illumination has on face images in the thermal infrared with other factors that contribute to changes in face imagery, such as variations in facial expression and more subtle variations due to camera noise.

Illumination invariance of the human face in the termal infrared can be qualitatively observed in Figure 6.5 for a coregistered LWIR and visible video camera sequence of a face under three different illumination conditions. For this sequence a single 60-W light bulb mounted in a desk lamp illuminates a face in an otherwise completely dark room and is moved into different positions. The top row of visible video imagery shows dramatic changes in the appearance of the face. The bottom row shows LWIR imagery which, unlike its coregistered visible counterpart, appears to be remarkably invariant across different illuminations, except in the image area corresponding to the glasses. As we will see, illumination invariance in the thermal infrared, while not being completely ideal, is nonetheless strongly approximate.

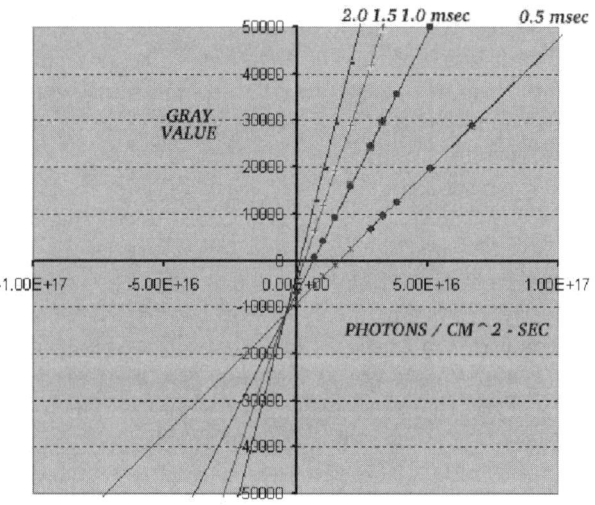

**Figure 6.6.** Responsivity curves for different integration times for the Indigo Merlin Series MWIR camera used for collecting face images.

Figure 6.7 shows simultaneously acquired MWIR and LWIR images of a subject from the Equinox database, together with corresponding gray value histograms of an individual under the three illumination conditions previously described. These images are the third image frame out of each respective 40-image frame sequence. Gray values in the histograms are represented as 16-bit integers with the high 12-bits being the actual image gray value. The gray level histograms are remarkably stable across different illuminations for both the MWIR and the LWIR images. Of the variations that are present in the respective histograms, which are due to change in illumination and which are due to other factors? For instance, note the darker mouth region in the MWIR image for right illumination as compared to the mouth region in the MWIR images for other illuminations. The darker mouth region is due to the subject breathing-in room temperature air at the moment, thereby cooling down the mouth. This has nothing to do with any illumination condition.

The histograms in Figure 6.7 can be compared with those in Figure 6.8, which shows gray value histograms corresponding to the fourth and twentieth image frame out of the 40-image frame sequence respective to the frontal illumination condition. In this case, illumination is the same but the fourth frame being consecutive with the third frame isolates changes due to camera noise, and the twentieth frame occuring just under two seconds later means the subject has changed facial expression. The variations in the gray level histogram due to camera noise and to different facial expression under the same illumination are of similar magnitude to variations occurring under different illumination.

A quantitative analysis of invariance in the framework of hypothesis testing was also performed. The following analysis is repeated for two different distance measures between images. Firstly we consider the $L^2$ distance between normalized images taken as vectors. Secondly, we use the Kullback–Leibler divergence[3] between the histograms of the normalized faces, given by

$$I(P,Q) = \int P \log \frac{P}{Q},$$

where $P$ and $Q$ are the respective normalized histograms.

For each video sequence of $40 + 3$ frames[4], we compute the $43 \cdot 42/2 = 903$ distances between normalized faces for distinct pairs of frames. Also, we compute the $43 \cdot 43 = 1849$ distances between normalized faces for sequences of the same subject and modality, one sequence with frontal illumination and the other with lateral illumination. From these computations we estimate (nonparametrically) the distribution of distances for images with the same illumination condition and with different illumination conditions. Figures 6.9, 6.10, 6.11, and 6.12 show the estimated distributions for the $L^2$ distance and $KL$-divergence for two subjects in our database. With an infinite supply of images, we would expect the distances to behave according to a $\chi$ distribution with the number of degrees of freedom matching the number of pixels in the normalized faces, and indeed the experimental estimates approximate $\chi$ distributions.

It is clear from Figures 6.9, 6.10, 6.11, and 6.12 that the distances between normalized visible faces with different illumination conditions are much larger

---

[3] The Kullback–Leibler divergence does not satisfy the triangle inequality, and thus is not strictly a distance. However, it provides an information-theoretic measure of similarity between probability distributions.

[4] Recall that 40 consecutive video frames were collected while subjects recited the vowels, and then three additional static frames were acquired while the subjects were asked to act out the expressions "smile," "frown," and "surprise."

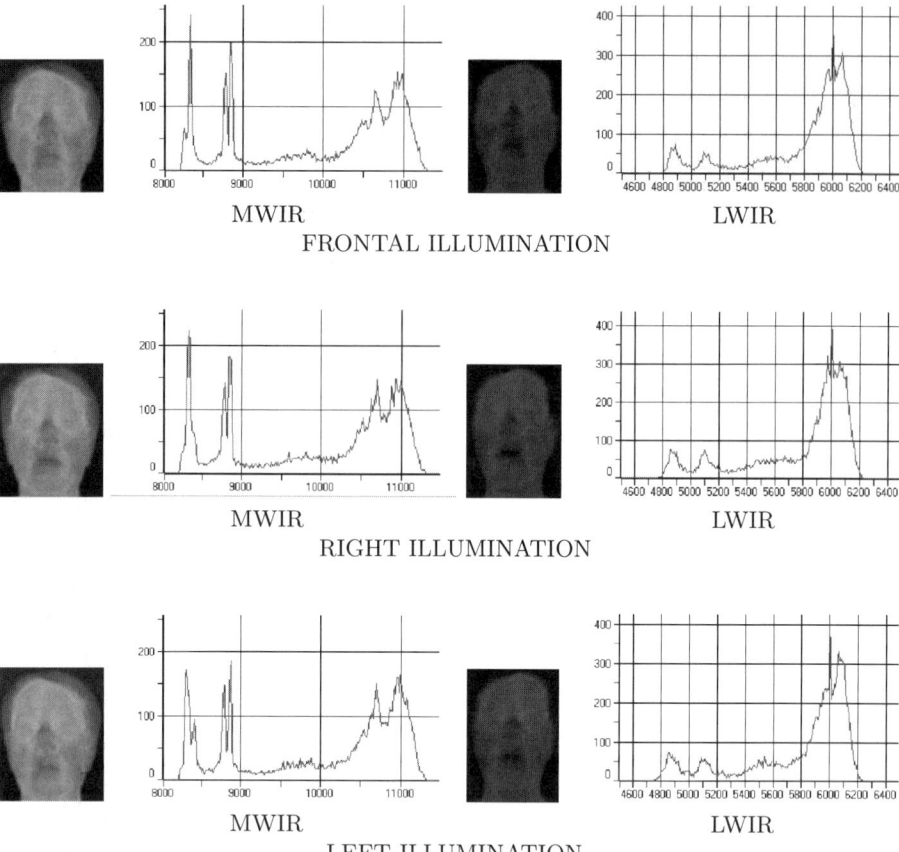

**Figure 6.7.** MWIR and LWIR imagery of a face for three illumination conditions and respective histograms of the third frame out of a sequence of 40 images.

than those for visible faces with the same illumination condition. This indicates that the variation in appearance due to change in illumination is much larger than that due to change in facial expression. The corresponding statement for LWIR imagery does not hold. That is, looking once again at Figures 6.9, 6.10, 6.11, and 6.12, one can see that the distribution of distances between normalized faces with different illumination conditions is comparable (but not equal; see below) to the distribution obtained by using images acquired with the same illumination condition. In other words, the variation in appearance introduced by changes in illumination and expression is comparable to that induced by changes in facial expression alone. Phrasing these statements as formal hypothesis, we can reject the null-hypothesis of illumination invari-

Chapter 6 Face Recognition in the Thermal Infrared    179

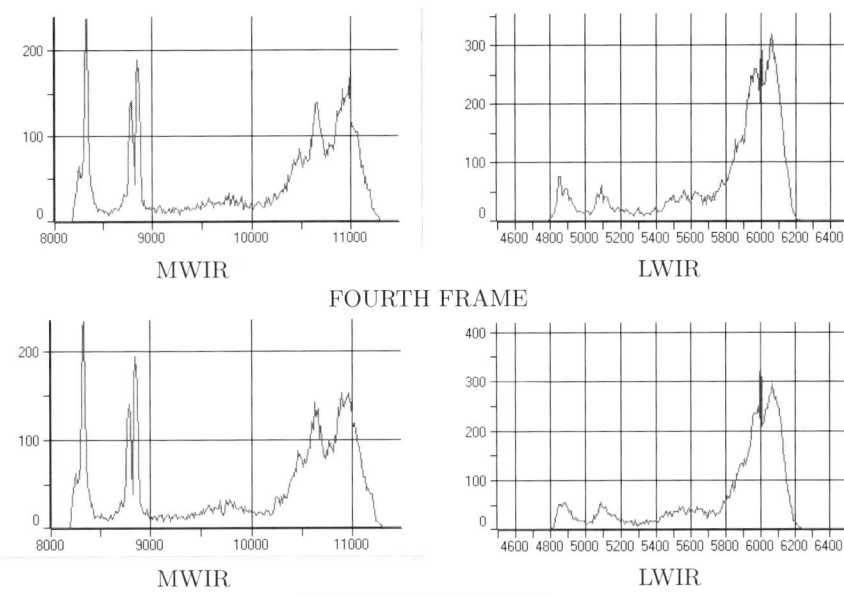

**Figure 6.8.** MWIR and LWIR imagery of the same face as Figure 6.7 respective to frontal illumination for the fourth frame (top row) and twentieth frame (bottom row) out of a sequence of 40 images.

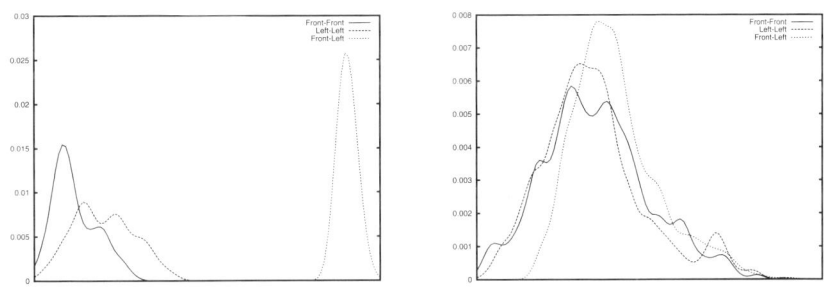

**Figure 6.9.** Distribution of $L^2$ distances for visible (left) and LWIR (right) images of subject 2344.

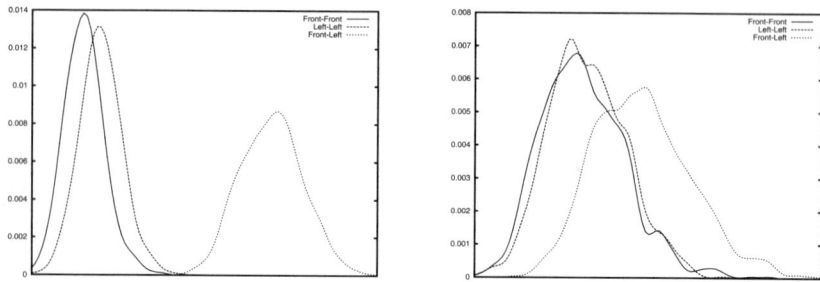

**Figure 6.10.** Distribution of Kullback–Leibler divergences for visible (left) and LWIR (right) images of subject 2344.

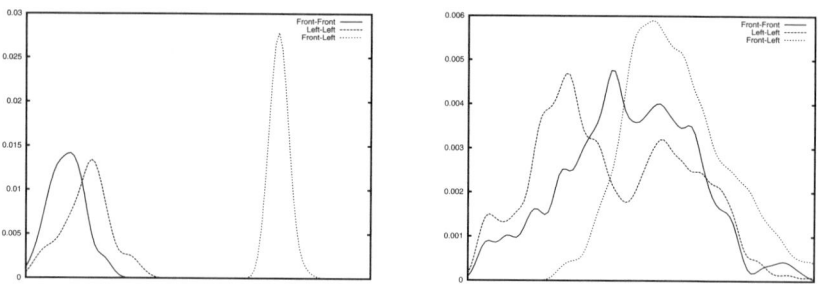

**Figure 6.11.** Distribution of $L^2$ distances for visible (left) and LWIR (right) images of subject 2413.

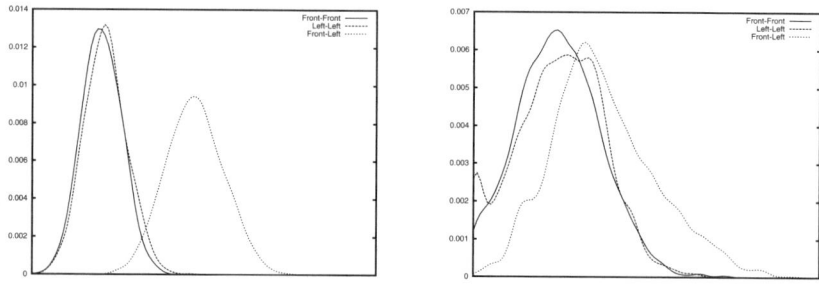

**Figure 6.12.** Distribution of Kullback–Leibler divergences for visible (left) and LWIR (right) images of subject 2413.

ance for visible imagery with a *p*-value smaller than $0.01^5$, whereas we are unable to reject the null-hypothesis for LWIR imagery with any significant confidence. The slight shift in the distributions to the right for variable illumination suggests that illumination invariance in the LWIR is not completely ideal.

## 6.5 Emissivity Of Human Facial Skin

Figure 6.4(a) shows that the amount of thermal emission from a common lightbulb is three to four orders of magnitude greater than the thermal emission from skin in both the 3–5 $\mu$m MWIR region and the 8–14 $\mu$m LWIR region. Empirical observation with our own MWIR and LWIR cameras showed that direct illumination from an incandescent filament through lightbulb glass and plastic diffuser is at least 300 times greater than thermal emission from human facial skin. This is a rather striking fact given that thermal IR imagery of faces is highly illumination invariant. Human skin must absorb a large quantity of radiation in both the MWIR and the LWIR implying that skin has very high emissivity.

Figure 6.13 shows a human subject in the same scene with a 6in ∗ 6in square blackbody (Mikron Model 345) imaged in the MWIR and LWIR spectra. Separate images are taken for the blackbody at two different temperatures: 32°C and 35° C. The corresponding histograms show gray value modes for the facial skin image region and for the blackbody image region. Prior to imaging, an Anritsu thermocouple was used to make contact temperature measurements on the forehead, on both cheeks and on the chin of the human subject. An average skin surface temperature of 32°C was observed. Note, however that the face thermally emits more energy than does a 32°C blackbody. Recall that a blackbody is, by definition, a perfect emitter at all temperatures and wavelengths. Therefore, we have a physical contradiction unless we can account for the extra radiation. Since the path self-emission from the atmosphere between the subject and the sensor is negligible compared to the emission from the subject, we conclude that the extra radiation must be originating below the skin surface (where body temperature is around 37°C) and shining through the translucent skin layer and onto the sensor. This may reveal an important aspect of how thermal emission arises from human anatomy and perhaps even a physical mechanism for why skin has such high absorption in the thermal IR.

Figure 6.14 illustrates a preliminary high-level model of human skin in terms of optical and thermal properties. Evidently, skin layers must be significantly transmissive to thermal emission from underlying internal anatomy

---
[5] This means that the likelihood of our rejecting the hypothesis of illumination invariance for visible imagery while at the same time the hypothesis being true is lower than 1%[16].

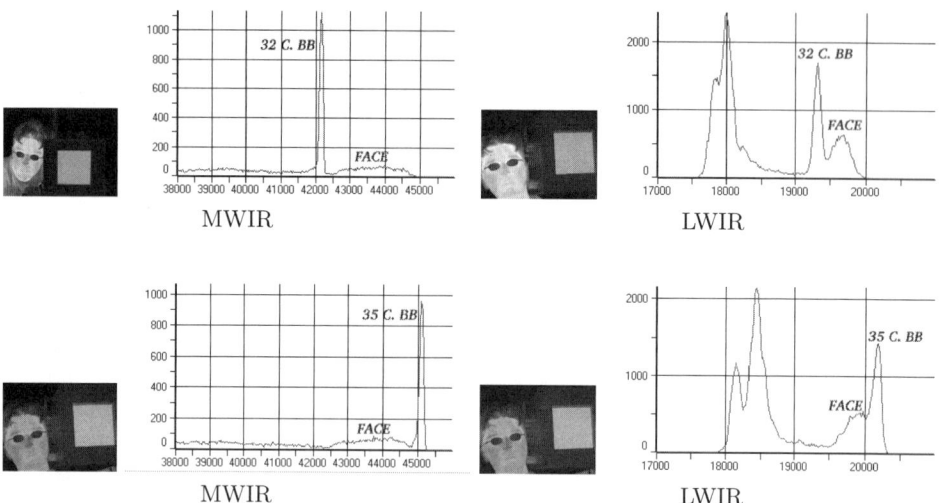

**Figure 6.13.** Direct comparison of MWIR and LWIR imagery of a face with a groundtruth blackbody at two different temperatures, 32°C and 35°C.

which is at a higher temperature. This is qualitatively evidenced from thermal observation of prominent vasculature beneath the skin particularly in the neck and forehead. Just how far below the skin surface thermal emission is transmitted is unclear and is an avenue for future research. If at least the outer layers of skin are transmissive, then incident thermal IR illumination must be first transmitted and then absorbed within deeper layers of skin or other anatomy. This may explain why the amount of thermal emission from skin seems to be independent of external skin color in the visible spectrum.

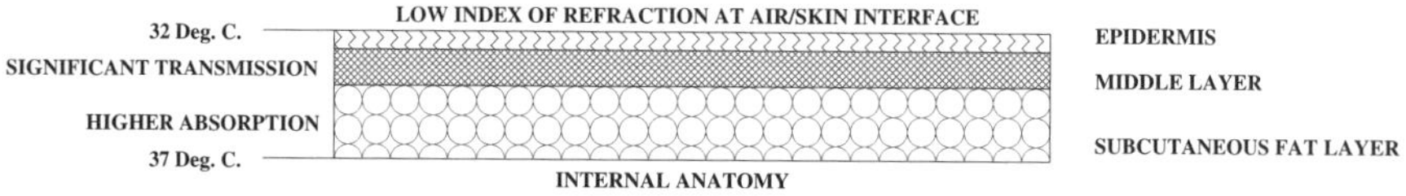

**Figure 6.14.** Preliminary thermal model for human skin.

We now proceed to compute a quantitative estimate of the average emissivity respective to the MWIR and the LWIR for human facial skin from the data in Figure 6.13. First we compute the mean thermally emitted energy of facial skin $Skin^{mean}_{energy}$. Since the thermal IR imagery used is radiometrically calibrated, we can compute the mean gray value in the histogram for the facial lobe and determine the corresponding energy by linearly interpolating

Chapter 6 Face Recognition in the Thermal Infrared 183

between the gray value peaks for the blackbody at 32°C (305 deg. K) and 35°C (308 K) and respective blackbody energies. For the MWIR this is

$$\text{Skin}_{\text{energy}}^{\text{mean}} = BB_{\text{energy}}^{305K} + [BB_{\text{energy}}^{308} - BB_{\text{energy}}^{305}] \frac{\text{Skin}_{\text{gray}}^{\text{mean}} - BB\text{gray}_{\text{max}}^{305K}}{BB\text{gray}_{\text{max}}^{308K} - BB\text{gray}_{\text{max}}^{305K}}, \quad (6.3)$$

where

$$BB_{\text{energy}}^{308K} = \int_3^5 Q(\lambda, 308K)d\lambda,$$

$$BB_{\text{energy}}^{305K} = \int_3^5 Q(\lambda, 305K)d\lambda.$$

For the LWIR replace $Q(\lambda, T)$ with $W(\lambda, T)$ and integration occurs over wavelengths from 8 to 14 microns.

We then make a conservative estimate of the lower bound for average emissivity, $\epsilon$, by comparing the mean thermally emitted energy of facial skin to a blackbody at internal body temperature 37°C. This yields:

$$\epsilon_{\text{mwir}}^{\text{skin}} > \frac{\text{Skin}_{\text{energy}}^{\text{mean}}}{\int_3^5 Q(\lambda, 310K)d\lambda} = 0.91,$$

$$\epsilon_{\text{lwir}}^{\text{skin}} > \frac{\text{Skin}_{\text{energy}}^{\text{mean}}}{\int_8^{14} W(\lambda, 310K)d\lambda} = 0.97.$$

These lower bounds are conservative as this effectively assumes that thermal emission is being sensed from a material that has a temperature of 37°C throughout. In reality there is a temperature gradient from the skin surface at 32°C through skin layers and blood vessels eventually to 37°C internal body temperature. The average temperature lies somewhere between 32 and 37°C. It is clear that skin at least has high emissivity in the MWIR and extremely high emissivity in the LWIR supporting a physical basis for excellent illumination invariance.

As the emissivity of skin is so close to 1.0, it is meaningful to quantify what is the average skin temperature due to the internal temperature gradient below the skin. This can be defined in terms of a blackbody equivalent temperature of skin, to be the temperature of a blackbody emitting equivalent energy as $\text{Skin}_{\text{energy}}^{\text{mean}}$. This temperature, $\text{Skin}BB^T$, can be computed by numerically solving the following integral equations:

$$\int_3^5 Q(\lambda, \text{Skin}BB_{\text{MWIR}}^T)d\lambda = \text{Skin}_{\text{MWIR energy}}^{\text{mean}},$$

$$\int_8^{14} W(\lambda, \text{Skin}BB_{\text{LWIR}}^T)d\lambda = \text{Skin}_{\text{LWIR energy}}^{\text{mean}}.$$

From the data presented in Figure 6.13 we compute:

$$\text{Skin}BB^T_{\text{MWIR}} = 34.3°C, \qquad \text{Skin}BB^T_{\text{LWIR}} = 34.7°C.$$

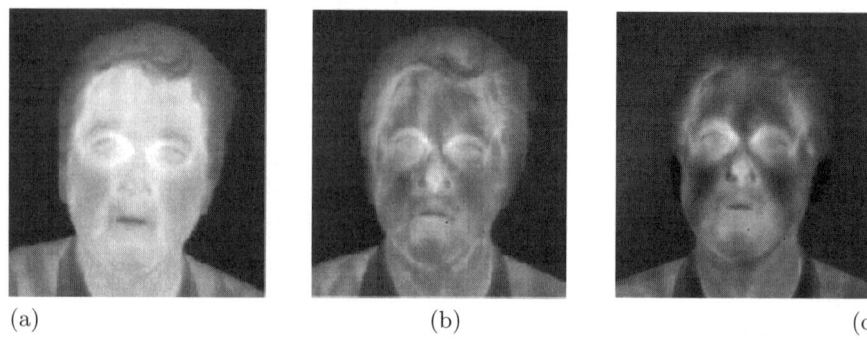

(a) (b) (c)

**Figure 6.15.** Face imaged in the LWIR (a) low activity, (b) after jogging, (c) after being outdoors at 0°C ambient temperature.

Modeling thermal emission from human faces is a good first step toward improving infrared face recognition performance. Much as understanding reflective phenomenology in the visible spectrum has led to development of algorithms that take explicit account of illumination variation [17, 18, 19], the same is true for understanding of underlying emissive phenomenology in the thermal context.

## 6.6 Comparison Of Face Recognition Performance In The Visible And Thermal Infrared

Over the course of the last two years, successively more comprehensive performance testing of existing appearance-based face recognition algorithms has occurred on the Equinox visible/infrared database [8, 9, 20]. The algorithms tested include principal component analysis (PCA) also known in the face recognition community as eigenfaces [12], local feature analysis (LFA) [21], linear discriminant analysis (LDA) — also known in the face recognition community as *Fisherfaces* [22] — and independent component analysis (ICA) [23]. Although we have available to us imagery from the visible spectrum and three different infrared spectra, we selected for direct comparison the visible and LWIR spectra since they are the most complementary respective to reflective versus emissive phenomenology. Also of key importance has turned out to be experimentation with precisely coregistered fusion of visible and LWIR imagery.

Prior to the mid-1990s, Wilder et al. [15] had directly compared performance on a smaller dataset of visible and thermal infrared imagery. In this case the thermal infrared was MWIR imagery taken with a platinum silicide sensor. Their study concluded that both modalities yielded approximately equal performance. No image fusion of visible and MWIR was tested, although it was suggested in conclusion that such fusion might be beneficial.

Only the most basic features of testing conducted in [8, 9, 20] will be reviewed presently, and these references should be consulted for further details. Prior to testing, each face image is preprocessed using standard geometric image normalization techniques by manually locating eye features and frenulum. These images are then subsampled and subsequently cropped to remove all but the inner face. Figure 6.16 shows examples of normalized visible and LWIR image pairs from the Equinox database. As discussed in [24] face recognition performance is analyzed using pairs of sets called gallery gallery and *probe* face image sets[6]. The gallery is an exemplar image set of known individuals, and the probe is an image set of unknown individuals that need to be classified. For testing correct identification performance of a face recognition algorithm, consecutive gallery images can be rank-ordered with respect to how well they match an unknown probe image with the closest gallery image match being the highest rank and consecutive lower rankings corresponding to consecutively worse matches with respect to a given metric. One way of quantifying correct identification by a given algorithm is by the percentage of probe images that correctly correspond to the matched individual who is highest ranked in the gallery. Table 6.1 shows a brief summary of performances for different algorithms on visible, LWIR, and fused imagery. This test set is particularly challenging for two reasons. First, the gallery and probe images were taken at different times, ranging from six months to two years apart. Secondly, while all the gallery images were acquired indoors, a portion of the probe images were acquired outdoors. We see that a PCA-based algorithm has very low performance on both visible and LWIR. Interestingly, in this case fused performance is actually lower than LWIR performance. This occurs only when recognition performance in one or both modalities is severely impaired, as is the case here. Performance for an LDA-based algorithm is much better, and exhibits improvement when visible and LWIR results are fused. Best performance on this set is obtained with Equinox's proprietary fused algorithm, which reduces the residual error by about 23% over the fused LDA-based result.

Figure 6.17 depicts distributions which compare the performance of an LDA-based algorithm with respect to a Monte Carlo simulation of 30,000 gallery-probe image set pairs for visible, LWIR and fused visible/LWIR modalities. Figure 6.17(a) shows the distribution of top-match recognition performance for an LDA-based algorithm when applied to visible, LWIR and fused imagery. It is easy to see that mean recognition rates are considerably higher for LWIR imagery than for visible imagery, and that when both modalities are fused, recognition performance climbs even higher. Not only is the mean

---

[6] A third set, the *training* set, is used to determine algorithm parameters, and is disjoint from gallery and probe sets.

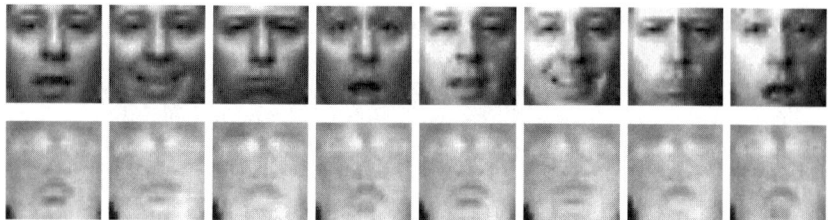

**Figure 6.16.** Example of visible (top) and LWIR (bottom) normalized face images.

**Table 6.1.** Top match recognition performance summary for different algorithms on visible, LWIR and fused imagery. Probe images are six months to two years older than corresponding gallery images. All gallery images are taken indoors, while some probes are taken outdoors.

|         | PCA Angle | LDA Angle | Equinox |
|---------|-----------|-----------|---------|
| Visible | 19.355    | 55.323    |         |
| LWIR    | 30.968    | 61.452    |         |
| Fused   | 23.548    | 74.451    | 80.323  |

correct identification highest for fused visible/LWIR but note the smaller standard deviation, indicating more stability over variations in gallery and probe sets. In Figure 6.17(b), we see paired performance differences for the same set of experiments. In this case for each random experiment, the performance difference between LWIR and visible and fused and visible becomes one data point. The distribution of these differences, shown in the figure, indicates that LWIR affords an average performance gain of 6 percentage points over visible imagery, while fusion of visible and LWIR increases that gain to 9 percentage points. Note that this constitutes a reduction of the residual error by 75% when using fused imagery versus visible imagery alone.

Figure 6.18 shows a set of receiver operating characteristic (ROC) curves for an LDA-based facial verification algorithm applied to visible, LWIR and fused visible/LWIR imagery. These curves are obtained by averaging the results from thousands of experiments generated by randomly selecting different nonoverlapping gallery and probe sets. Recall that an ROC curve shows the trade-off between correct verification versus false acceptance, as the security setting of the system is varied from low to high. The equal-error-rate (EER) is the point on the curve at which false acceptance equals correct verification, and is often used as a scalar summary of the entire ROC curve, with lower values indicating higher performance. In this case, we see that the use of fused visible/LWIR imagery cuts the EER by more than 50% versus visible imagery alone.

An end-to-end face recognition system based upon coregistered visible and LWIR imagery has been developed. The bottom row of images in Figure 6.5 illustrates a visualization of fused visible/LWIR imagery. This system, called

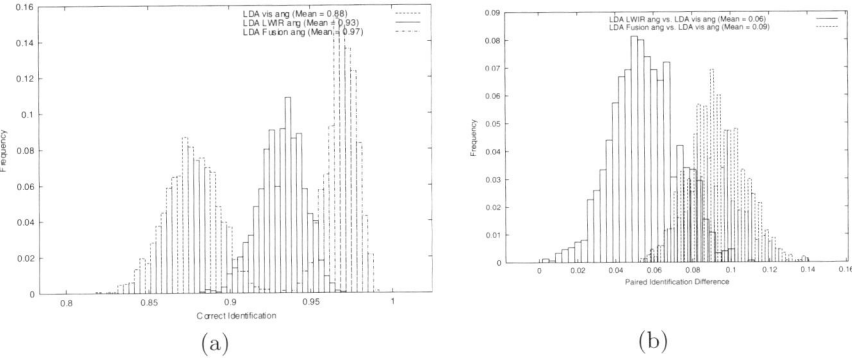

**Figure 6.17.** Performance comparisons of linear discriminant analysis (LDA) for visible, LWIR and fused visible/LWIR modalities. (a) Performance distributions, (b) paired performances. Taken from [20].

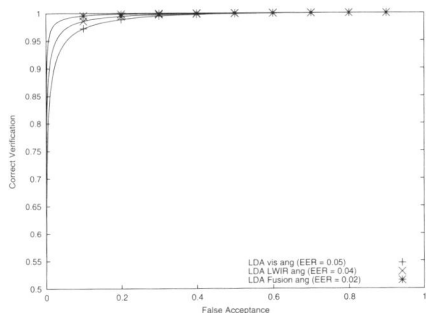

**Figure 6.18.** Receiver operator characteristic (ROC) curve comparing the same LDA-based algorithm for visible, LWIR, and fused visible/LWIR modalities. Taken from [20].

the Equinox access control environment (ACE), is capable of enrolling individuals into a database, and then for unknown individuals automatically detecting their faces in an image and recognizing whether they belong to the database. Face detection is another involved technical aspect separate from the recognition stage, and is beyond the scope of this article [25]. A recent demonstration of the Equinox ACE system over three days in July 2002 enrolled 105 individuals into this system at a trade show. Approximately two-thirds of these individuals returned to be recognized of which well over 90% were correctly identified as a top ranked match in the database. This included subjects that purposely attempted to fool the system by partially obstructing the view of their face, or attempting to mask their thermal appearance by applying ice cubes to their face. Plate I shows one of the interfaces for this system explained in the caption. Note that while the system exploits fused visible and thermal imagery, the interface may be used in visible-only mode for ease of interpretation by the operator.

## 6.7 Conclusions, Challenges Ahead, and On-Going Work

This article provides a broad overview of research that has been proceeding at Equinox Corporation for the past two years on using thermal infrared imagery for enhancing face recognition. The following key aspects were described:

- Collection of a comprehensive database of thermal infrared imagery of human faces incorporating radiometric calibration, multiple illumination conditions, and imagery of duplicated individuals over time.
- Extensive experimental testing of the performance of appearance-based face recognition algorithms, directly comparing performance on visible, LWIR, and fused visible/LWIR modalities.
- Quantification of illumination invariance and examination of physical phenomenology responsible for illumination invariance of human faces in thermal infrared imagery.
- Development of a working end-to-end face recognition system using a novel sensor configuration that precisely coregisters visible and LWIR image modalities.

Statistically significant evidence was presented indicating that appearance-based face recognition algorithms applied to thermal infrared, particularly LWIR imaging, have consistently better performance than when applied to visible imagery. Application of these algorithms to fused visible/LWIR consistently showed even better improvement in performance.

To date, the largest issue not yet addressed by face recognition using thermal infrared is performance analysis under different extreme activity levels and extreme ambient temperature. It should be carefully noted that examples of extreme varying activity levels, which pose a potentially serious disadvantage to recognition using thermal infrared, have not yet been incorporated into the Equinox visible/infrared database. By virtue of the fact that the Equinox database is comprised of imagery taken at intervals of six months for over two years means that some range of normal activity level must be inherently incorporated for duplicate individuals. So far, this does not appear to have an adverse effect on recognition performance. More precisely, performance degradation over time is similar for visible and thermal imagery, with a possible advantage toward the thermal modality. However, it is important to note that data collections under different deliberate changes in activity level do need to be performed. Figure 6.15 shows the large difference in thermal infrared signatures for the same face at rest, after jogging and after coming in from outdoors at winter time. Additionally, the effect of other confusers such as heavy makeup application must be evaluated. Existing data is not sufficient to perform a valid evaluation, but we intend to collect data specifically for this task. Evidence from small-scale experiments performed in-house indicates that recognition performance using fused visible/LWIR imagery remains high in the presence of differences in facial hair or glasses between gallery and probe images. These variations in appearance do not noticeably hinder our system's ability to recognize faces. We should, however, point out that these results may not be statistically significant, due to the small sample size. We do expect to see some performance degradation in a large-scale experiment. It has already been noted that thermal infrared imagery has the potential for being used to identify an individual's activity state and even state of inebriation [14, 26]. Unfortunately, this benefit may counterbalance to some degree the performance accuracy of unique face recognition capability. On-going work is incorporating more thermal infrared face imagery in outdoor environments, and will also shortly include varying activity and ambient temperature conditions as well.

# References

[1] Dowdall, J., Pavlidis, I., Bebis, G.: A face detection method based on multib and feature extraction in the near-IR spectrum. In: Proceedings IEEE Workshop on Computer Vision Beyond the Visible Spectrum: Methods and Applications, Kauai, Hawaii (2002)

[2] Pavlidis, I., Symosek: The imaging issue in an automatic face/disguise of detection system. In: Proceedings IEEE Workshop on Computer Vision Beyond the Visible Spectrum: Methods and Applications, Hilton Head (2000)

[3] Horn, B.: Understanding image intensities. Artificial Intelligence (1977) 1–31

[4] Horn, B., Sjoberg, R.: Calculating the reflectance map. Applied Optics **18** (1979) 1770–1779

[5] Siegal, R., Howell, J.: Thermal Radiation Heat Transfer. McGraw-Hill, New York (1981)

[6] Prokoski, F.: Method for identifying individuals from analysis of elemental shapes derived from biosensor data. In: U.S. Patent 5,163,094, November 10 (1992)

[7] DARPA Human Idenification at a Distance (HID) Program, Equinox Corporation, P.W.: Using Visible and Thermal Infrared Imagery for Human Identification. DARPA/AFOSR Contract# F49620-01-C-0008 (2000-2003)

[8] Socolinsky, D., Wolff, L., Neuheisel, J., Eveland, C.: Illumination invariant face recognition using thermal IR imagery. In: Proceedings of IEEE Conference on Computer Vision and Pattern Recognition (CVPR), Kauai, Hawaii (2001)

[9] Socolinsky, D.A., Selinger, A.: A comparative analysis of face recognition performance with visible and thermal infrared imagery. In: Proceedings IAPR International Conference on Pattern Recognition, Quebec, Canada (2002)
[10] Wyszecki, G., Stiles, W.S.: Color Science : Concepts and Methods, Quantitative Data and Formulae. Wiley Series in Pure and Applied Optics, John Wiley & Sons (1981)
[11] Dereniak, E., Boreman, G.: Infrared Detectors and Systems. John Wiley & Sons (1996)
[12] Turk, M., Pentland, A.: Eigenfaces for recognition. J. Cognitive Neuroscience **3** (1991) 71–86
[13] Adini, Y., Moses, Y., Ullman, S.: Face rcognition: The problem of compensating for changes in illumination direction. IEEE Transactions on Pattern Analysis and Machine Intelligence **19** (1997) 721–732
[14] Prokoski, F.: History, current status, and future of infrared identification. In: Proceedings IEEE Workshop on Computer Vision Beyond the Visible Spectrum: Methods and Applications, Hilton Head (2000)
[15] Wilder, J., Phillips, P., Jiang, C., Wiener, S.: Comparison of visible and infrared imagery for face recognition. In: Proceedings of 2nd International Conference on Automatic Face & Gesture Recognition, Killington, VT (1996) 182–187
[16] P. J. Bickel, K. A. Doksum: Mathematical Statistics. Prentice-Hall, Englewood Cliffs, NJ (1977)
[17] Shashua, A.: On photometric issues in 3D visual recognition from a single 2D image. IJCV **21** (1997) 99–122
[18] Shashua, A., Raviv, T.R.: The quotient image: Class-based re-rendering and recognition with varying illuminations. IEEE TPAMI **23** (2001) 129–139
[19] Zhao, W., Chellappa, R.: Robust Face Recognition using Symmetric Shape-from-Shading. Technical report, Center for Automation Research, University of Maryland, College Park, MD (1999) Available at http://citeseer.nj.nec.com/zhao99robust.html".
[20] Socolinsky, D.A., Selinger, A., Neuheisel, J.: Face recognition with visible and thermal infrared imagery. Computer Vision and Image Understanding (CVIU) Special Issue on Face Recognition (2003) Submitted.
[21] Penev, P., Attick, J.: Local feature analysis: A general statistical theory for object representation. Network: Computation in Neural Systems **7** (1996) 477–500
[22] Belhumeur, P., Hespanha, J., Kriegman, D.: Eigenfaces vs. Fisherfaces: Recognition using class specific linear projection. IEEE Transactions PAMI **19** (1997) 711–720
[23] Comon, P.: Independent component analysis: a new concept? Signal Processing **36** (1994) 287–314

[24] Phillips, P., Moon, H., Rizvi, S., Rauss, P.: The FERET Evaluation Methodology for Face-Recognition Algorithms. Technical Report NIS-TIR 6264, National Institiute of Standards and Technology (1999)
[25] Eveland, C., Socolinsky, D., Wolff, L.: Tracking faces in infrared video. In: Proceedings IEEE Workshop on Computer Vision Beyond the Visible Spectrum: Methods and Applications, Kauai, Hawaii (2002)
[26] Prokoski, F.: Method and apparatus for recognizing and classifying individuals based on minutiae. In: U.S. Patent 6,173,068, January 9 (2001)

# Chapter 7

# Cardiovascular MR Image Analysis

Milan Sonka[1], Daniel R. Thedens[1], Boudewijn P. F. Lelieveldt[2], Steven C. Mitchell[1], Rob J. van der Geest[2], and Johan H. C. Reiber[2]

[1] The University of Iowa, Iowa City, USA
 {milan-sonka, dan-thedens, steven-mitchell}@uiowa.edu;
[2] Leiden University Medical Center, Leiden, The Netherlands
 {b.p.f.lelieveldt, rvdgeest, j.h.c.reiber}@lumc.nl

**Summary.** Magnetic resonance (MR) imaging allows 2D, 3D, and 4D imaging of living bodies. The chapter[1] briefly introduces the major principles of magnetic resonance image generation, and focuses on application of computer vision techniques and approaches to several cardiovascular image analysis tasks. The enormous amounts of generated MR data require employment of automated image analysis techniques to provide quantitative indices of structure and function. Techniques for 3D segmentation and quantitative assessment of left and right cardiac ventricles, arterial and venous trees, and arterial plaques are presented.

## 7.1 Introduction

Cardiovascular disease is the number one cause of death in the western world. Cardiac imaging is an established approach to diagnosing cardiovascular disease and plays an important role in its interventional treatment. Three-dimensional imaging of the heart and the cardiovascular system is now possible with x-ray computed tomography, magnetic resonance, positron emission tomography, single photon emission tomography, and ultrasound, to name just the main imaging modalities. While cardiac imaging capabilities are developing rapidly, the images are mostly analyzed visually and therefore qualitatively. The ability to quantitatively analyze the acquired image data is still not sufficiently available in routine clinical care. Large amounts of acquired data are not fully utilized because of the tedious and time-consuming character of manual analyses. This is even more so when 3D image data need to be processed and analyzed.

In this chapter, we will concentrate on cardiac image analysis that uses magnetic resonance (MR) to depict cardiovascular structure. After briefly describing capabilities of MR to image the heart and vascular system, we will

---

[1] Portions reprinted, with permission, from *IEEE Transactions on Medical Imaging*, Volume 21, pp. 1167–1178, September 2002. © 2002 IEEE.

## 7.2 Capabilities of MRI

**Principles of MR Imaging**

Magnetic resonance imaging (MRI) relies on the phenomenon of *nuclear magnetic resonance* to generate image contrast [1, 2]. The hydrogen atom (along with other species having an odd number of protons or neutrons, such as sodium and phosphorous) possess a spin angular momentum. The single proton of the hydrogen atom (often referred to in this context as a *spin*) is by far the most abundant, and thus is considered in the vast majority of imaging applications. Most importantly, for the purposes of imaging, the spins will give rise to a magnetic moment and will act like microscopic bar magnets. As a result, when the protons are placed in a strong static magnetic field, at equilibrium they tend to line up in the same direction as the external field. The net effect of all the spins lined up in this way generates a small but measurable magnetization along the longitudinal direction of the large external field. The magnitude of this magnetization increases as the strength of the external field is increased.

By itself, this magnetization does not give much useful information about the distribution of the protons within the object. The application of a second small (relative to the primary strong field) magnetic field oscillating in the radiofrequency (RF) range sets up a resonance condition and will perturb the spins away from their equilibrium state, "tilting" them away from their alignment with the main field into the transverse plane. Much like a gyroscope, this will *excite* the spins causing them (and their magnetic fields) to precess about the direction of the main field, and the rate at which the spins precess is directly proportional to the strength of the main magnetic field. Figure 7.1 shows the relationship between the two magnetic fields and the resulting perturbation of the magnetization vector. A fundamental principle of electromagnetics is that a time-varying magnetic field can induce an electric current in an appropriately placed coil of wire, generating a signal that can measure the distribution of the spins within the object. Since the rate of precession depends on the magnetic field strength, slightly varying the strength across the bore of the magnetic with *gradient* fields yields a spatially varying rate of precession [3]. When the RF field is removed, the spins begin to return towards their equilibrium state aligned with the strong static magnetic field.

The rate of return of spins to their equilibrium state is governed by two time constants intrinsic to different tissue types, T1 and T2. T2 determines how long it will take for the signal generated by the "tipped" spins to decay away. T1 measures the amount of time it takes for the spins to completely return to their equilibrium alignment with the main magnetic field (see Figure

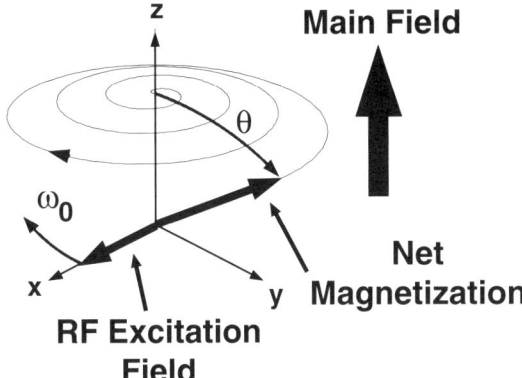

**Figure 7.1.** Generation of MRI signal with RF field. Application of a small rotating magnetic field at the resonant frequency $\omega_0$ causes the magnetization vector to tilt and precess into the transverse $(x-y)$ plane by an angle $\theta$. The precessing transverse component generates the MR signal, which can be detected by a receiver coil.

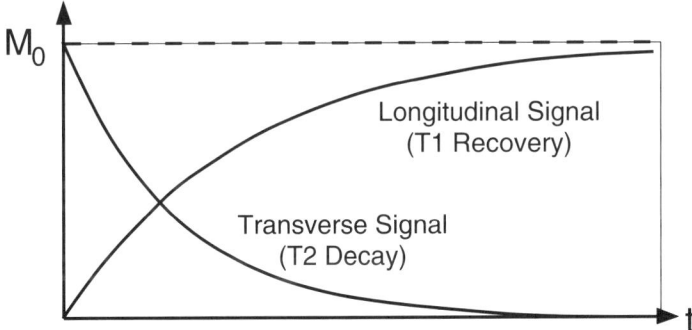

**Figure 7.2.** Relaxation of transverse and longitudinal magnetization. After excitation, the measurable transverse signal decays away with time constant T2, while the longitudinal magnetization is recovered with a time constant T1. Both constants are dependent on tissue type and strength of the main field.

7.2. Because of this signal decay, an MR imaging experiment generally must consist of several cycles of signal generation followed by signal measurement or acquisition.

Hence, the signal measured from a tissue will depend on its density of protons as well as its T1 and T2 relaxation parameters. Motion and flow also contribute to the final signal generated. The remarkable ability of MRI to generate a wide variety of tissue contrast arises from the fact that the imaging experiment can be designed to vary the relative weight of each of these parameters in the measured signal. For example, muscle and fat have very different T1 and T2 parameters and by varying the timing of the applied RF excitation pulses, maximum contrast between the two can be achieved. Other strategies may enhance or suppress flowing blood compared to stationary tissues.

## Challenges of Cardiovascular MRI

Cardiovascular imaging presents formidable challenges for any imaging modality as well as some unique to MRI. Because the signal generated by a single RF excitation decays away quite rapidly (on the order of milliseconds), formation of a complete image nearly always requires repeated excitations and data acquisition. Motion of tissues between acquisitions or during a single acquisition causes blurring and other artifacts that compromise image quality. The heart is in constant motion throughout the cardiac cycle, and respiration also causes the heart and other organs of the chest and abdomen to shift positions throughout duration of an image acquisition. Both types of motion must be suppressed in some way if high-resolution images of the cardiovascular anatomy are to be generated.

The problem of cardiac motion is overcome by synchronizing each excitation and data acquisition to the heart cycle, as measured by electrocardiography [4]. Each data acquisition is acquired during a narrow window of time at the same point or phase of the heart cycle, yielding a correctly registered set of measurements that can be reconstructed with minimal motion artifact. The drawback is that only a fraction of each heart cycle can be utilized for data collection, causing a concomitant increase in the time required to acquire an image. This time can be used to acquire images from the same or other locations at different points within the heart cycle without an additional time penalty, however.

Respiratory motion is generally not as predictable as cardiac motion and occurs over a different time scale. Three basic strategies have been used to overcome image degradations. For relatively rapid two-dimensional techniques that can be acquired in fewer than 15 to 20 heartbeats, subjects are instructed to hold their breath for the duration of the scan. Longer scans can be achieved with repeated breath-holding, though this requires a consistent breath-hold position on the part of the subject to achieve optimal results. Respiratory triggering is another option, whereby data are acquired only during a limited part of the respiratory cycle (analogous to cardiac triggering). A more sophisticated method for lengthy 2D and 3D acquisitions is navigator gating [5, 6]. For each image data acquisition, an additional set of pulses determines the position of the diaphragm and adjusts the imaging plane and data acquisition to minimize or track this motion. This permits high-resolution and artifact-free images while the subject breathes freely.

Recently developed real-time imaging strategies can acquire a complete image in a matter of milliseconds, eliminating the need for either type of motion suppression [7]. Such techniques still remain limited in the resolution and signal-to-noise ratio (SNR) that can be achieved and cannot yet replace the more time-consuming scan protocols.

**Cardiac Morphology**

The most basic use of cardiac MRI is to depict the structure or morphology of the heart. Two general classes of imaging techniques are widely used for cardiac imaging, commonly referred to as black-blood and bright-blood techniques.

**Black-Blood Imaging.** Black-blood images are produced by T2-weighted spin-echo (SE) imaging sequences, in which two RF excitations (an excitation pulse and an inversion pulse) are applied to the imaged volume [4]. After the excitation pulse, the excited spins begin to lose coherence due to slight variations in their resonant frequencies, resulting in a rapid loss of overall signal. The inversion pulse "flips" the magnetization about one of the axes permitting these spins to regain their coherence and generate an *echo* when the signal has been restored. When the two pulses are separated by a sufficient interval, flowing blood experiences only one of these pulses and thus does not produce a restored signal echo, leaving a flow void in the chambers of the heart. The timing of the two RF pulses sets the echo time (TE) at which the signal refocuses (and data are acquired) and determines the precise signal and contrast features of the image. For black blood imaging, a TE of at least 20 msec is usually used. A longer TE yields greater contrast based on T2 characteristics of the tissues, which may be useful to identify such lesions as acute myocardial infarction or myocardial scar. This comes at the expense of reduced overall signal due to signal decay. Standard SE sequences show excellent contrast among myocardium (medium intensity), epicardial fat (high intensity), and flowing blood (low intensity). The signal void created by SE sequences generates images with especially good contrast in endocardial regions, valves, and vessel walls.

The main limitation of standard SE sequences is the acquisition time required in a cardiac-triggered exam, which results in poor temporal resolution and the prospect of significant respiratory motion artifact. Fast SE (FSE) sequences overcome this limitation by applying multiple inversion pulses and additional signal readouts during a single cardiac cycle. Speedups of an order of magnitude are possible in this way. However, the longer readout times degrade the image contrast due to the more complex dependence on relaxation times.

The currently preferred black-blood technique for imaging cardiac morphology is a T2-weighted inversion recovery (IR) pulse sequence [8]. This sequence applies additional RF excitation pulses to effectively null the signal from blood (and possibly fat as well) based on its T1 relaxation parameters. This is usually followed by a FSE sequence that can be acquired in 15 to 20 heartbeats, suitable for a breath-held acquisition and yielding a robust black-blood sequence with T2 contrast.

**Bright-Blood Imaging.** Bright-blood images originate from gradient echo (GRE) imaging sequences which only use a single RF excitation, relying on the gradient hardware instead of an inversion pulse to refocus the signal for data acquisition. Much shorter TE times (1–10 msec) are used, and the excitation and data readouts can be repeated more frequently (every 10–20 msec). Because blood need only experience the single RF pulse to generate a signal, it appears brighter than myocardium on GRE acquisitions. The short TE between excitation and data readout enhances this effect since there is less time for signal decay due to relaxation. Additional flow-compensation pulses can also be applied to further enhance blood signal and improve contrast with nearby myocardium. As with FSE imaging, the fastest imaging sequences utilize multiple excitations and data readouts over an extended interval (80 msec is a typical duration) synchronized to the cardiac cycle to generate images that can be acquired within a breath-holding interval [9]. Contrast between blood and myocardium is generally not as good as with SE imaging, as varying flow profiles may result in heterogeneous blood pool.

The availability of faster gradient hardware has seen a resurgence of techniques based on steady-state free precession (SSFP) [10]. SSFP maximizes the use of signal from blood by applying rapid excitations repeated at very short intervals. The resulting contrast is a function of relaxation parameters as $T1/T2$. The short repetition times greatly reduce flow effects and show a more homogeneous depiction of myocardial blood pool, which in turn improves contrast with myocardium and visualization of papillary muscles. Rapid excitations also permit better temporal resolution [11, 12], or the time savings can be traded off for higher resolution at the same time resolution. As state-of-the-art MR gradient hardware proliferates, SSFP will likely become even more common.

The rapid repetition of readouts in both GRE and SSFP mean that several images at the same location can be taken at different time points within the heart cycle. Alternatively, the imaging time can be used to acquire multiple slices at a reduced temporal resolution. Using segmented acquisitions, a multi-slice multiphase view of the cardiac morphology can be acquired within a single breath-hold of 15 to 20 heartbeats.

### Cardiac Function

Many of the techniques mentioned above for imaging of cardiac morphology, including both black-blood and bright-blood imaging, are also suitable for measuring cardiac function indices as well. Compared to other modalities, MRI has the advantage that completely arbitrary image orientations can be chosen, guaranteeing that true long-axis or short-axis views serve as the basis for quantitative measurements. The availability of three-dimensional information in the form of multiple parallel slices eliminates the need for any geometric assumptions about ventricular anatomy when estimating masses and volumes, a significant advantage over x-ray and ultrasound.

Bright-blood GRE imaging is more commonly used for evaluation of ventricular function. The shorter acquisition time permits a greater number of slices to be acquired during the cardiac cycle, which can be used for higher temporal resolution (more frames per cycle) or for a greater volume coverage (more slice locations). The acquisition of images at multiple phases of the cardiac cycle is known as cine MRI (example shown later in Figure 7.8) [13]. With present system hardware, a complete multislice multiphase cine data set suitable for quantitative analysis can be acquired in a single breath-hold interval. The limiting factor with standard GRE imaging is the contrast between medium-intensity myocardium and the bright blood pool. Areas of slower flowing blood will demonstrate reduced intensity making delineation of the endocardial contours difficult.

The recent advances in SSFP imaging cited above may solve this problem to some degree with its more robust contrast. The faster repetition time used in SSFP also increases the frame rates possible in a cine study. With state-of-the-art gradient hardware, truly 3D cine MRI with no gaps between slices is now possible within a single breath-hold interval [14, 15].

Improving gradient and computing hardware has now made real-time imaging feasible for functional imaging. Rates of 16 frames per second or more can be continuously obtained much like x-ray fluoroscopy [7]. The scan plane can be modified directly on the real-time images, dramatically reducing the time required for "scout" scans to find the proper short-axis orientation. At such rates, cardiac gating and breath-holding are unnecessary, which permits imaging of patients with arrhythmias. Presently, spatial resolution of real-time studies remain comparatively limited, but a number of ongoing developments in image reconstruction techniques are improving this. Two such strategies exploit the widespread use of multiple receiver coils. Simultaneous acquisition of spatial harmonics (SMASH) [16] and sensitivity encoding (SENSE) [17] use the spatially varying response of a group of coils as an additional means of spatial encoding to reduce the time needed to acquire a given resolution image. Other techniques analyze the temporal dimension of the acquisition to reduce the acquisition of redundant information and enhance either temporal or spatial resolution [18].

Each of these forms of cine and real-time MRI data are useful for computing several global measures of cardiac function. Accurate and reproducible quantitative measurements of ventricular volumes at both systole and diastole, masses, and ejection fraction (difference between the diastolic and systolic ventricular volumes) are all computable with multislice or volume data sets. In each case, myocardial border identification is necessary to extract quantitative results. Compared to x-ray and ultrasound, MRI also accurately depicts epicardial borders, again eliminating the geometric assumptions that often must be made in competing modalities. As a result, regional myocardial function assessments can also be made with cine techniques. This may be done subjectively viewing cine or real-time "loops" or through quantitative measurements of regional wall thickness and strain.

Regional measurements of three-dimensional strain is possible using myocardial tagging. This imaging method excites myocardium with a pattern of lines or grids whose motion can then be tracked over the heart cycle, providing a precise depiction of the deformations occurring within the myocardial tissues. Analysis of these deformations in short- and long-axis views gives 3D strain measurements useful for determining local myocardial function. A promising rapid technique is harmonic phase (HARP) imaging which has potential as a real-time technique [19].

**Myocardial Perfusion**

Another important indicator that can be assessed by MRI is regional blood flow (or perfusion) in the myocardium. This may indicate areas of damage to myocardium from a cardiac event or insufficient blood flow resulting from a significant arterial stenosis. Determination of blood flow within the myocardium depends on the use of contrast agents (usually gadolinium-based) that change the relaxation characteristics of blood, particularly the T1 relaxation time [20]. Gadolinium causes a considerable shortening of the T1 relaxation time, meaning that magnetization returns to equilibrium much more rapidly. As shown in Figure 7.3, when RF excitation pulses are applied in rapid succession, tissues with short T1 relaxation will still have time to recover and generate greater signal for subsequent excitations. Longer T1 relaxation times means that little magnetization has returned to the equilibrium state, so later excitations result in much less signal. Appropriate timing of a pair of RF pulses can maximize the signal difference between two tissues with known T1 relaxation times.

Perfusion is mostly measured during the "first pass" into the myocardium after injection of contrast agent [21, 22]. Areas of myocardium with adequate blood flow will have enhanced intensity from the shortened T1 of the inflowing blood. Perfusion deficits will not receive this material and remain at lower intensity. The time of the imaging window is limited as contrast material may soon begin to diffuse from normal to deficit regions, and the contrast agent will recirculate with the blood within 15 seconds. Hence, rapid GRE sequences are used to image quickly and permit multiple slices to be obtained over a volume. T1 contrast is maximized by applying an RF "preparation" pulse that initially excites or saturates all of the blood and tissues. After a delay time that causes contrast-enhanced material to return towards equilibrium while the longer T1 tissues recover much less magnetization to yield strong T1 contrast, a standard fast GRE imaging sequence is applied. The result is bright signal in normal tissue and low-intensity regions of perfusion deficit. Acquisition of several time frames during this process permits quantitative measurements of the severity of these perfusion abnormalities.

Further myocardial tissue characterization is possible using gadolinium contrast agents by waiting a considerable duration (20 min or more) before imaging [22, 23]. Gadolinium contrast will eventually move to the extracellular

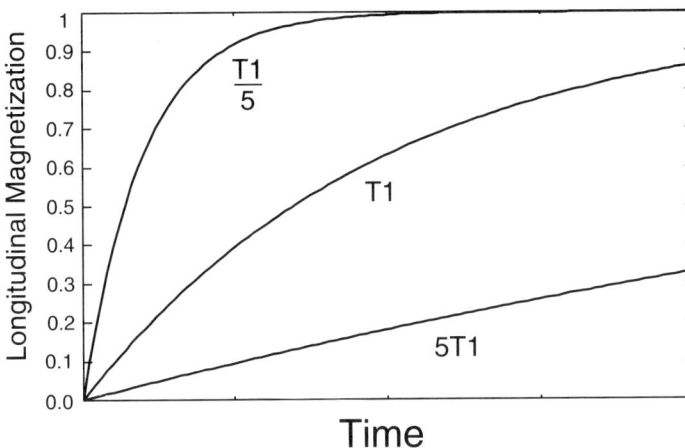

**Figure 7.3.** Generation of T1-based contrast. A typical RF excitation causes the longitudinal magnetization to go to zero, followed by recovery via T1 relaxation. The amount of recovery depends on the T1 time constant and determines the amount of signal available for imaging during the next excitation. When the repetition time is brief, short T1 species will recover and generate much greater signal than longer T1 species.

space and accumulate more in areas of nonviable myocardium, resulting in enhanced signal in these areas on T1 weighted images compared to normal tissue.

## Angiography

In addition to imaging of the heart, MRI has also been widely applied to imaging vessels throughout the body. Its advantages over conventional X-ray angiography go beyond the fact that it is much less invasive. MRI can also collect true 3D data, permitting arbitrary selection of views and slices in post-processing to optimize the visualization of vessels. This is especially helpful in complex vascular trees where tracing the vessel of interest may be difficult. Contrast for MR angiography can be developed in two ways. Pulse sequences may exploit the different signal properties of flowing and stationary tissues to produce images. Other sequences rely on the relaxation characteristics of arterial and venous blood, usually enhanced by T1-shortening contrast agents as described for myocardial perfusion. In both cases, the goal is to generate images of the vessel lumen suitable to detect and evaluate stenoses.

Two flow-based imaging techniques are in common use for MR angiography and both effectively produce "bright-blood" images of the vessel lumen. Phase-contrast (PC) imaging takes advantage of the fact that flowing blood will move during the data acquisition readout. Since spatial information is encoded by a spatially varying magnetic field gradient, flowing spins experience a changing magnetic field as they move, resulting in a phase change in their signal compared to stationary tissues. By applying an appropriate encoding gradient pattern prior to imaging, flowing blood can be selectively viewed. PC imaging can also quantitatively measure flow velocities. Time-of-flight (TOF) imaging uses the continuous replacement of flowing blood in the imaged slice to differentiate it from static tissue. Rapid repetition of excitation pulses covering the imaged slice saturates and eventually eliminates signal from stationary material because there is not enough time to regain any equilibrium magnetization. Flowing blood retains a signal since fresh unsaturated blood is constantly flowing into the slice to be excited and flows away again before saturation can be complete. The result produces high signal from flowing blood against the low intensity of background structures.

Reliance on flow for image contrast may introduce artifacts where flow patterns are not ideal. Such anomalies will affect both PC and TOF sequences. Areas of slow flow may have reduced signal, either due to reduced phase changes for PC or saturation in TOF. Complex flow patterns and turbulence can also cause reduced intensities within the vessel lumen in both cases. The consequences could include stenoses that are overestimated or a false appearance of an occlusion of the vessel.

The limitations of flow-based angiography have made flow-independent techniques more prevalent. It is possible to create high-contrast angiographic images using only the intrinsic T1 and T2 relaxation characteristics of blood using a variety of "prepared contrast" pulses that saturate or suppress one or more background tissues [24]. However, injectable contrast agents such as those based on gadolinium compounds have proven to be safe and well tolerated and are widely available. These contrast agents dramatically reduce the T1 relaxation time of blood and greatly enhance its signal on TOF images. Much of MR angiography is now dominated by contrast-agent-based protocols.

Once again, the main limiting factor in contrast studies is the time before the contrast agent leaks outside the blood vessels and begins to enhance the signal in tissues other than blood. Successful contrast angiography therefore requires careful timing of contrast injection and image acquisition and a rapid acquisition technique to minimize artifacts due to contrast dispersion and respiratory motion. Fast 3D GRE imaging is most commonly used to acquire the T1-based contrast to yield bright contrast-enhanced blood pool. Subtraction of a non-contrast-enhanced volume may also be used to further suppress background structures. A variety of strategies have been employed to reduce the imaging time to acquire a 3D data set even further and assure accurate timing of the acquisition. Partial acquisition methods which acquire 60–75% of a full data set and synthesize the rest based on mathematical assumptions can help reduce imaging times. More extreme versions of this have been applied to radial sampling patterns to reduce acquisition time even further, trading the shortened time for some increased and coherent background noise [25].

The timing of the acquisition relative to the injection of contrast agent is also crucial. If the data acquisition occurs too early, the signal will not yet be enhanced, while a late acquisition will show poor contrast because of a heightened signal from other tissues or veins. For many applications, a fixed time delay based on previous experience may be sufficient, although increased doses of contrast often accompany this technique to increase the window of enhancement. A much smaller dose of contrast may be given and tracked with a sequence of rapid 2D images used to pinpoint the transit time prior to a full 3D acquisition. Automatic monitoring of the signal at a predefined location upstream from the desired location has also been implemented. The use of real-time imaging to monitor contrast passage is another possibility.

The limited volume imaging time available because of the dispersion of contrast agent into other tissues is currently being addressed. New intravascular contrast agents that do not leak into tissues during the course of a typical MR exam are being perfected by a number of researchers [26, 27]. As a result, their T1 shortening properties can be utilized for longer or multiple exams without the enhancement of background tissues. MR angiograms in higher-resolution 3D or over the whole body then become possible. The longer persistence in the blood pool does mean that both arteries and veins will be displayed for longer 3D scan durations. Some means of separating the two may be needed for diagnostic use of such images.

Coronary artery imaging may be a particular beneficiary of such contrast agents, as the necessity of high-resolution, 3D coverage, and motion correction requires longer scan times than are feasible with standard contrast material. The flow and saturation effects that often compromise 3D techniques are also improved with such contrast agents [28]. Perfection of a minimally invasive coronary MR imaging is of particular interest because of the number of highly invasive x-ray angiography procedures that are performed that show no clinically significant disease.

In summary, MR imaging shows tremendous promise to assess virtually all areas of cardiovascular health. While the much-heralded ascendance of MRI as the "one-stop shop" for noninvasive cardiovascular imaging has not yet come to pass, the current state-of-the-art and continued advances in cardiac MRI still point towards such a possibility.

## 7.3 Cardiac MR Segmentation

The rapid development of cardiac magnetic resonance (MR) acquisition techniques as described in the previous section has created a vast diagnostic potential, and within one patient examination, several aspects of cardiac function can be evaluated. A major bottleneck for cardiac MR methods in routine clinical practice however, is the prohibitively large amount of data involved in a comprehensive patient examination (typically between 2000 and 5000 images). Therefore, to utilize the full diagnostic potential of cardiac MR, highly automated quantitative analysis is essential; hence image segmentation is of primary importance to further advance the clinical utility of cardiac MR.

Though much effort has been directed to automated segmentation of cardiac MRI image data, there are three main reasons why existing methods frequently exhibit a lower success rate in comparison with human expert observers, especially when applied to clinical-quality images — existing methods do not incorporate a sufficient amount of a priori knowledge about the segmentation problem; do not consider 3D or temporal context as an integral part of their functionality; and position the segmentation boundaries at locations of the strongest local image features not considering true anatomical boundary locations.

### 7.3.1 Cardiac Segmentation Approaches

A number of 3D medical image analysis approaches occurred recently, many of them addressing one or more of the above-mentioned shortcomings of available segmentation techniques. A detailed review of existing 3D cardiac modeling approaches is provided in [29]. In the context of our work and considering the goal of segmenting 3D volumetric and temporal cardiac images and image sequences, statistical modeling of 3D shape and 3D image properties is crucial. Vemuri and Radisavljevic concentrated on a 3D model that combines deformed superquadric primitives with a local displacement field expressed on an orthonormal wavelet basis [30]. As a result of this orthonormal basis, the shape parameters become physically meaningful, and thus a preferred shape can be imposed based on parameter distributions in a set of training samples. Similarly, Staib and Duncan developed a 3D balloon model [31]. The model is parameterized on an orthonormal Fourier basis such that the statistics of the Fourier coefficients in a training set allow a constrained image search. Model fitting in these two methods is performed by balancing an internal energy term with an external, gradient derived, scalar field. Metaxas et al. introduced physics-based deformable models for modeling rigid, articulated, and deformable objects, their interactions with the physical world, and the estimate of their shape and motion from visual data [32, 33].

Cootes and Taylor and colleagues developed a statistical point distribution model (PDM) and demonstrated its utility for 2D image segmentation [34, 35]. One of the primary contributions was an ease of automated learning of the model parameters from sets of corresponding points as well as the PDM's ability to incorporate shape and boundary gray level properties and their allowed variations. Applications to segmentation of echocardiographic data [36] and deep neuroanatomical structures from MR images of the brain may serve as examples [37]. Following the point distribution model ideas, Kelemen et al. built a statistical model of 3D shapes using parametric surface representations [38]. Similar to PDMs, shape and gray level information in the boundary vicinity was incorporated in the model. The method's performance was demonstrated on 3D segmentation of neuroanatomical structures. A multiscale 3D shape modeling approach called M-reps was developed by Pizer et al. [39]. M-reps support a coarse-to-fine hierarchy and model shape variations via probabilistically described boundary positions with width- and scale-proportional tolerances. Three-dimensional echocardiographic image segmentation using core atoms was reported by Stetten and Pizer[40]. Davatzikos et al. presented a deformable model in which geometric information is embedded via a set of affine-invariant attribute vectors; these vectors characterize the geometric structure around a model point from a local to global scale, forming an adaptive focus deformable statistical shape model [41]. The methodology was applied to segmentation of neuroanatomical structures.

In all the above-referenced approaches, the models primarily hold information about shape and its allowed variations. The information about image appearance is only considered in a close proximity to the object borders. A powerful, model driven segmentation technique called active appearance model (AAM) was recently introduced by Cootes and coworkers [42, 43, 44]. An AAM describes the image appearance and the shape of an object in a set of examples as a statistical shape-appearance model. AAMs can be applied to image segmentation by minimizing the difference between the model and an image along statistically plausible shape/intensity variations (analysis by synthesis). AAMs have been shown to be highly robust in the segmentation of routinely acquired single-phase, single-slice cardiac MR [45] and echo images [46], because they exploit prior knowledge about the cardiac shape, image appearance, and observer preference in a generic way. For a detailed background on active appearance models and their application to image segmentation, the reader is referred to [43].

Two-dimensional active appearance motion models [45, 47, 46] have demonstrated the ability of time-continuous segmentation by exploiting temporal coherency in the data. However, these 2D + time AAMs do not represent a true 3D approach. Their segmentation ability is limited to cases with fixed numbers of preselected frames; they rely on a priori knowledge of image frame correspondences within each cardiac cycle. The 3D model presented below is the first such to date capable of successful segmentation of cardiac MR images [48]. The model's behavior is learned from manually traced segmentation examples during an automated training stage. The shape and image appearance of the cardiac structures are contained in a single model. This ensures a spatially and/or temporally consistent segmentation of three-dimensional cardiac images.

**Point Distribution Model Concept**

Point distribution models describe populations of shapes using statistics of sets of corresponding landmarks of the shape instances [34, 35, 49]. By aligning $N$ shape samples (consisting of $n$ landmark points) and applying a principal component analysis (PCA) on the sample distribution, any sample **x** within the distribution can be expressed as an average shape $\bar{\mathbf{x}}$ with a linear combination of eigenvectors $P$ superimposed

$$\mathbf{x} = \bar{x} + P\mathbf{b}. \tag{7.1}$$

In 2D models, $p = \min(2n, N-1)$ eigenvectors $P$ form the principal basis functions, while in a 3D model; $p = \min(3n, N-1)$ eigenvectors are formed. (The minimum operator is needed since we frequently have more corresponding shape points than training set samples.) In both cases the corresponding eigenvalues provide a measure for compactness of the distribution along each axis. By selecting the largest $q$ eigenvalues, the number of eigenvectors can be reduced, where a proportion $k$ of the total variance is described such that

$$\sum_{i=1}^{q} \lambda_i \geq k \cdot \text{Total} \quad \text{where} \quad \text{Total} = \sum_{i=1}^{p} \lambda_i. \tag{7.2}$$

### 7.3.2 Representing the Shape of 3D Cardiac Ventricles

Extending the 2D PDM to three dimensions is a nontrivial task. In order to create a compact and specific model, point correspondences between shapes are required. Even if landmark points are easily identifiable in both models, specifying a uniquely corresponding boundary surface built from points in between these landmarks is difficult in 3D. In a 2D case [45], a boundary sequence of points may be identified by evenly sampling points spanning from one landmark to the next. In a 3D case, defining a unique sampling of the object surfaces is ill-posed but the problem can be solved in simplified geometries. An approach like that was used for left-ventricular segmentation.

For the purpose of ventricular segmentation, a normalized cylindrical coordinate system is defined with its primary axis aligned with the long axis of the heart, and the secondary axis aligned with the posterior junction of the right and left ventricles in the basal slice. The cardiac ventricles resemble a cylindrical or paraboloid shape. First, contours are sampled slice-by-slice at even angle increments. To transform the rings in the normalized cylindrical coordinate system, each point on the ring is connected by a straight line to the next adjacent corresponding point on the rings above and below. Starting from the apex slice to the basal slice, a fixed number of slicing planes are placed evenly along the long axis. Apex slice was defined as the most inferior slice with a visible left ventricular cavity, slices with merely a small muscle cap were excluded. New points are interpolated where the planes intersect the lines. This yields a set of corresponding boundary points for each sampled left ventricle across the population of ventricles (Figure 7.4).

### 7.3.3 Three-Dimensional Point Distribution Models

Aligning shape samples to a common scale, rotation, and translation is important for a compact model to be generated during the PCA stage. Procrustes

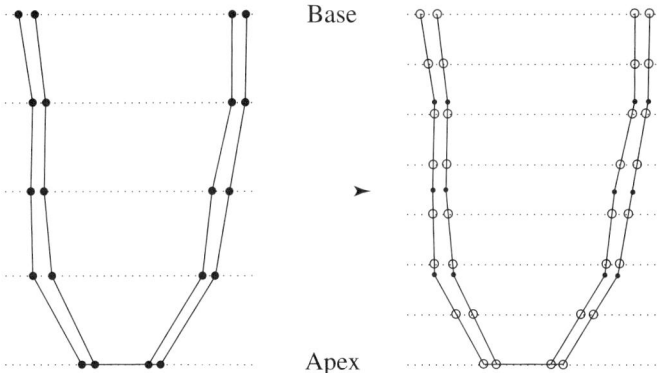

**Figure 7.4.** A cross-sectional depiction of transforming a cardiac MR stack with manually placed landmarks to a normalized cylindrical coordinate system.

analysis [50, 51] is used whereby an arbitrary shape is selected as the initial average shape estimate. All the other shapes are aligned to this average using a least-squares minimization. A new average is computed by a simple mean across the corresponding points, and the algorithm repeats until convergence.

For the 2D case, aligning one shape to another can be solved analytically by minimizing scale, rotation, and translation terms. Extending to 3D, the minimization of scaling, translations, and rotation differences along the three axes may lead to singularities known as gimbal lock. Assuming that 3D translation is represented by a separate translation vector $\mathbf{t}$, a quaternion $\mathbf{q}$ representation of scaling and rotation avoids such behavior [52].

A quaternion $\mathbf{q}$ is defined as the linear combination of a scalar term $q_0 \geq 0$, and three right-handed orthonormal vectors ($\mathbf{i}$, $\mathbf{j}$, and $\mathbf{k}$) :

$$\mathbf{q} = q_0 + q_1\mathbf{i} + q_2\mathbf{j} + q_3\mathbf{k}. \tag{7.3}$$

Together, the position and orientation of a 3D object can be represented as a seven-element pose vector $(\mathbf{q}|t) = [q_0, q_1, q_2, q_3, t_i, t_j, t_k]$.

The alignment of two 3D shape instances is accomplished using a well-known procedure given by Besl and McKay [53] to optimize for $\mathbf{q}$ and $\mathbf{t}$. Aligning all the shapes is a matter of employing the Procrustes analysis using Besl's procedure to calculate the pose parameters. Once shape alignment is finished, principal component analysis is applied to the 3D models in a way that is no different from the conventional 2D application [35].

### 7.3.4 Modeling Volume Appearance

The first part of creating an appearance model of volume is to warp all the sample volumes to the average shape to eliminate shape variation and bring

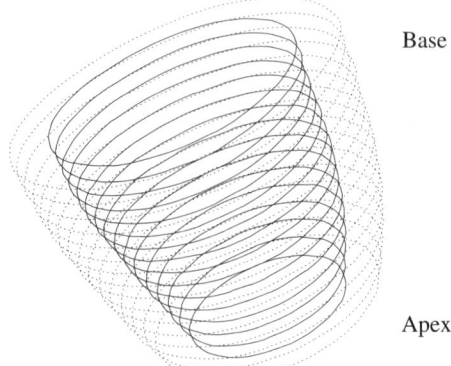

**Figure 7.5.** A wireframe representation of the mean LV shape in the normalized cylindrical coordinate system.

voxelwise correspondence across all the training samples, such that the voxel intensities can be represented as a shape-free vector of intensity values. Warping an image **I** to a new image **I**′ involves creating a function which maps control points $\mathbf{x}_i$ to $\mathbf{x}'_i$ as well as the intermediate points in between. For the 2D case, either piecewise affine warping or thin-plate spline warping is adequate. In our models piecewise warping is preferred because it is significantly faster than thin-plate spline warping.

In 2D piecewise affine warping, landmark points are used to construct the shape area as a set of triangles. The well-known Delaunay triangulation algorithm is suitable for computing such a triangular mesh and can be found in many computational geometry references. Individual triangular patches are locally warped using barycentric coordinates. Given a triangle with the three corners, $\mathbf{x}_1$, $\mathbf{x}_2$, and $\mathbf{x}_3$, we can represent any point $\mathbf{x}$ within the triangle as $\mathbf{x} = \alpha\mathbf{x}_1 + \beta\mathbf{x}_2 + \gamma\mathbf{x}_3$ where $\gamma = 1 - (\alpha + \beta)$ and $\alpha + \beta + \gamma = 0$. In order for a point $\mathbf{x}$ to fall inside a triangle, $0 \leq \alpha, \beta, \gamma \leq 1$ must be true.

Piecewise affine warping is implemented as follows:

- For each pixel location $\mathbf{x}'$ in $\mathbf{I}'$:
  - Find the triangle $t'$ which contains $\mathbf{x}'$ by solving $\alpha$, $\beta$, and $\gamma$ for each triangle and finding the triangle where $0 \leq \alpha, \beta, \gamma \leq 1$.
  - Find the equivalent pixel location $\mathbf{x}$ by computing $\mathbf{x} = \alpha\mathbf{x}_1 + \beta\mathbf{x}_2 + \gamma\mathbf{x}_3$ where $\mathbf{x}_1$, $\mathbf{x}_2$, and $\mathbf{x}_3$ are the triangle points from the original image.
  - Copy the pixel value in **I** located by $\mathbf{x}$ into the warped image **I**′ located at $\mathbf{x}'$. Some form of pixel interpolation such as bilinear may be used at this stage.

In our 3D models, piecewise affine warping is extended to tetrahedrons with four corners, $\mathbf{x}_1$, $\mathbf{x}_2$, $\mathbf{x}_3$, and $\mathbf{x}_4$. Any point within the tetrahedron is represented as $\mathbf{x} = \alpha\mathbf{x}_1 + \beta\mathbf{x}_2 + \gamma\mathbf{x}_3 + \delta\mathbf{x}_4$. In a general case, creating a tetrahedral representation of volume is solved using a 3D Delaunay triangulation

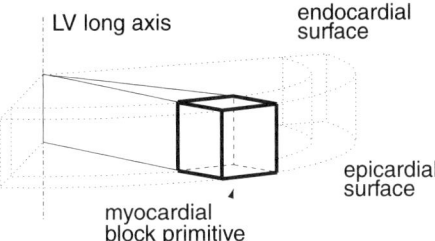

**Figure 7.6.** Definition of myocardial block primitives from concentric wedges.

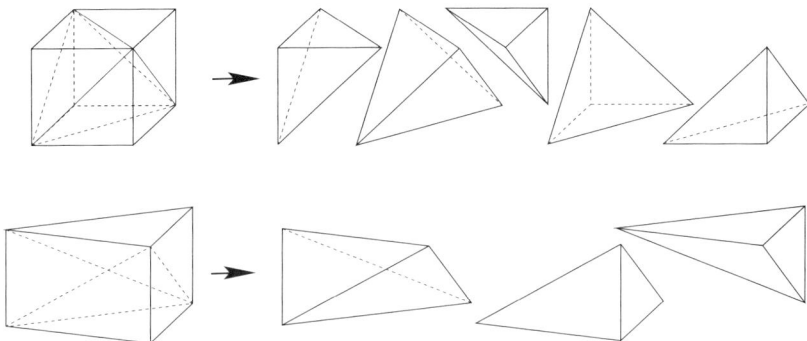

**Figure 7.7.** Decomposition of a cube (above) and a wedge (below) into tetrahedrons.

algorithm. However due to the cylindrical nature of the LV shape, a manually defined volume partitioning in regular tetrahedrons was utilized. Each slice level is constructed of pie-shaped wedges built on three tetrahedrons with exterior profile cubes built with five tetrahedrons (Figures 7.6, 7.7). Piecewise affine warping is implemented in a similar fashion as the 2D case. Because all volumes are warped to the average volume, barycentric coordinates, $\alpha, \beta, \gamma, \delta$ are precomputed for each fixed voxel point eliminating the time-consuming process of searching for the enclosing tetrahedron for each voxel point during the matching. Due to the regular geometry of the tetrahedrons in our volume partitioning, the barycentric coordinate computation did not become ill posed.

After the warping phase, the shape-free intensity vectors are normalized to an average intensity of zero and an average variance of one to remove the effects of brightness and contrast variations across scans. Next, PCA is applied to the shape-free intensity vectors to create an intensity model. In agreement with the AAM principle, shape information and intensity information are combined into a single active appearance model. Lastly, another PCA is applied to the coefficients of the shape and intensity models to form a combined appearance model [54].

In the equations below, the subscript $s$ corresponds to shape parameters while the subscript $g$ represents intensity (gray-level) parameters. To summarize, the 3D AAM is created as follows:

1. Let $\mathbf{x}_i$ denote a vector of 3D landmark points for a given sample $i$. Compute a 3D PDM and approximate each shape sample as a linear combination of eigenvectors, where $\mathbf{b}_s = P_s^T(\mathbf{x} - \bar{\mathbf{x}})$ represents the sample shape parameters.
2. Warp each image to the mean shape using a warping such as piecewise affine or thin plate spline warping to create shape-free intensity vectors.
3. Normalize each intensity vector, applying a global intensity transform with parameters $\mathbf{h}_i$, to match the average intensity vector $\bar{\mathbf{g}}$.
4. Perform a PCA on the normalized intensity images.
5. Express each intensity sample as a linear combination of eigenvectors, where $\mathbf{b}_g = P_g^T(\mathbf{g} - \bar{\mathbf{g}})$ represents the sample shape parameters.
6. Concatenate the shape vectors $\mathbf{b}_s$ and gray-level intensity vectors $\mathbf{b}_g$ in the following manner

$$\mathbf{b} = \begin{pmatrix} W\mathbf{b}_s \\ \mathbf{b}_g \end{pmatrix} = \begin{pmatrix} WP_s^T(\mathbf{x} - \bar{\mathbf{x}}) \\ P_g^T(\mathbf{g} - \bar{\mathbf{g}}) \end{pmatrix}, \qquad (7.4)$$

the weighting matrix $W$ is a diagonal matrix relating the different units of shape and intensity coefficients.

7. Apply a PCA to the sample set of all $\mathbf{b}$ vectors, yielding the appearance model

$$\mathbf{b} = Q\mathbf{c}. \qquad (7.5)$$

### 7.3.5 Active Appearance Models: 3D Matching

Matching an appearance model to image data involves minimizing the root-mean-square intensity difference between the image data and appearance model instance by modifying the affine transformation, global intensity parameters, and the appearance coefficients. A gradient descent method is used that employs the relation between model coefficient changes and changes in the voxel intensity difference between the target image and synthesized model [54]. This relation is derived during a training stage.

Let $\mathbf{t}$ and $\mathbf{q}$ represent the translation and quaternion transformation parameters, and $\mathbf{h}$ the intensity transform parameters. As shown above, shape $\mathbf{x}$ is derived in the target image from the appearance coefficient $\mathbf{c}$ and the affine transformation vectors $\mathbf{t}$ and $\mathbf{q}$. Then, shape intensity vector $\mathbf{g}_s$ is sampled from the target volume data after warping the space defined by $\mathbf{x}$ to the mean shape $x$. The model intensity vector $\mathbf{g}_m$ is derived from the appearance coefficients $\mathbf{c}$ with the global intensity corrected via $\mathbf{h}$. The error function, $E$, is the root-mean-square difference of $\mathbf{g}_s - \mathbf{g}_m$.

Gradient descent optimization requires the partial derivatives of the error function defined by the intensity of the target and synthesized model volume. While it is not possible to create such a function analytically, these derivatives may be approximated using fixed matrices computed by randomly perturbing model coefficients for a set of known training images and observing the resulting difference in error images [54]. Using a set of training images, their corresponding modeling parameters **c**, **t**, **q**, and **h** are randomly displaced, thus creating a difference between $\mathbf{g}_s$ and $\mathbf{g}_m$. From the parameter displacements and the resulting difference intensity vectors, gradient approximating matrices $A_c$, $A_t$, $A_q$, and $A_h$ can be determined using reduced-rank multivariate linear regression. Alternatively, the gradient matrices may be built one column at a time by averaging the Gaussian weighted differences between the target and synthesized image of each individual model perturbation. The latter method is preferred for 3D AAM matching due to lower memory requirements, better representation of high-order eigenmodes, and faster computation. This iterative refinement technique of precomputed fixed matrices versus brute force gradient descent optimization was formulated by Cootes [43] as well as by Baker and Matthews [55]. Formally, the gradient matrices are created as follows:

1. Select an object from the training set with known appearance model parameters $\mathbf{c}_0$, $\mathbf{t}_0$, $\mathbf{q}_0$, and $\mathbf{h}_0$.
2. For each element in the model parameters, **c**, **t**, **q**, or **h**, perturb a single element by a fixed $\delta \mathbf{p}$ with the rest of $\delta \mathbf{c}$, $\delta \mathbf{t}$, $\delta \mathbf{q}$, and $\delta \mathbf{h}$ assigned to zero. Typically, **c** is perturbed within $\pm 1.5$ standard deviation, **t** by 3-5 voxels, and **q**, **h** by 10% of their original value.
3. Let $\mathbf{c} = \delta \mathbf{c} + \mathbf{c}_0$. Compute shape **x** and texture $\mathbf{g}_m$.
4. Apply an affine transformation to **x** by first transforming **x** using $\delta \mathbf{t}$ and $\delta \mathbf{q}$, then transforming the result by $\mathbf{t}_0$ and $\mathbf{q}_0$. This cascaded transform is required to maintain linearity.
5. Create the image patch $\mathbf{g}_s$ warped from the target image to the mean shape using shape **x**.
6. Apply global intensity scaling to $\mathbf{g}_s$ by using $\delta \mathbf{h}$ first and then scaling the result by $\mathbf{h}_0$.
7. Compute $\delta \mathbf{g} = \mathbf{g}_s - \mathbf{g}_m$.
8. Compute the slope, $\delta \mathbf{s} = \delta \mathbf{g} / \delta \mathbf{p}$. Weight the slope by a normalized Gaussian function with the $\pm 3$ standard deviation is set to the maximum and minimum model perturbation values.
9. Accumulate the slope with previous slopes for that given element.
10. Go to step (2) and repeat until all elements and perturbations of each element are sufficiently covered. Place the average slope into the appropriate column in the gradient matrices $A_c$, $A_t$, $A_q$, or $A_h$.
11. Go to step (1) and repeat until there is sufficient coverage of displacement vectors.

The corresponding model correction steps are computed as

$$\delta \mathbf{c} = A_c \left( \mathbf{g}_s - \mathbf{g}_m \right), \tag{7.6}$$

$$\delta \mathbf{t} = A_t \left( \mathbf{g}_s - \mathbf{g}_m \right), \tag{7.7}$$

$$\delta \mathbf{q} = A_q \left( \mathbf{g}_s - \mathbf{g}_m \right), \tag{7.8}$$

$$\delta \mathbf{h} = A_h \left( \mathbf{g}_s - \mathbf{g}_m \right). \tag{7.9}$$

Matching the AAM to the image data is accomplished as follows:

1. Place the mean appearance model ($\mathbf{c}, \mathbf{h} = 0$; $\mathbf{t}, \mathbf{q}$ defined by the initial model position) roughly on the object of interest and compute the difference image $\mathbf{g}_s - \mathbf{g}_m$.
2. Compute the root-mean-square (RMS) error of the difference image, $E$.
3. Compute the model corrections $\delta \mathbf{c}$, $\delta \mathbf{t}$, $\delta \mathbf{q}$, and $\delta \mathbf{h}$ from the difference image [Equations (7.6)–(7.9)].
4. Set $k = 1$.
5. Compute new model parameters as $\mathbf{c} := \mathbf{c} - k\delta \mathbf{c}$, $\mathbf{t} := \mathbf{t} - k\delta \mathbf{t}$, $\mathbf{q} := \mathbf{q} - k\delta \mathbf{q}$, and $\mathbf{h} := \mathbf{h} - k\delta \mathbf{h}$.
6. Based on these new parameters, recompute $\mathbf{g}_s - \mathbf{g}_m$ and find the RMS error.
7. If the RMS error is less than $E$, accept these parameters and go to step (2).
8. Else try setting $k$ to 1.5, 0.5, 0.25, 0.125, etc., and go to step (5). Repeat steps (5)–(8) until the error cannot be reduced any further.

### 7.3.6 Case Study

To investigate the clinical potential of the 3D active appearance model under clinically realistic conditions, AAM's were trained and tested in multislice short-axis cardiac magnetic resonance images collected from 38 normal subjects and 18 patients yielding a total of 56 short-axis 3D cardiac MR data sets. Patients were selected suffering from different common cardiac pathologies (amongst others, different types of myocardial infarction, hypertrophic cardiomyopathy, arrhythmia). Images were acquired using standard ECG gated fast field echo MR pulse sequences on a Philips Gyroscan NT 15 scanner. Slices were acquired in a per-slice manner, under breathhold in end-expiration. End-diastolic images were used in this study. Image resolution was $256 \times 256$ pixels, with a field of view of 400–450 mm, slice thickness 8–11 mm. Between 8 and 14 slices were scanned to at least cover the entire left ventricle, depending on LV dimensions and slice spacing.

In midventricular short-axis MR images, the left ventricle can usually be identified as an approximately circular object (Figure 7.8a). This fact is used for automated initialization of the 3D AAM. A previously validated Hough transform based method determines a 2D centroid of the LV long axis for each MR image slice [56]. A 3D centroid of a line segment fitted through the 2D centroids of individual MR slices defines the initial position of the 3D AAM.

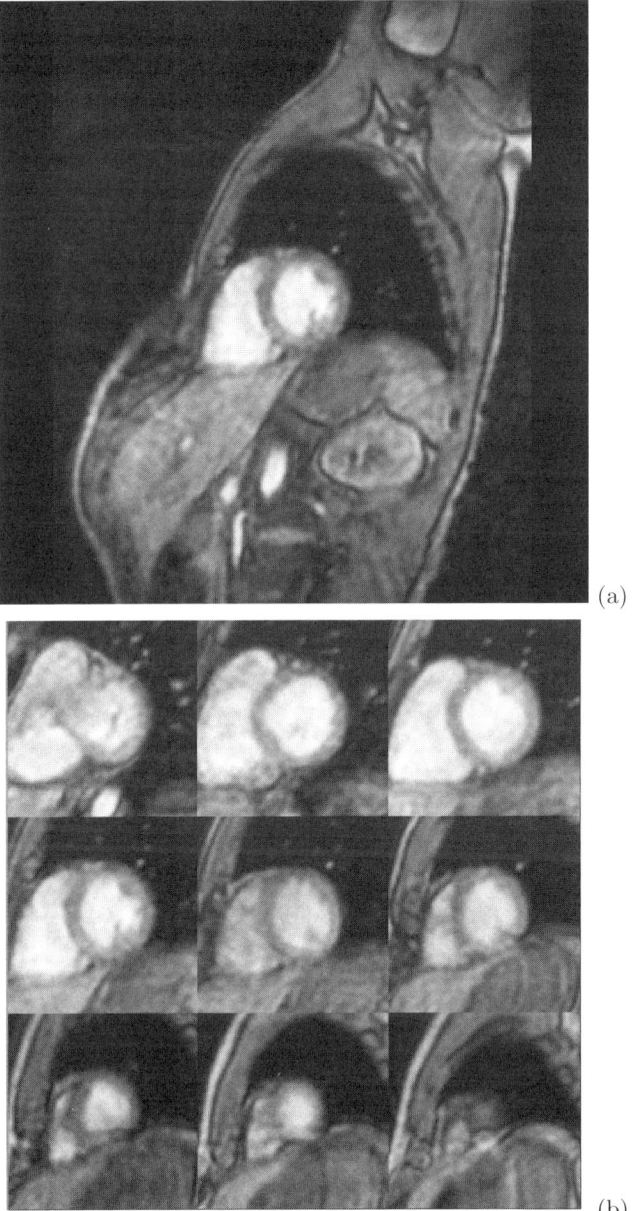

**Figure 7.8.** Example cardiac MR images used for validation. (a) Left-ventricular segmentation was performed in volumetric images consisting of 8–12 full-size MR images like the one shown here. (b) Subimages depicting LV detail in all 9 images of this volumetric data set. See Plate I for the segmentation results.

To make the 3D segmentation procedure completely independent from any user interaction regarding the rotation and scale of the heart in the short-axis plane, the matching process was repeatedly performed for a range of five orientations and three scales. This multiple initialization is important because AAM matching may be dependent on initial positioning since gradient descent may contain local minima. The matching result yielding the smallest quadratic intensity error was selected as the final match. The matching procedure resulted in a set of endo- and epicardial contours for each volumetric MR image.

Left ventricular endocardium and epicardium were manually traced by an expert observer who was blinded to the results of the computer analysis, these borders defined the independent standard. To quantitatively assess the performance of the 3D AAM approach, surface positioning errors were determined comparing the automatically detected endo- and epicardial surfaces with the independent standard. The average signed and unsigned surface positioning errors were defined by measuring the distances between points along rays perpendicular to the centerline between the respective manual contours and the computer-determined surfaces; 100 rays were used for each contour. Surface positioning errors are expressed in millimeters as mean ± standard deviation. Negative sign of the signed error value means that the automatically-determined surface was inside of the observer-defined surface.

Three clinically important measures were calculated and used for performance assessment: LV cavity volume, LV epicardial volume, and LV myocardial mass. The volumetric indices were determined using all slices for which both manually traced contours and computer-determined surfaces were available and were expressed in cm$^3$. The LV mass measurements are reported in $g$. Regression analysis was used to compare the computer measurements with the independent standard.

Plate I shows an example of an automatically analyzed volumetric MR data set. PLATES II demonstrates several stages of the model matching process, starting with the initial model position and ending with the final fit. Mean signed endo- and epicardial surface positioning errors were $-0.46 \pm 1.33$ mm and $-0.29 \pm 1.16$ mm, respectively, showing a slightly negative border detection bias. The mean unsigned positioning errors were $2.75 \pm 0.86$ mm for the endocardial contours and $2.63 \pm 0.76$ mm, for the epicardium, demonstrating small absolute differences from the independent standard (voxel sizes ranged from $1.56 \times 1.56 \times 8$ mm to $1.76 \times 1.76 \times 11$ mm). A very good correlation of the manually-identified and 3D AAM-determined LV endo- and epicardial volumes as well as correlation of computer-determined LV wall mass with the independent standard were achieved – ($y = 0.88x + 8.4, R^2 = 0.94; y = 0.91x + 12.1, R^2 = 0.97; y = 0.80x + 17.9, R^2 = 0.82$, respectively).

### 7.3.7 Extension to 4D Analysis

The heart is a dynamic system making time-independent segmentation inadequate. Applying 3D AAM segmentation to the full cardiac cycle would require multiple models for different phases because any temporal knowledge of the interrelationship between frames would be lost. Several existing methods have been developed for 3D + time cardiac segmentation taking into account the temporal relationship between frames. For example, one technique by Jacob et al. [57] solves temporal coherency between active shape models in echocardiograms through the use of a Kalman filter creating a motion model to predict the cardiac cycle in addition to a shape model. An alternative method by Montagnat *et al.* [58] segments cylindrical echocardiographic images using deformable models. Here temporal coherence is accomplished by reinitializing the deformable model using the previous segmentation, while incorporating a 4D anisotropic diffusion filter that significantly improves the spatial and temporal information between frames.

To extend the 3D AAM framework to 3D + time, we propose to incorporate a time element to the model by phase normalizing objects to a common time correspondence and concatenating shape and texture vectors of individual phases into a single shape and texture vector. Such a technique has been found efficient in 2D + time AAM and ASM models [47, 46, 59] and is promising as a future extension of 3D AAMs.

## 7.4 Vascular MR Image Analysis

### 7.4.1 Magnetic Resonance Angiography

Magnetic resonance angiography (MRA) is a powerful clinical tool that challenges the preeminence of conventional contrast angiography — the gold standard of vascular imaging. MRA offers combined imaging of vascular and soft tissues during a single comprehensive examination.

The automatic segmentation and labeling of the vascular structures are motivated by the clinical desire for quantitative information about a patient's vascular anatomy and function. Various cardiovascular problems, including aneurysms and stenoses, can be more accurately assessed using volumetric information than from x-ray angiographic projections. It is time consuming and impractical to manually segment the vessel structures to be analyzed. This indicates the need for robust and quick methods to perform accurate separation and identification of vascular structures within the anatomy with a minimal amount of user interaction. This segmentation can then be used to perform clinically useful tasks, including selective visualization (region of interest display), and quantitation of an individual's vascular function.

### 7.4.2 Methods for Vascular MR Segmentation

The volumetric segmentation techniques were preceded by numerous attempts to visualize vascular structures in 3D from 2D projections [60, 61, 62, 63, 64]. These approaches continue to be used in x-ray angiography. After the arrival of MR in sufficient speed and resolution, Fessler and Macovski developed a detailed object-based approach to the reconstruction of the arterial trees using projections from magnetic resonance angiograms [62]. Although they were using MR, which is inherently volumetric, they utilized concepts from earlier research and used planar images as the basis for their reconstruction technique. Garreau et al. further refined the mapping and biplane angiography reconstruction issues by introducing an expert knowledge base to give a map of the topology of the vessel paths [63]. This knowledge base was used as a basis for structure and feature labeling of the vessel tree.

More recent work in the area of vessel segmentation uses 3D MR or CT data sets. Clearly, similar analysis methodologies can be utilized for MRA and CTA (computed tomography angiography). Some of the less complex approaches involve direct segmentation of the data without modeling of the vessel structure or explicit determination of the vessel path. An example of such an approach can be found in [65] where abdominal aortic aneurysms are segmented and quantified directly from the image data. The user selects two starting points in the distal iliac arteries and the segmentation algorithm travels in a proximal direction along the center of the vessels, determining the lumen outline of iliac arteries and the abdominal aorta in CTA volumes. More comprehensive segmentation of the cerebral vessels in CTA and MRA has been accomplished using an iterative dilation approach [66]. A bounded space dilation operation was used to build up a vessel tree. Cerebral arteries were well segmented and the algorithm avoided inclusion of neighboring bone structures and thin connections to adjacent regions by additional restrictions on the growth process. This work is notable as a method which has the ability to indicate where bifurcations of the vessels occur, as the growth front algorithm can readily determine when a newly grown region is not connected in the 3D space. Bifurcation detection is a crucial portion of the overall vessel segmentation process, especially in algorithms that utilize topology analysis as an aid to segmentation.

In addition to direct 3D segmentation approaches, several preprocessing methodologies exist that are worth discussing, most notably finding central axes of the elongated vascular structures using 3D skeletonization [67, 68, 69] or by identifying vessel medial axes and cores [70]. A complete segmentation and analysis package for coronary angiograms was described by Higgins et al. [69] which used skeletonization to determine the central axes of vessel paths in 3D CTA data sets. Segmentation and quantitation of the coronary arteries were performed in an integrated system. The central paths determined by a 3D skeletonization algorithm were used as the guiding topological map for the segmentation.

Finding the central axes of vessels using a medial axis transform is another powerful approach. The concepts of height ridges and medial axes were extended to the concept of core atoms by Pizer *et al.* [71, 72, 70]. They extended the medial axis concept to gray-scale images and defined a quantifiable notion of medialness — the core . A core is a locus in a space whose coordinates are position, radius, and associated orientations [70]. Vessel central axes are characterized by high medialness, thus relating them to cores. Core transform finds the center of the vessel through a specific computation of the medial axis and follows the local maxima along the path of vessel propagation as a step in image segmentation. The concept of core atoms encapsulates information about edge direction, radius, and shape. Each core atom holds information about two edge elements and is located midway in between these two edge elements. Using statistical methods, the core atoms can indicate a vessel path or other anatomic shape [72]. Section 7.4.3 discusses this approach in more detail.

A codimension 2 geodesic active contour approach is under development by Lorigo et al. [73]. A mathematical modeling technique is used to represent complicated curvelike structures of vasculature as seen in 3D MRA image data. The segmentation task is defined as an energy minimization problem over all 3D curves. Mean curvature evolution techniques that were previously developed and implemented with level set methods [74, 75, 76] were extended to a higher codimension and applied to segmentation of brain vessel vasculature. While this approach needs to mature before it reaches clinical applicability, it represents an interesting and promising direction.

Despite the fact that most of the methods employed to segment vessels from 3D image data sets are very different in implementation, two main concepts are commonly utilized. The first concept is that of determining the center of the vessel paths. The following of a particular vessel path is the crucial step in understanding the vessel topology that is important for segmentation. The second concept is the use of some a priori knowledge about the segmentation task. This knowledge can be used to provide a road map to either guide the segmentation process or to identify structures with physically relevant names. It is the application of these two major concepts that form the basis for most of the recently developed methods that are outlined in the following discussion.

Separation of arteries and veins is an emerging challenge in MRA analysis. With the development of new MR contrast agents that have longer persistence in the blood, there is the ability to image the vasculature fully enhanced and at high resolution. These high-resolution "steady state" images have simultaneous enhancement of both the artery and vein blood pools (Figure 7.9). This enhancement can be useful. However, it can also obscure critical detail when analyzing the vessels using maximum intensity projection and other visualization strategies. Artery–vein separated images have unobstructed artery visualization comparable to dynamic MRA scans. Nevertheless, they do not suffer from the limited resolution that is necessary to achieve a dynamic im-

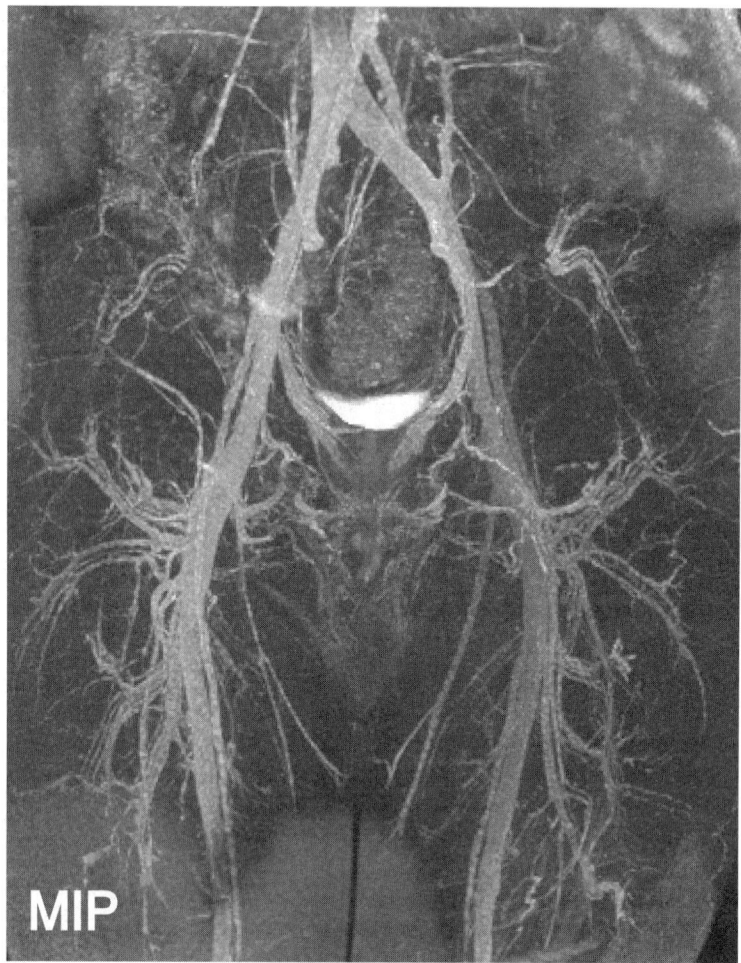

**Figure 7.9.** Maximum intensity projection of a typical blood-pool contrast enhanced MRA data set of the abdomen and lower extremities.

age sequence. To meet the incompatible goals of high resolution with the ability to view unobstructed arterial vessels, artery–vein separation and selective visualization techniques must be developed to achieve acceptance of contrast-enhanced MRA as an alternative to x-ray angiography. Currently, separation of arteries and veins in MRA images is limited to research applications and manual segmentations. Recently, there has been considerable progress and several different approaches have shown potential as a method for artery–vein separation.

Generally, artery–vein separation methods can be divided into two categories — acquisition based methods [77, 78, 79] and postprocessing techniques [62, 66, 67, 68, 69, 70]. Acquisition-based techniques seek to exploit various flow and physical properties. Phase behavior and time series acquisitions image the vessels attempting to provide information about the vessel identity directly from the imaging protocol itself. Postprocessing techniques seek to provide the arterial–venous separations by analyzing the acquired data after the scan is performed. These methods have to rely on information present in the scan itself, while having the advantage of not being tied to a particular protocol or scan type. These methods vary in complexity and all of them employ user interaction to some extent. Another postprocessing approach seeks to segment the artery and vein using image intensity properties to determine a membership in an artery or vein. Two representative methods of this research are gray-scale connectivity [80] and fuzzy connectivity [81].

### 7.4.3 Vasculature Assessment via Tubular Object Extraction and Tree Growing

One of the prominent characterizations of vessel shape is undoubtedly their tubular character. As mentioned above, Pizer et al. developed a generalized methodology for determination of central axes of tubular structures via calculation of intensity ridges, medialness, and cores in gray level images [82, 70]. This concept was applied to time-of-flight MRA images by Aylward et al. [83] and further extended to a vascular tree representation by Bullitt et al. [84].

The main steps of the method are given in the following algorithm:

1. Vasculature assessment via tubular object extraction and tree growing.
2. Geometry-based semi-automated segmentation of the MRA volume to extract individual vessel segments in the region of interest.
3. Comparison of the extracted vessel segments with the maximum intensity projection of the original MRA data in the region of interest. If vessel segments are missing, repeat step (1).
4. Iterative construction of a vessel tree.
5. 3D visualization and interactive editing of the resulting vessel tree.

Extraction of each vessel segment starts from a user-supplied seed point. A vessel segment is a nonbranching 3D portion of the vasculature. Using the seed point, image intensity ridges are automatically extracted utilizing user-supplied information about the approximate width of the segmented vessel. As a result, the medial axis (skeleton) of the vessel segment is formed. The vessel segment's width is determined at each point of the skeleton under the assumption that the vessel is approximately circular. The width is calculated to be proportional to the scale that produces maximal response from a cylindrical medialness measure. The vessel segmentation process is repeated for all vessel segments in the region of interest, yielding a set of unbranched, directed 3D skeleton curves with a width associated with each point [83].

The segmentation step is complete when the maximum intensity projection of the original MRA data agrees with the visualization of the vessel segments resulting from the segmentation. If vessel segments are missing in the region of interest, the segmentation process is continued from newly identified seed points.

Before the vasculature tree is constructed, several potential problems of the identified vessel segments must be considered since the result of vessel extraction may not be ideal:

- Spurious vessel segments must be eliminated.
- Excessively long vessel segments extending past one or more branches must be divided.
- Adjacent vessel skeletons belonging to the same vessel must be connected.
- Directionality of blood flow must be determined.

All these problems are addressed during the tree construction step.

The tree construction process utilizes linear distance properties of individual vessel segment skeletons and the image intensity of the original MRA data. During the segment-connecting process, I and Y connections are allowed while X connections are not. Consequently, at least one end-point of the two segments to be connected is engaged. Tree construction starts from one or more interactively identified tree roots. The maximum distance to be considered for establishing a connection between two segments is specified by the user. Starting with the root nodes, a tree is constructed by iterative addition of segments that satisfy both the distance and intensity criteria.

The sequence of forming tree segment connections is controlled by the *connection cost CC* with the best possible connection being realized at each iteration. For each two segments, one is considered a potential parent (the one already connected to the tree root) and one is a potential child. Three pairs of possible connections are determined and a line is constructed connecting each pair of points in the 3D image. These lines form three axes along which hollow cylinders are constructed with radii slightly greater than that of the potential child segment so that the cylinder surface is positioned outside of the child vessel. The *intensity ratio* is defined as a ratio of the average image intensities of a cylinder surface and its axis. A low ratio (bright central axis and dark cylinder surface) is considered to be evidence of a valid connection. The connection cost is defined as a weighted sum of the *linear distance LD* of the connected segments and the intensity ratio $IR$:

$$CC = LD + 4\,IR \ . \tag{7.10}$$

The connection of a minimum cost $CC$ is identified and formed. After the best connection from all possible ones is found, flow direction in the connected segment may be reversed to agree with that of the already formed tree. The spurious segments mentioned earlier are to a large extent removed automatically since they fail to meet the connectivity requirements of distance and intensity.

As can be expected, the process described above may lead to missing or incorrect connections caused by MR imaging artifacts, a limited size of the imaged region of interest, or other ambiguities present in MR data. Therefore, the resulting tree is carefully inspected and editing tools are employed to include missing connections and remove inappropriate ones. The user is allowed to delete proximal or distal segments and associated subtrees, delete an entire vessel and the associated subtree, disconnect a subtree from a parent and reconnect it to another user-specified parent or parent point, and reverse blood flow in any segment (causing automated update of the parent-child data structure).

### 7.4.4 Knowledge-Based Approach to Vessel Detection and Artery–Vessel Separation

The following algorithm was developed to perform arteriovenous separation in the peripheral vasculature, specifically the iliac and femoral vessels. When analyzing intravascular contrast agent-enhanced MRA data sets that image this area, the most challenging aspect to overcome is the partial volume effects brought on by limited spatial resolution and the proximity of the vessels. These effects cause the artery and vein segments to become aliased within some voxels of the data, causing incorrect connections between the artery and vein pathways, when in reality there is only close proximity between the two. These incorrect connections cause simpler methodologies such as region growing to fail in separating the arteries and veins into two distinct objects.

To cope with this problem, a knowledge-based method was developed and tested by Stefancik and Sonka that consists of the following main steps [85], Figure 7.10.

1. Knowledge-based segmentation of arterial and venous trees in lower extremities
2. Binary mask generation — the contrast-enhanced MRA data are segmented in a 3D connected combined vessel tree (consisting of arteries and veins) and nonvessel regions.
3. Tree-structure generation — the combined vessel tree is topologically described as a tree structure using vessel-bounded space dilation for identification of bifurcations.
4. Optimal vessel path calculation — vascular central axes are determined using 3D dynamic programming in a vessel-bounded space.
5. Vessel segment labeling — vessel segments are labeled as belonging to arteries or veins.
6. Conflict resolution — if any branch segment belongs to more than one path through the tree, individual voxels within that segment are assigned to their appropriate paths and their anatomic labels.

These main steps will now be described in more detail.

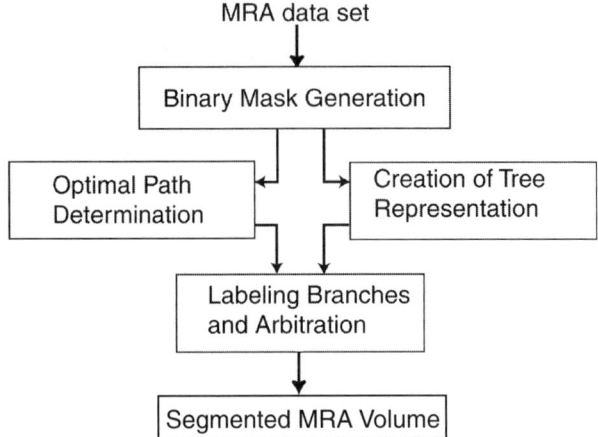

**Figure 7.10.** Graphical overview of the segmentation process described in the text.

### Binary Mask Generation

The combined vessel tree consisting of arteries and veins and forming a binary volumetric mask is acquired by a two-step process — percentage-based gray-scale threshold calculation is followed by a seeded region growing. For the abdominal region and lower extremities, the vessel structures occupy approximately 5% of the data set by volume. Therefore, 95% thresholding is employed with a threshold value derived from a gray-level histogram [86]. As a result, virtually all vessels, some MRI artifacts, and some subcutaneous fat regions are segmented. To remove the imaging artifacts and subcutaneous fat from the binary mask, a seeded region growing operation follows. A seed point is identified within a vessel structure, and all connected voxels over the threshold value are labeled. All nonconnected voxels are then discarded. An example of the resulting combined tree is shown in Figure 7.13(a).

### Tree-Structure Generation

For a vascular structure, it is reasonable to segment the vessel volume into vessel segments, with each segment representing a section between two subsequent bifurcations. The vessel segments forming a vessel tree serve as an aid to topology analysis and as a method of grouping voxels in structural primitives for subsequent processing.

The conditional bounded-space dilation operation of mathematical morphology [86] is used in the growth front algorithm tree generation [66]. The tree contains relevant information about continuous branch segments, and higher-level parameters such as length and volume can be calculated from these segments (Figure 7.11).

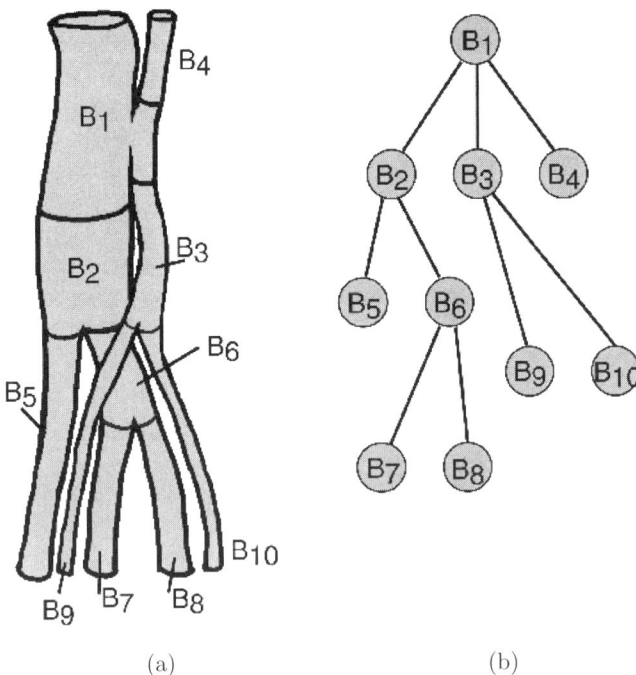

**Figure 7.11.** Tree structure representation. (a) Vessel tree with labeled branch segments. (b) Node topology created by the bounded-space dilatation growth algorithm.

### Optimal Vessel Path Calculation

The complex topology of the combined tree along with multiple false bifurcations that are always present lead to an over-segmented tree which is not practical to label directly. Additional information about the spatial path of individual vessel segments is needed. The optimal vessel path is determined using a dynamic programming path cost maximization applied to 3D cost volumes corresponding to the analyzed 3D data set.

### Vessel Segment Labeling

The optimal path calculation yields discrete paths through the volume that tend to follow the center of the vessel. Since the artery or vein label is known for each proximal seed point used for dynamic programming path search, this label is propagated to all vessel segments along the identified paths. When this labeling is performed for each path, segments may have conflicting labels — some of the segments are labeled as belonging to both arteries and veins. Such cases are solved by the conflict resolution step.

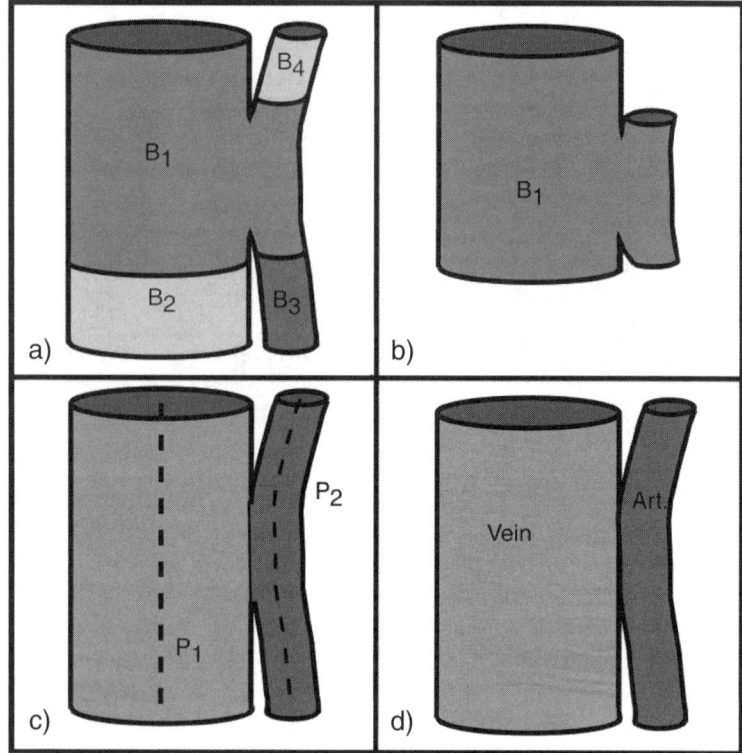

**Figure 7.12.** Conflict resolution: (a) Example area of combined artery and vein membership in one branch ($B_1$). (b) Branch $B_1$ is labeled as a contention branch. (c) Optimal paths and their radii of influence dictate voxel membership. (d) Artery and vein segmented.

## Conflict Resolution

When a vessel segment has conflicting labels, a spatial decision conflict resolution function is invoked to resolve the labeling conflict. The employed decision function allows a specific vessel segment to label its voxels by different labels according to their proximity to the conflicting paths that run through that vessel segment (Figure 7.12). This effectively separates the artery and vein connections that occur due to the partial volume effects.

The algorithm presented above was tested in artificially generated data sets and in in vivo acquired MRA volumes that were acquired 5 minutes post-injection of a blood pool contrast agent[2] [87]. The original image data are shown in Figure 7.9. Automatically and manually segmented data were compared to assess the performance of the automated method in separation of main vessels in the abdominal and upper leg area. The artery–vein labeling error ranged from 1 to 15%. Complete quantitative validation results can be found in [87]. Figures 7.13(a) (b) show representative segmentation of two different data sets. Figures 7.13(c) (d) show selective visualization of the segmented structures from Figure 7.13(b). As indicated above, this approach focused on the separation part and may benefit from improving the initial binary tree segmentation. The first step of the fuzzy connectivity approach described in the next paragraphs seems to be especially attractive for that purpose.

### 7.4.5 Fuzzy Connectivity Approach to Vessel Detection and Artery–Vessel Separation

Udupa et al. developed a fuzzy connectivity–based methodology for vessel extraction and artery–vein separation. The method is based on a principle of fuzzy connected image segmentation that was applied, e.g., to MR brain segmentation, multiple sclerosis lesion detection in MR data sets, and separation of bone and soft tissues from skin in CT images [88, 89]. This concept was further enriched by the recent addition of scale-based fuzzy connectivity [90].

The basic principle of this approach is the notion that image intensity information itself is insufficient for segmenting heterogeneous objects with varying image intensities within the objects. Variations of intensities may be caused by a plethora of reasons — imaging artifacts, noise, inhomogeneity of the object itself, etc. All these causes are common in medical images. Udupa et al. stated an important concept that voxels belonging to the same objects tend to hang together, thus defining objects by a combination of spatial relationship of their elements (pixels, voxels), at the same time considering the local image intensity properties. The spatial relationships should be determined for *each* pair of image elements in the entire image. To accomplish that, local and global image properties are considered.

The local fuzzy relation is called affinity and represents a strength of hanging togetherness of nearby image elements and has a value in [0,1]. The affinity is a function of the spatial distance between the two nearby image elements as well as of their image intensities or other intensity-derived features (e.g., edges). Fuzzy connectedness is then a global fuzzy relation that assigns every pair of image elements $E_1$ and $E_2$ a value in [0,1] based on the hanging-togetherness values along all possible paths between these two image elements. Note that the elements $E_1$ and $E_2$ are not expected to be nearby. For each path, its strength is defined as the minimum affinity value for all pairwise elements of the path. In other words, the strength of the entire path is defined by the strength of its weakest local connection. Then, the value of fuzzy

---
[2] AngioMARK, EPIX Medical, Cambridge MA.

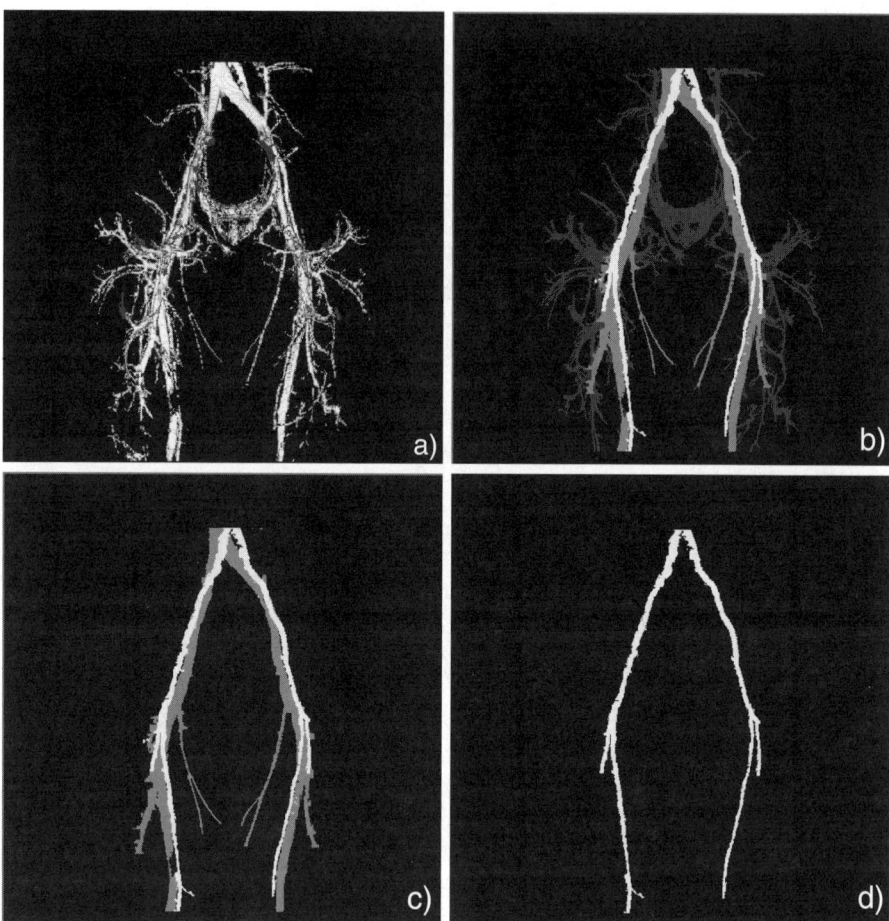

**Figure 7.13.** Segmentation of main arteries and veins in an MRA data set of abdomen and lower extremities. See Fig. 7.9 for maximum intensity projection of this data set. (a) Combined tree (volume rendering). (b) Main arteries and veins. (c) Selective visualization of major arteries and veins from (b). (d) Segmented arterial tree from (b).

connectedness (global hanging togetherness) of $E_1$ and $E_2$ is determined as the maximum of the strengths of all possible paths between $E_1$ and $E_2$. The strength of connectedness of all possible pairs of elements defining a fuzzy connected object is determined via dynamic programming.

An approach for arterial and venous tree segmentation and artery–vein separation was developed based on fuzzy connectivity principles [81] and applied to blood-pool contrast enhanced images[3] of abdomen and lower extremi-

---

[3] AngioMARK, EPIX Medical, Cambridge MA.

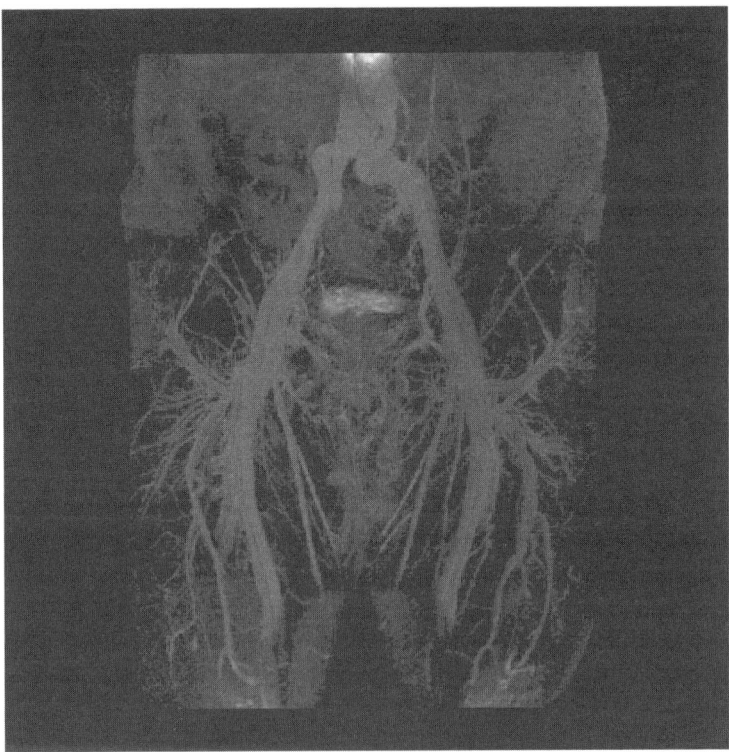

**Figure 7.14.** Maximum intensity projection image of the original data used in Plate III. Courtesy of J. K. Udupa, University of Pennsylvania.

ties. First, an entire vessel tree is segmented from the MRA data sets utilizing absolute fuzzy connectedness. Next, arteries and veins are separated using relative fuzzy connectedness. For the artery–vein separation step, seed image elements are interactively determined inside an artery and inside a vein. Large-aspect arteries and veins are separated, smaller-aspect separation is performed in the following iterations, four iterations being typically sufficient. To separate the arteries and veins, a distance transform image is formed from the binary image of the entire vessel structure. Separate centerlines of arterial and venous segments between two bifurcations are determined using a cost function reflecting the distance transform values. All image elements belonging to the arterial or venous centerlines are then considered new seed elements for the fuzzy connectivity criterion, thus allowing artery–vein separation.

The vessel tree segmentation and artery–vein separation procedure is quite fast, requiring only several minutes to complete. Figure 7.14 and Plate III show the functionality of the method.

## 7.5 MR Assessment of Atherosclerotic Plaque

### 7.5.1 Vessel Wall Imaging

Recent studies indicate that the composition of atherosclerotic plaque lesions may be more important than the morphology. The depiction of the vessel lumen generated by most angiography techniques will be insufficient to determine the significance of lesions, such as whether they are stable or are likely to rupture.

Magnetic resonance imaging is capable of distinguishing between the different components of atherosclerotic plaque, identifying its composition in addition to its morphology and potentially providing a more clinically relevant assessment of the severity of disease. Direct imaging of the vessel wall depicts components of atherosclerotic plaque. Calcium, lipids, thrombus, and fibrous tissues each have distinct relaxation characteristics that can be measured by MRI using a multicontrast approach for tissue characterization.

Imaging of the vessel wall for plaque characterization is considerably more difficult than vessel lumen imaging. Signal generated by the vessel wall limits the signal-to-noise ratio that can be achieved, while the needed high resolution further reduces the available SNR. Comprehensive tissue characterization requires multiple image acquisitions to generate enough contrast weightings to distinguish between the major plaque components. In addition, blood signal must be thoroughly suppressed to achieved adequate contrast. This may be especially difficult in areas prone to slow or turbulent flow (such as the carotid bifurcation) where this signal may mimic that of a legitimate plaque lesion. Despite these obstacles, considerable progress has been made in finding feasible imaging protocols for imaging vessel wall in the aorta, carotid, and coronary arteries.

The most successful protocols to date in all three of these locations incorporate fast spin echo imaging coupled with cardiac gating (and respiratory gating in the case of the coronary arteries) and double inversion recovery (double IR) for blood suppression. The double IR preparation applies a nonselective inversion to the entire volume, followed by a second selective inversion to restore the magnetization in the imaged slice or slab. After waiting an appropriate time interval for blood signal to be nulled, a standard FSE sequence is applied to depict vessel wall and any associated plaque. This protocol may be used with either 2D multislice acquisitions or true 3D volumes. By acquiring T1-, T2-, and proton-density-weighted images, the major components of atherosclerotic plaque (lipid, calcification, fibrous cap) can be segmented. Two-dimensional acquisitions have been shown with resolution down to 0.4 × 0.4 mm in-plane with 2 mm slices, while true 3D acquisitions can reduce slice thickness to 0.5 mm. The need for high-quality blood suppression in the double IR sequence results in rather long acquisition times, and in the case of coronary imaging, respiratory motion suppression is necessary as well.

Several ongoing developments in atherosclerotic plaque imaging with MRI promise to further improve its capabilities. Imaging at higher field strengths

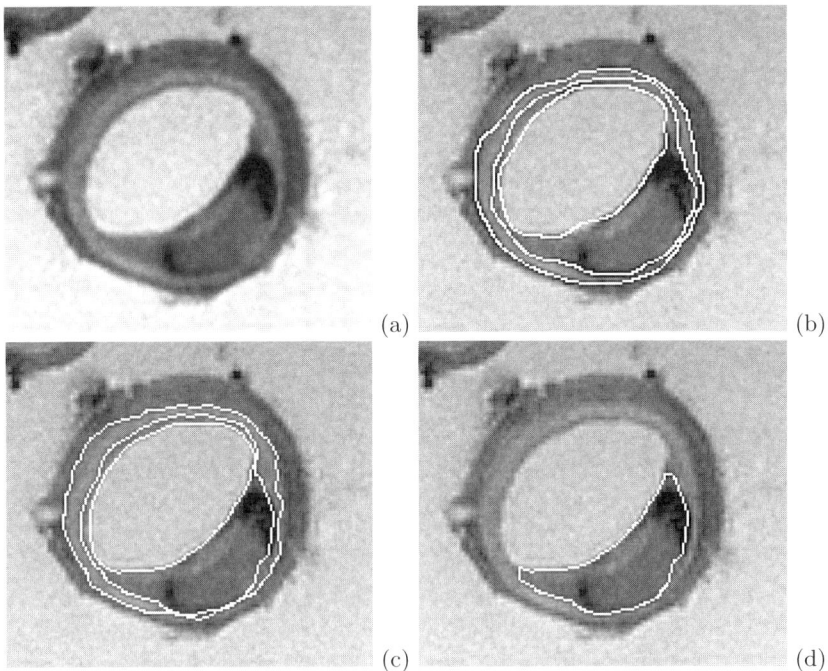

**Figure 7.15.** Example of segmented T2-weighted MR cross-section. (a) Original in vitro MR image of an excised artery, (b) observer-defined segmentation of lumen, intima-media, and adventitia–media borders, (c) fully automated segmentation, (d) identified region of plaque. MR image data courtesy of Drs. Stollberger and Holzapfel, Karl-Franzens University of Graz and Graz University of Technology, Austria.

(3T and beyond) will help improve SNR as such systems proliferate [91, 92]. Development of specialized receiver coils (including intravascular coils) may also dramatically improve image quality and resolution potential [93]. Contrast agents that are selectively absorbed by plaque components are also showing promise for plaque characterization [94, 95, 96, 93, 97, 98, 99, 100, 101, 102, 103].

### 7.5.2 Plaque Assessment via MR Wall Imaging

To determine plaque properties, MR images must have sufficient resolution to allow segmentation of the vessel wall. The segmentation results in determination of the morphology of the vessel wall and plaque. After that, plaque composition must be determined. This overall approach requires performing a sequence of complex steps [104].

Quantification of plaque morphology in high-resolution volumetric multicontrast data (Plate IV) requires automated segmentation. To perform this task, a previously developed and successfully method for automated design of border detection criteria [105] was employed that substantially simplifies implementation of new border-based image segmentation applications. All information necessary to perform image segmentation is automatically derived from a training set that is presented in the form of expert-traced segmentation examples. Therefore, borders of lumen, internal elastic lamina (intima–media), and external elastic lamina (media–adventitia) were manually traced in a set of training images. From these examples, border detection criteria using edge- and gray-level pattern information [105] were derived and used for segmentation of a testing set of images. Border detection errors were assessed in the testing set.

Five atherosclerotic iliac arterial segments were imaged in vitro as described above. The arterial wall layers and plaque were clearly visible as is shown in Plate IV . The corresponding histology demonstrated good agreement with the MR images. An example of an MR image Figure 7.15 (a) with automatically identified lumen, intima–media, and media–adventitia borders is shown in Figure 7.15 (b) (c). After training as described above, the borders automatically detected in a disjoint testing set were quantitatively compared with a manually-identified independent standard. The automated method yielded promising accuracy and minimal bias: root-mean-square and signed mean border positioning errors of the media–adventitia border detection were $1.03\pm0.28$ and $0.36\pm0.35$ pixel, respectively. Corresponding border positioning errors were $0.96\pm0.12$ and $0.34\pm0.31$ pixel for the intima–media borders and $1.04\pm0.27$ and $0.08\pm0.45$ pixel for the lumen border detection. Regions of plaque were identified as any portion of arterial wall with thickening between lumen–intima and media–adventitia borders (Fig. 7.15 d). This thickness can be determined directly from the boundaries (surfaces) identified in the volumetric MR images — an increased thickness indicates presence of plaque.

The achieved quality of MR images and the segmentation results are encouraging and facilitate further development of plaque vulnerability indices based on plaque mechanics [106]. Once noninvasive assessment of plaque vulnerability becomes available, the way in which atherosclerotic disease is diagnosed, monitored, and treated may change dramatically. Determination of plaque composition and its mechanical stress consequences will allow assessment of cardiovascular risks on a per-subject basis.

## 7.6 Conclusions

Computer vision methods are being employed in medical image analysis more and more frequently. MR imaging provides enormous flexibility to probe living bodies and deliver image information about the morphology, structure, and function of individual organs and their parts. Magnetic resonance imaging

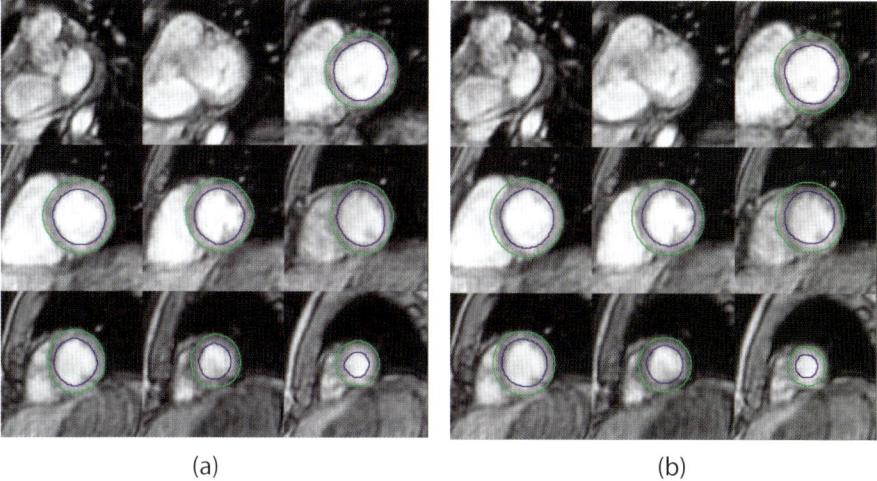

**PLATE I.** Segmentation results in testing-set image data. (a) Manually-identified contours forming an independent standard. (b) 3D AAM determined segmentation of the left ventricle. The 3D AAM segmentation was performed in full-size image volumes.

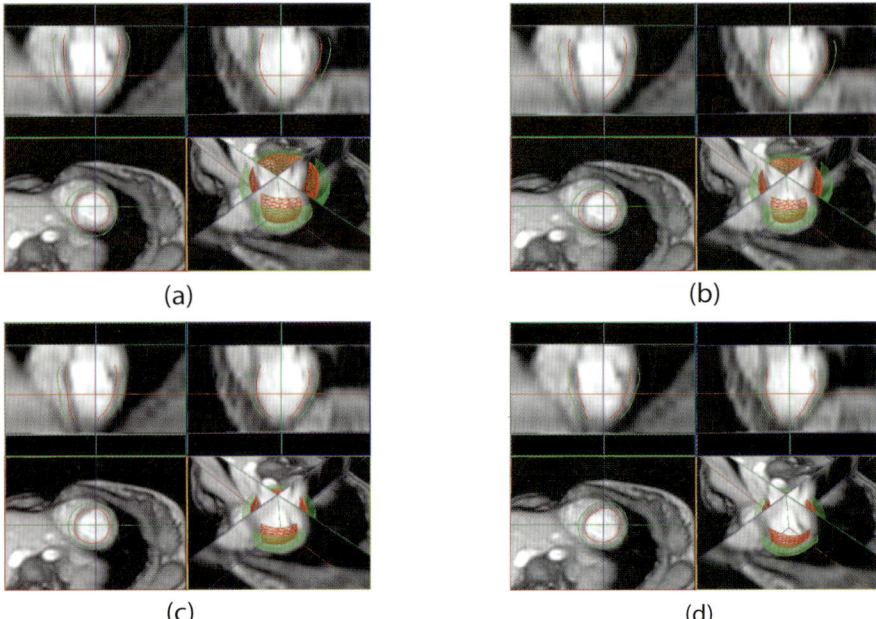

**PLATE II.** 3D AAM matching process. Note the color coding of all frames and the coordinate axes. The color-coded straight lines show position of frames in the other two cutting planes. (a) The initial position of the model in the volumetric data set. (b) & (c) Stages during the iterative model matching process. (d) The final match.

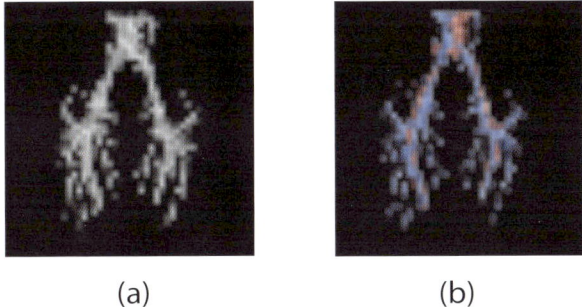

(a)                  (b)

**PLATE III.** Performance of the fuzzy connectivity approach to vessel detection and artery-vessel separation. (a) Segmentation of the entire vessel tree using absolute fuzzy connectivity. (b) Artery-vein separation using relative fuzzy connectivity. Courtesy of J. K. Udupa, University of Pennsylvania.

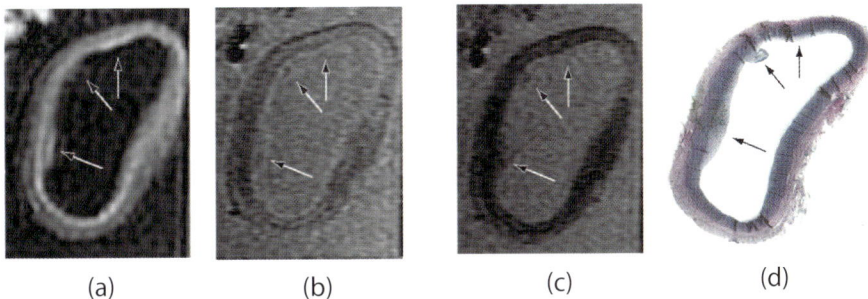

(a)        (b)        (c)        (d)

**PLATE IV.** MR imaging of a diseased human cadaver iliac artery. Panels (a) and (b) show T1 and PD MR images respectively. Panel (c) shows T2 MR image. Panel (d) shows a histological section at the corresponding arterial location. Arrows depict identified plaque in MR and histology images.

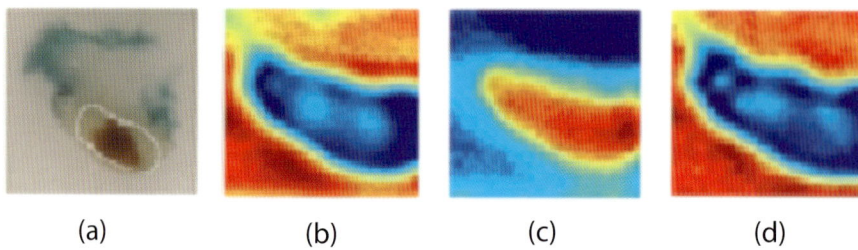

**PLATE V.** (a) Photograph of a wax embedded histo-pathological melanoma section. (b) Pulse amplitude relative to reference pulse amplitude, in time domain. (c) Time delay between transmitted pulse peak and peak of reference. (d) Transmittance (ratio of transmitted and incident intensities after Fourier transformation of pulses) at 2 THz.

and analysis of MRT images are indeed extremely powerful examples of how image processing techniques can be applied beyond the visible spectrum.

The chapter focused on describing basic principles of MR imaging and described several application areas in which automated image analysis will play a major role in the near future — segmentation and analysis of cardiac MR images, vascular MR angiography, and MR images of atherosclerotic plaque. In all of the covered application areas, the imaging is performed three-dimensionally and the analysis is three-dimensional as well. Needless to say, the step from 2D to 3D is not an easy one and many new solutions must be found to perform truly 3D medical image analysis. MR images is capable to deliver four dimensional information as well (3D + time).

To summarize, the field of MR imaging and MR image analysis is an extremely exciting one. The information about living bodies is provided in real time, in 4D, with consequences almost beyond human imagination. The amounts of data generated exceed the visual analysis capabilities of the diagnostic radiologists and call for development of reliable, accurate, precise, and fast quantitative techniques for 3D and 4D MR image analysis. The authors have no doubts that the general and broad field of computer vision has much to contribute to automate the medical image analysis in general, and MR image analysis in particular.

## 7.7 Acknowledgments

This work was supported in part by NIH grants-R01 HL63373, R37 HL61857, American Heart Association grant 0160452Z (Heartland Affiliate), and Innovative Research Incentive 2001 grant from the Netherlands Organization for Scientific Research.

## References

[1] Bloch, F.: Nuclear induction. Phys. Rev. **70** (1946) 460–473
[2] Purcell, E.M., Torrey, H.C., Pound, R.V.: Resonance absorption by nuclear magnetic moments in a solid. Phys. Rev. **69** (1946) 37–38
[3] Lauterbur, P.C.: Image formation by induced local interactions: examples employing nuclear magnetic resonance. Nature **242** (1973) 190–191
[4] Lanzer, P., Barta, C., Botvinick, E.H., Wiesendanger, H.U., Modin, G., Higgins, C.B.: ECG-synchronized cardiac MR imaging: method and evaluation. Radiology **155** (1985) 681–686
[5] Danias, P., McConnell, M., Khasgiwala, V., Chuang, M., Edelman, R., W.J., M.: Prospective navigator correction of image position for coronary MR angiography. Radiology **203** (1997) 733–736

[6] Wang, Y., Rossman, P.J., Grimm, R.C., Riederer, S.J., Ehman, R.L.: Navigator echo-based real-time respiratory gating and triggering for reduction of respiration effects in three-dimensional coronary MR angiography. Radiology **198** (1996) 55–60
[7] Yang, P., Kerr, A., Liu, A., Hardy, C., Meyer, C., Macovski, A., Pauly, J., Hu, B.: New real-time interactive cardiac magnetic resonance imaging system complements echocardiography. J. Am. Coll. Cardiol **32** (1998) 2049–2056
[8] Simonetti, O., Finn, J., White, R., Laub, G., Henry, D.A.: "Black blood" T2-weighted inversion recovery MR imaging of the heart. Radiology **199** (1999) 49–57
[9] Atkinson, D.J., Edelman, R.R.: Cineangiography of the heart in a single breath hold with a segmented turboFLASH sequence. Radiology **178** (1991) 357–360
[10] Oppelt, A., Graumann, R., Barfuss, H., Fischer, H., Hartl, W., Shajor, W.: FISP: a new fast MRI sequence. Electromedia **54** (1986) 15–18
[11] Barkhausen, J., Ruehm, S.G., Goyen, M., Buck, T., Laub, G., Debatin, J.F.: MR evaluation of ventricular function: true fast imaging with steady-state precession versus fast low-angle shot cine MR imaging: feasibility study. Radiology **219** (2001) 264–269
[12] Plein, S., Bloomer, T.N., Ridgway, J.P., Jones, T.R., Bainbridge, G.J., Sivananthan, M.U.: Steady-state free precession magnetic resonance imaging of the heart: comparison with segmented $k$-space gradient-echo imaging. J. Magn. Reson. Imaging **14** (2001) 230–236
[13] Buser, P.T., Auffermann, W., Holt, W.W., Wagner, S., Kircher, B., Wolfe, C., Higgins, C.B.: Noninvasive evaluation of global left ventricular function with use of cine nuclear magnetic resonance. J Am Coll Cardiol **13** (1989) 1294–1300
[14] Foo, T.K.F., Ho, V.B., Kraitchman, D.: Single breath-hold single-phase and CINE 3D acquisition using variable temporal $k$-space sampling. In: Proc. ISMRM, 9th Annual Meeting, Glasgow (2001) 112
[15] Scheffler, K.: 3D cardiac cine imaging in a single breath-hold using elliptically reordered 3D trueFISP. In: Proc. ISMRM, 9th Annual Meeting, Glasgow (2001) 113
[16] Sodickson, D., Manning, W.: Simultaneous acquisition of spatial harmonics (SMASH): fast imaging with radiofrequency coil arrays. Magn. Reson. Med. **38** (1997) 591–603
[17] Weiger, M., Pruessmann, K., Boesiger, P.: Cardiac real-time imaging using SENSE. Magn. Reson. Med. **43** (2000) 177–184
[18] Madore, B., Glover, G., Pelc, N.: Unaliasing by Fourier-encoding the overlaps using the temporal dimension (UNFOLD), applied to cardiac imaging and fMRI. Magn. Reson. Med. **42** (1999) 813–828
[19] Osman, N., Kerwin, W., McVeigh, E., Prince, J.: Cardiac motion tracking using CINE harmonic phase (HARP) magnetic resonance imaging. Magn. Reson. Med. **42** (1999) 1048–1060

[20] Lima, J.A., Judd, R.M., Bazille, A., Schulman, S.P., Atalar, E., Zerhouni, E.A.: Regional heterogeneity of human myocardial infarcts demonstrated by contrast-enhanced MRI: Potential mechanisms. Circulation **92** (1995) 1117–1125
[21] Wu, K.C., Zerhouni, E.A., Judd, R.M., Lugo-Olivieri, C.H., Barouch, L.A., Schulman, S.P., Blumenthal, R.S., Lima, J.A.: Prognostic significance of microvascular obstruction by magnetic resonance imaging in patients with acute myocardial infarction. Circulation **97** (1998) 765–772
[22] Kim, R.J., Fieno, D.S., Parrish, T.B., Harris, K., Chen, E.L., Simonetti, O., Bundy, J., Finn, J.P., Klocke, F.J., Judd, R.M.: Relationship of MRI delayed contrast enhancement to irreversible injury, infarct age, and contractile function. Circulation **100** (1999) 1992–2002
[23] Rehwald, W.G., Fieno, D.S., Chen, E.L., Kim, R.J., Judd, R.M.: Myocardial magnetic resonance imaging contrast agent concentrations after reversible and irreversible ischemic injury. Circulation **105** (2002) 224–229
[24] Brittain, J., Hu, B., Wright, G., Meyer, C., Macovski, A., Nishimura, D.: Coronary angiography with magnetization-prepared T2 contrast. Magn. Reson. Med. **33** (1995) 689–696
[25] Peters, D., Korosec, F., Grist, T., Block, W., Holden, J., Vigen, K., C.A., M.: Undersampled projection reconstruction applied to MR angiography. Magn. Reson. Med. **43** (2000) 91–101
[26] Taylor, A.M., Panting, J.R., Keegan, J., Gatehouse, P.D., Amin, D., Jhooti, P., Yang, G.Z., McGill, S., Burman, E.D., Francis, J.M., Firmin, D.N., Pennell, D.J.: Safety and preliminary findings with the intravascular contrast agent NC100150 injection for MR coronary angiography. J Magn Reson Imaging **9** (1999) 220–227
[27] Stillman, A.E., Wilke, N., Li, D., Haacke, E.M., McLachlan, S.: Evaluation of ultrasmall superparamagnetic iron oxide to enhance magnetic resonance angiography of the renal and coronary arteries: work in progress. J. Comput. Assist. Tomogr. **20** (1996) 51–55
[28] Li, D., Dolan, B., Walovitch, R., Lauffer, R.: Three-dimensional MR imaging of coronary arteries using an intravascular contrast agent. Magn. Reson. Med. **39** (1998) 1014–1018
[29] Frangi, A.F., Niessen, W.J., Viergever, M.A.: Three-dimensional modeling for functional analysis of cardiac images: A review. IEEE Trans. Med. Imaging **20** (2001) 2–25
[30] Vemuri, B.C., Radisavljevic, A.: Multiresolution stochastic hybrid shape models with fractal priors. ACM Trans. on Graphics **13** (1994) 177–207
[31] Staib, L.H., Duncan, J.S.: Model-based deformable surface finding for medical images. IEEE Trans. Med. Imaging **15** (1996) 720–731
[32] Metaxas, D.N.: Physics-Based Deformable Models. Kluwer, Boston MA (1996)

[33] Metaxas, D.N., Kakadiaris, I.A.: Elastically adaptive deformable models. IEEE Trans. Pattern Anal. and Machine Intelligence **24** (2002) 1310–1321
[34] Cootes, T.F., Cooper, D.H., Taylor, C.J., Graham, J.: Trainable method of parametric shape description. Image and Vision Computing **10** (1992) 289–294
[35] Cootes, T.F., Taylor, C.J., Cooper, D.H., Graham, J.: Active shape models-their training and application. Computer Vision and Image Understanding **61** (1995) 38–59
[36] Hill, A., Cootes, T.F., Taylor, C.J., Lindley, K.: Medical image interpretation: A generic approach using deformable templates. Medical Informatics **19** (1994) 47–59
[37] Duta, N., Sonka, M.: Segmentation and interpretation of MR brain images: An improved active shape model. IEEE Trans. Med. Imaging **17** (1998) 1049–1062
[38] Kelemen, A., Szekely, G., Gerig, G.: Elastic model-based segmentation of 3D neurological data sets. IEEE Trans. Med. Imaging **18** (1999) 828–839
[39] Pizer, S.M., Joshi, S., Fletcher, T., Styner, M., Tracton, G., Chen, J.Z.: Segmentation of single-figure objects by deformable M-reps. In Niessen, W., Viergever, M., eds.: MICCAI 2001, Berlin, Springer Verlag (2001) 862–871
[40] Stetten, G.D., Pizer, S.M.: Medial-node models to identify and measure objects in real-time 3D echocardiography. IEEE Trans. Med. Imaging **18** (1999) 1025–1034
[41] Shen, D., Herskovits, E.H., Davatzikos, C.: An adaptive focus statistical shape model for segmentation and shape modeling of 3D brain structures. IEEE Trans. Med. Imaging **20** (2001) 257–270
[42] Cootes, T.F., Beeston, C., Edwards, G.J., Taylor, C.J.: A unified framework for atlas matching using active appearance models. In Kuba, A., Samal, M., eds.: Information Processing in Medical Imaging. Lecture Notes in Computer Science, Visegrad, Hungary, Springer Verlag, Berlin (1999) 322–333
[43] Cootes, T.F.: Statistical models of appearance for computer vision. Technical report, available at http://www.isbe.man.ac.uk/~bim/Models/app_model.ps.gz (1999)
[44] Edwards, G., Taylor, C., Cootes, T.: Interpreting face images using active appearance models. In: 3rd International Conference on Automatic Face and Gesture Recognition 1998, Nara, Japan (1998) 300–305
[45] Mitchell, S.C., Lelieveldt, B.P.F., van der Geest, R.J., Bosch, H.G., Reiber, J.H.C., Sonka, M.: Cardiac segmentation using active appearance models. IEEE Trans. Med. Imaging **20** (2001) 415–423
[46] Bosch, J.G., Mitchell, S.C., Lelieveldt, B.P.F., Nijland, F., Kamp, O., Sonka, M., Reiber, J.H.C.: Automatic segmentation of echocardio-

graphic sequences by active appearance models. IEEE Trans. Med. Imaging **21** (2002) 1374–1383
[47] Lelieveldt, B., Mitchell, S., Bosch, J., van der Geest, R., Sonka, M., Reiber, J.: Time-continuous segmentation of cardiac image sequences using Active Appearance Motion Models. In: Information Processing in Medical Imaging: Lecture Notes in Computer Science, vol. 2082, New York, Springer-Verlag (2001) 446–452
[48] Mitchell, S.C., Bosch, J.G., Lelieveldt, B.P.F., van der Geest, R.J., Reiber, J.H.C., Sonka, M.: 3D Active Appearance Models: Segmentation of cardiac MR and ultrasound images. IEEE Trans. Med. Imaging **21** (2002) 1167–1178
[49] Sonka, M., Hlavac, V., Boyle, R.: Image Processing, Analysis, and Machine Vision. 2nd edn. PWS, Pacific Grove, CA (1998)
[50] Goodall, C.: Procrustes methods in the statistical analysis of shape. J. Royal Stat. Soc. B **53(2)** (1991) 285–339
[51] Bookstein, F.L.: Morphometric Tools for Landmark Data. Cambridge University Press, Cambridge, MA (1991)
[52] Altmann, S.: Rotations, Quaternions and Double Groups. Clarendon Press, Oxford, England (1986)
[53] Besl, P.J., McKay, N.D.: A method for registration of 3D shapes. IEEE Trans. Pattern Anal. and Machine Intelligence **14** (1992) 239–256
[54] Cootes, T.F., Edwards, G.J., Taylor, C.: Active appearance models. IEEE Trans. Pattern Anal. and Machine Intelligence **23** (2001) 681–685
[55] Baker, S., Matthews, I.: Equivalence and efficiency of image alignment algorithms. In: Computer Vision and Pattern Recognition Conference. Volume 1. (2001) 1090–1097
[56] van der Geest, R.J., Buller, V.G.M., Jansen, E., Lamb, H.J., Baur, L.H.B., van der Wall, E.E., de Roos, A., Reiber, J.H.C.: Comparison between manual and semiautomated analysis of left ventricular volume parameters from short-axis MR images. Journal of Computer-Assisted Tomography **21** (1997) 756–765
[57] Jacob, G., Noble, A., Mulet-Parada, M., Blake, A.: Evaluating a robust contour tracker on echocardiographic sequences. Medical Image Analysis **3** (1999) 63–75
[58] Montagnat, J., Sermesant, M., Delingette, H., Malandain, G., Ayache, N.: Anisotropic filtering for model-based segmentation of 4D cylindrical echocardiographic images. Pattern Recognition Letters **24** (2003) 815–828
[59] Hamarneh, G., Gustavsson, T.: Deformable spatio-temporal shape models: Extending ASM to 2D+Time. In: 12th British Machine Vision Conference. (2001) 13–22
[60] Barillot, C., Gibaud, B., Scarabin, J., Coatrieux, J.: 3D reconstruction of cerebral blood vessels. IEEE Computer Graphics Applications (1985) 13–19

[61] Vignaud, J., Rabischong, P., Yver, J.P., Pardo, P., Thurel, C.: Multidirectional reconstruction of angiograms by stereogrammetry and computer: Application to computed tomography. Neuroradiology **18** (79) 1–7
[62] Fessler, J., Macovski, A.: Object-based 3D reconstruction of arterial trees from magnetic resonance angiograms. IEEE Trans. Med. Imaging **10** (1991) 25–39
[63] Garreau, M., Coatrieux, J., Collorec, R., Chardenon, C.: A knowledge-based approach for 3D reconstruction and labeling of vascular networks from biplane angiographic projections. IEEE Trans. Med. Imaging **10** (1991) 122–131
[64] Suetens, P.: 3D reconstruction of the blood vessels of the brain from a stereoscopic pair of subtraction angiograms. Image and Vision Computing **1** (1983) 43–51
[65] Wink, O., Niessen, W., Viergever, M.: Fast quantification of abdominal aortic aneurysms from cta volumes. In: Medical Image Conference and Computer-Assisted Interventions (MICCAI'98), Berlin, Springer-Verlag (1998) 138–145
[66] Masutani, Y., Schiemann, T., Hohne, K.: Vascular shape segmentation and structure extraction using a shape-based region-growing model. In: MICCAI '98, First International Conference on Medical Image Computing and Computer-Assisted Intervention, Boston, MA, Springer0-Verlag, New York (1998) 1242–1249
[67] Lobregt, S., Verbeek, P.W., Groen, F.C.A.: Three-dimensional skeletonization. IEEE Transactions on Pattern Analysis and Machine Intelligence **2** (1980) 75–77
[68] Saha, P.K., Chaudhuri, B.B., Chandra, B., Majumder, D.D.: Topology preservation in 3D digital space. Pattern Recognition **27** (1994) 295–300
[69] Higgins, W., Spyra, W., Ritman, E.: System for analyzing high-resolution three-dimensional coronary angiograms. IEEE Trans. Med. Imaging **15** (1996) 377–385
[70] Pizer, S.M., Eberly, D.H., Morse, B.S., Fritsch, D.S.: Zoom invariant vision of figural shape: The mathematics of cores. Computer Vision and Image Understanding **69** (1998) 55–71
[71] Morse, B.S., Pizer, S.M., Liu, A.: Multiscale medial analysis of medical images. Proceedings on Information Processing in Medical Images **687** (1993) 112–131
[72] Furst, J., Pizer, S.M., Eberly, D.: Marching cores: A method for extracting cores from 3d medical images. In: Proceedings of the Workshop on Mathematical Models in Biomedical Image Analysis, IEEE Computer Society Technical Committee on Pattern Anaysis and Machine Intelligence (1996) 124–130
[73] Lorigo, L.M., Faugeras, O.D., Grimson, W.E.L., Keriven, R., Kikinis, R., Nabavi, A., Westin, C.F.: CURVES: Curve evolution for vessel segmentation. Medical Image Analysis **5** (2001) 195–206

[74] Sethian, J.A.: Level Set Methods. Cambridge University Press, Cambridge, MA (1996)
[75] Sapiro, G.: Vactor-valued actrive contours. In: Proc. of Computer Vision and Pattern Recognition, Los Alamitos, CA, IEEE Comp. Soc. (1996) 680–685
[76] Caselles, V., Kimmel, R., Sapiro, G.: Geodesic active contours. Int. J. of Comp. Vision **22** (1997) 61–79
[77] Carpenter, J.P., Owen, R.S., Holland, G.A., Baum, R.A., Barker, C.F., Perloff, L.J., Golden, M.A., Cope, C.: Magnetic resonance angiography of the aorta, iliac, and femoral arteries. Surgery **116** (1994) 17–23
[78] Evans, A.J., Sostman, H.D., Knelson, M.H., Spritzer, C.E., Newman, G.E., Paine, S.S., Beam, C.A.: Detection of deep venous thrombosis: Prospective comparison of MR imaging with contrast venography. Am J Radiology **161** (131–139) 1993
[79] Spritzer, C.E.: Venography of the extremities and pelvis. Magn Reson Imag Clinics of North Amer **1** (1993) 239–251
[80] Smedby, O.: Visualizations of arteries and veins with 3D image processing methods. In: XI International Workshop on Magnetic Resonance Angiography: New Aspects on Visualisation of Macro- and Microcirculation. (1999) 102
[81] Lei, T., Udupa, J.K., Saha, P.K., Odhner, D.: Artery-vein separation via MRA-An image processing approach. IEEE Trans. Med. Imaging **20** (2001) 689–703
[82] Fritsch, D.S., Eberly, D., Pizer, S.M., McAuliffe, M.J.: Stimulated cores and their applications in Medical imaging. In: Information Processing in medical Imaging, Berlin, Springer-Verlag (1995) 365–368
[83] Aylward, S., Pizer, S.M., Bullitt, E., Eberly, D.: Intensity ridge for tubular object segmentation and description. In: Proc. Workshop on Mathematical Methods in Biomedical Image Analysis, IEEE Catalog 96TB100056, IEEE (1996) 131–138
[84] Bullitt, E., Aylward, S., Liu, A., Stone, J., Mukherji, S.K., Coffey, C., Gerig, G., Pizer, S.M.: 3D graph description of the intracerebral vasculature from segmented MRA and tests of accuracy by comparison with x-ray angiograms. In Kuba, A., Samal, M., Todd-Pokropek, A., eds.: Information Processing in Medical Imaging. Springer-Verlag, Berlin (1999) 308–321
[85] Stefancik, R.M., Sonka, M.: Highly automated segmentation of arterial and venous trees from three-dimensional MR angiography. Int. J. Card. Imag. **17** (2001) 37–47
[86] Sonka, M., Hlavac, V., Boyle, R.: Image Processing, Analysis, and Machine Vision. 2nd edn. PWS, Pacific Grove, CA (1998) (1st edition Chapman and Hall, London, 1993).
[87] Stefancik, R.M.: Segmentation of arteries and veins in contrast-enhanced magnetic resonance angiography using a graph-search technique. Master's thesis, The University of Iowa (1999)

[88] Udupa, J.K., Wei, L., Samarasekera, S., Miki, Y., van Buchem, M.A., Grossman, R.I.: Multiple sclerosis lesion quantification using fuzzy-connectedness principles. IEEE Trans. Med. Imaging **16** (1997) 598–609

[89] Rice, B.L., Udupa, J.K.: Fuzzy-connected clutter-free volume rendering for MR angiography. International Journal of Imaging Systems and Technology **11** (2000) 62–70

[90] Saha, P.K., Udupa, J.K., Odhner, D.: Scale-based fuzzy connected image segmentation: Theory, algorithms, and validation. Computer Vision and Image Understanding **77** (2000) 145–174

[91] Noeske, R., Seifert, F., Rhein, K.H., Rinneberg, H.: Human cardiac imaging at 3 T using phased array coils. Magnetic Resonance in Medicine **44** (2000) 978–982

[92] Yang, Y., Gu, H., Zhan, W., Xu, S., Silbersweig, D.A., Stern, E.: Simultaneous perfusion and BOLD imaging using reverse spiral scanning at 3T: characterization of functional contrast and susceptibility artifacts. Magnetic Resonance in Medicine **48** (2002) 278–289

[93] Fayad, Z.A., Fuster, V.: Characterization of atherosclerotic plaques by magnetic resonance imaging. Annals of the New York Academy of Sciences. **902** (2000) 173–86

[94] Fuster, V., ed.: The Vulnerable Atherosclerotic Plaque: Understanding, Identification, and Modification. American Heart Association. Futura publishing company, Armonk, New York (1999)

[95] Touissant, J.F., LaMuraglia, G.M., Southern, J.F.: Magnetic resonance images lipid, fibrous, calcified, hemorrhagic, and thrombotic components of human atherosclerosis in vivo. Circulation **94** (1996) 932–938

[96] Shinnar, M., Fallon, J., Wehrli, S., Levin, M., Dalmacy, D., Fayad, Z., Badimon, J., Harrington, M., Harrington, E., Fuster, V.: The diagnostic accuracy of ex vivo MRI for human atherosclerotic plaque characterization. Arteriosclerosis, Thrombosis & Vascular Biology **19** (1999) 2756–2761

[97] Yuan, C., Mitsumori, L.M., Ferguson, M.S., Polissar, N.L., Echelard, D., Ortiz, G., Small, R., Davies, J.W., Kerwin, W.S., Hatsukami, T.S.: In vivo accuracy of multispectral magnetic resonance imaging for identifying lipid-rich necrotic cores and intraplaque hemorrhage in advanced human carotid plaques. Circulation. **104** (2001) 2051–6

[98] Yuan, C., Hatsukami, T.S., Obrien, K.D.: High-resolution magnetic resonance imaging of normal and atherosclerotic human coronary arteries ex vivo: discrimination of plaque tissue components. Journal of Investigative Medicine. **49** (2001) 491–9

[99] Yuan, C., Zhang, S.X., Polissar, N.L., Echelard, D., Ortiz, G., Davis, J.W., Ellington, E., Ferguson, M.S., Hatsukami, T.S.: Identification of fibrous cap rupture with magnetic resonance imaging is highly associated with recent transient ischemic attack or stroke. Circulation. **105** (2002) 181–5

[100] Yuan, C., Kerwin, W.S., Ferguson, M.S., Polissar, N., Zhang, S., Cai, J., Hatsukami, T.S.: Contrast-enhanced high resolution MRI for atherosclerotic carotid artery tissue characterization. J. Mag. Res. Imaging. **15** (2002) 62–7
[101] Serfaty, J.M., Chaabane, L., Tabib, A., Chevallier, J.M., Briguet, A., Douek, P.C.: Atherosclerotic plaques: classification and characterization with t2-weighted high-spatial-resolution MRI imaging: an in vitro study. Radiology. **219** (2001) 403–10
[102] Coombs, B.D., Rapp, J.H., Ursell, P.C., Reilly, L.M., Saloner, D.: Structure of plaque at carotid bifurcation: High-resolution MRI with histological correlation. Stroke **32** (2001) 2516–2521
[103] Hatsukami, T.S., Ross, R., Polissar, N.L., Yuan, C.: Visualization of fibrous cap thickness and rupture in human atherosclerotic carotid plaque in vivo with high-resolution magnetic resonance imaging. Circulation **102** (2000) 959–964
[104] Sonka, M., Thedens, D.R., Schulze-Bauer, C., Holzapfel, G., Stollberger, R., Bolinger, L., Wahle, A.: Towards MR assessment of plaque vulnerability: Image acquisition and segmentation. In: 10th Scientific Meeting of the International Society for Magnetic Resonance in Medicine, Berkeley, CA, ISMRM (2002) 1570
[105] Brejl, M., Sonka, M.: Object localization and border detection criteria design in edge-based image segmentation: Automated learning from examples. IEEE Trans. Med. Imaging **19** (2000) 973–985
[106] Holzapfel, G.A., Schulze-Bauer, C.A.J., Stadler, M.: Mechanics of angioplasty: Wall, balloon and stent. In: Mechanics in Biology. J Casey and G Bao (Eds.), New York, The American Society of Mechanical Engineers (ASME) (2000) AMD-Vol. 242/BED-Vol. 46, 141-156.

# Chapter 8

# Visualization and Segmentation Techniques in 3D Ultrasound Images

Aaron Fenster, Mingyue Ding, Ning Hu, Hanif M. Ladak, Guokuan Li, Neale Cardinal, and Dónal B. Downey

Robarts Research Institute, 100 Perth Drive London, ON,N6A 5K8, Canada, {afenster,mding,nhu,hladak,mli,cardinal,ddowney}@imaging.robarts.ca

**Summary.** Although ultrasonography is an important cost-effective imaging modality, technical improvements are needed before its full potential is realized for accurate and quantitative monitoring of disease progression or regression. 2D viewing of 3D anatomy, using conventional ultrasonography limits our ability to quantify and visualize pathology and is partly responsible for the reported variability in diagnosis and monitoring of disease progression. Efforts of investigators have focused on overcoming these deficiencies by developing 3D ultrasound imaging techniques using existing conventional ultrasound systems, reconstructing the information into 3D images, and then allowing interactive viewing of the 3D images on inexpensive desktop computers. In addition, the availability of 3D ultrasound images has allowed the development of automated and semi-automated segmentation techniques to quantify organ and pathology volume for monitoring of disease. In this chapter, we introduce the basic principles of 3D ultrasound imaging as well as its visualization techniques. Then, we describe the use of 3D ultrasound in interventional procedures and discuss applications of 3D segmentation techniques of the prostates, needles, and seeds used in prostate brachytherapy.

## 8.1 Introduction

Ultrasonography is an inexpensive and safe imaging modality that is widely used for different applications such as material defect detection, and diagnosis and staging of human disease. Conventionally, ultrasound images are two-dimensional (2D) making comprehension of complex three-dimensional (3D) structures and related applications including volume measuring, 3D anatomy display and animation difficult. In order to overcome this problem, 3D ultrasound imaging techniques have been developed in the past decade, which can reconstruct 3D ultrasound images of organs and tissues from acquisition of multiple conventional 2D images.

In Sections 8.2 and 8.3 we address the problems of acquisition and visualization of 3D ultrasound images. In Section 8.4, we introduce the application of 3D US techniques in interventional procedures, such as image-guided surgery and therapy. Finally, we discuss the use of 3D ultrasound for segmentation techniques used to segment the prostate (Section 8.5), needles (Section 8.6) and brachytherapy seeds (Section 8.7).

## 8.2 Basic Principles of 3D Ultrasound

Three-dimensional visualization of the interior of the human body has been a goal of diagnostic radiology since the discovery of x-rays. In the 1970s and 1980s, computed tomography (CT), ultrasound (US), positron emission tomography (PET), and Magnetic Resonance Imaging (MRI) have revolutionized diagnostic radiology by providing true 3D information about the interior of the human body. However, 3D visualization techniques were slower to develop, primarily because of the demanding computational requirements for 3D reconstruction and manipulation of the large amount of data in the 3D images. Thus, early systems presented the acquired 3D information as 2D images, requiring the physician to view multiple cross-sections of the anatomy and assemble the 3D information in his or her mind.

Medical ultrasound (US) imaging is a versatile and inexpensive imaging modality available in most hospitals in the world. Current US imaging produces images of high quality, making it an indispensable tool in the management of many diseases, as well as for providing image guidance for interventional procedures. Nevertheless, conventional US imaging still suffers from disadvantages, related to its 2D nature, which 3D imaging attempts to address. Despite decades of exploration, it is only in the past five years that 3D US imaging has advanced sufficiently to move out of the research laboratory and become a commercial product for routine clinical use.

### 8.2.1 Limitations of 2D US Imaging

The development of 3D US addresses the disadvantages of 2D US imaging that are related to the flexibility and subjectivity of the conventional 2D US exam. Specifically, 3D ultrasound developments address the following limitations:

- Because conventional ultrasound images are 2D, the operators must mentally transform multiple 2D images to develop a 3D impression of the anatomy and pathology during the diagnostic examination or during an image-guided interventional procedure. This imaging approach is time-consuming, inefficient, requiring an experienced operator, and can potentially lead to incorrect diagnostic or therapeutic decisions.
- Staging and planning of interventional procedures often requires accurate estimation of organ or tumor volumes. Current 2D US volume measurement techniques assume an idealized shape, and use only simple measures of the width and length in a few views. This practice leads to inaccuracy and operator variability in volume estimation.

- During interventional procedures and in monitoring the results of therapy, it is important to obtain the same views repeatedly. However, it is difficult to localize the thin 2D US image plane to a particular feature in the organ, and more difficult to reproduce the same image location and orientation later, making conventional 2D US imaging nonoptimal for the quantitative monitoring of interventional techniques or for follow-up studies for monitoring the effects of therapy.
- The patient's anatomy or position sometimes restricts the image angle attainable with the US transducer, resulting in inaccessibility of the optimal image plane necessary for diagnosis or image-guided therapy.

## 8.2.2 Requirements for 3D US Imaging

The most common 3D US approach makes use of conventional ultrasound machines and transducers with 1D arrays. These transducers are manipulated in various ways to produce multiple 2D images, which are then reconstructed into 3D images. Because the 3D images are produced from a series of 2D images, their relative positions and orientations must be accurately known, so that the reconstructed 3D image is not distorted.

Three approaches have been used to produce 3D US images with conventional 2D US systems: tracked freehand, untracked freehand, and mechanical assemblies. A fourth approach has been developed in which 2D arrays are used to produce 3D US images directly. In the following sections, we briefly describe these approaches and describe segmentation techniques for 3D US based planning and guidance of interventional procedures. For detailed descriptions of 3D ultrasound imaging approaches, the reader can refer to recent review articles and a book [1, 2, 3].

## 8.2.3 3D US Scanning Mechanisms

### 2D Arrays

To produce 3D images in real time, systems using 2D arrays keep the transducer stationary, and use electronic scanning to sweep the US beam over the anatomy. The system developed at Duke University for real-time 3D echo cardiography has shown the most promise, and has been used for clinical imaging [4, 5]. The transducer is composed of a 2D array of elements, which are used to transmit a broad beam of ultrasound diverging away from the array and sweeping out a pyramidal volume. The returned echoes are then detected by the 2D array and processed to display multiple planes from the volume in real time. These planes can then be interactively manipulated to allow the user to view the desired image region.

## Mechanical Assemblies

Instead of using 2D arrays to produce 3D images, conventional transducers with 1D arrays can be used to scan the desired anatomical volume. As the transducer is swept across the volume, the series of 2D images produced by the conventional US system is recorded rapidly. If mechanical assemblies are used to move the transducer in a precisely predefined manner, the relative position and orientation of each 2D image can be accurately known, and the acquired series of 2D images can therefore be reconstructed into a 3D image.

Numerous mechanical scanning assemblies have been developed, in which the transducer is made to rotate or translate by a motor. The mechanical assemblies vary in size from small, integrated probes that house the mechanical mechanism, to larger external mounting mechanical assemblies.

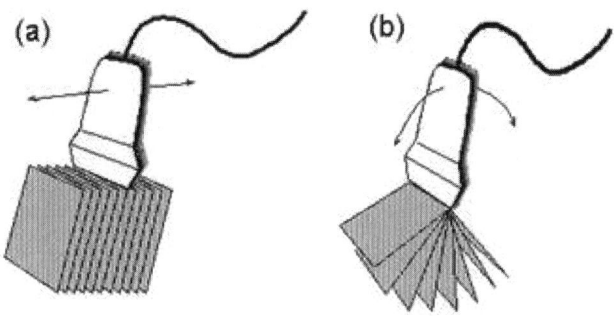

**Figure 8.1.** Diagram showing two 3D US mechanical scanning approaches: (a) linear motion, and (b) tilt motion. In both cases, the 2D images are acquired with constant spatial or angular spacing.

The integrated mechanical 3D scanning systems are easy for the operator to use, but are larger and heavier than conventional transducers. In addition, they require a special ultrasound machine, and cannot be used with any other conventional ultrasound systems. The external mounting assemblies are generally bulkier, but can be adapted to any conventional US machine. Different types of mechanical assemblies have been used to produce 3D images, as shown schematically in Figure 8.1.

Linear Motion: The transducer is mounted in a mechanical assembly which is translated linearly over the patient's skin, as shown in Figure 8.1(a), so that the set of 2D images acquired parallel to one another at a known spatial interval. Because the scanning geometry is known, the 3D reconstruction parameters can be precomputed, resulting in the 3D image being available for viewing immediately after the scan is performed [6].

Because the resolution in the acquired 2D images is nonisotropic, the resolution in the reconstructed 3D image produced by linear scanning will also be nonisotropic. Because conventional transducers have poor elevational resolution, the resolution in the scanning direction will be the worst. However, the resolution will be unchanged in the planes corresponding to the original 2D images.

This approach has been used successfully in many vascular applications, including B-mode [6, 7, 8], color Doppler imaging of the carotid arteries [7, 9, 10, 11, 12], tumor vascularity [13, 6], test phantoms [9, 14], and power Doppler imaging [6, 7, 9].

To avoid distortions in the reconstructed 3D images, Cardinal et al. [15] have shown that the distance between the acquired 2D images, and the tilt angle of the 2D image planes with respect to the scanning direction must be known accurately. To ensure that the error in volume measurements is less than 5%, the distance between acquired images must be less than 0.05 mm for a spacing of 1 mm.

Tilt Motion: The transducer is mounted in a mechanical assembly and tilted about an axis parallel to the face of the transducer, as shown in Figure 8.1(b). The tilting axis can be either at the face of the transducer, producing a set of 2D images that intersect at the face, or above it, causing the set of 2D planes to intersect above the skin. This approach allows the transducer face or the 3D probe housing to be placed at a single location on the patient's skin, making it useful for a wide range of abdominal ultrasound imaging applications [16, 7, 17].

This scanning approach has also been used successfully with endocavity transducers, such as transesophageal (TE) and transrectal (TRUS) transducers. In these applications, a side-firing linear array is used with either an external fixture or an integrated 3D transducer. The transducer is rotated about its long axis while 2D images are acquired. After a rotation of about 100, the acquired 2D images are arranged in fanlike geometry similar to that shown in Figure 8.1(b). This approach has been used successfully to image the prostate for diagnostic applications (Figure 8.2) [6, 7, 18, 19], and for 3D US-guided cryosurgery and brachytherapy [20, 6]. The main advantage of the tilt scanning approach is that the 3D scanning mechanism can be made compact to allow easy hand-held manipulation. In addition, using a suitable choice of scanning angle and angular interval, the scanning time can be short. Because the set of planes is acquired with a predefined angular interval, geometric 3D reconstruction parameters can be precalculated, allowing immediate viewing of the 3D image.

Because the acquired 2D images are arranged in a fanlike geometry, the distance between the planes will increase with distance from the trans-

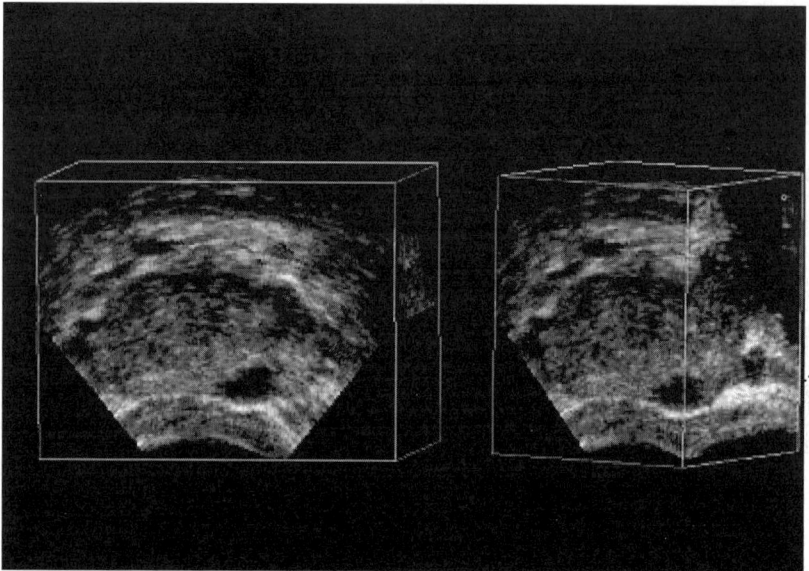

**Figure 8.2.** Two multiplanar rendering views of a 3D US image of a prostate with a carcinoma obtained with a mechanical tilt scanning approach.

ducer (rotation axis). However, the change with distance in the sampling interval can be approximately matched to the change with distance of the elevational resolution, thereby minimizing the effect of spatial sampling degradation. In combination, these two effects make the resolution worst in the scan (tilt) direction, and degrade with distance from the transducer. Tong et al. [21] have reported on an analysis of the linear, area, and volume measurement errors for the mechanical tilting transducer approach. They showed that if the location of the axis of rotation is known exactly, then the percentage error in volume will equal the percentage error in the rotation angle. To ensure a volume error of less than 5% for a typical scanning angle of 100, the total accumulated error in the scan angle must be less than 5. Thus, for a scan containing 100 2D images, the systematic error in the angular interval between images must be less than 0.05.

**Tracked Free-Hand Scanning**

Although mechanical scanning approaches result in geometrically accurate 3D images, the mechanical assemblies are generally bulky and at times inconvenient to use. To overcome this problem, free-hand scanning approaches have been developed that allow the user to manipulate the transducer freely by hand without significant constraints. While the anatomy is being scanned, the positions and orientations of the acquired 2D images are tracked and recorded, so that the 3D image can be reconstructed. Because the scanning geometry is not predefined, the operator must ensure that the set of acquired 2D images has no significant gaps. Several free-hand scanning approaches have been developed, which use four basic position-sensing techniques: acoustic tracking, articulated arms, magnetic field tracking, and image-based information.

- Acoustic Tracking: In this approach, sound-emitting devices (spark gaps) were mounted on the transducer, and an array of fixed microphones was mounted above the patient. As the operator moved the transducer over the patient's skin in the usual manner, the 2D ultrasound images were continuously acquired, and the acoustic pulses from the sound emitters were continuously recorded by the microphones. Using the speed of sound in air and the time-of-flight of the acoustic pulses from the emitters to the microphones, the positions and orientations of the acquired 2D images were then determined [22, 23].
- Articulated Arms: A partially constrained free-hand scanning approach was achieved by scanning the patient with the transducer mounted on a multijointed mechanical arm system. The relative rotation of the arms was measured with potentiometers located at each joint, and the relative positions and orientations of the acquired 2D images were calculated. While the transducer was manipulated over the patient's anatomy, a computer recorded the acquired 2D images and the relative orientation of all the arms, which were then used to reconstruct the 3D image.
- Magnetic Field Tracking: The most successful free-hand scanning approach makes use of a six degree-of-freedom magnetic field sensor to track the ultrasound transducer. In this approach, a transmitter is used to produce a spatially varying magnetic field, and a small receiver containing three orthogonal coils mounted on the transducer is used to sense the magnetic field strength. By measuring the strength of three components of the local magnetic field, the US transducer's position and orientation can be continuously monitored.

  Although magnetic field sensors are small and unobtrusive, their accuracy can be compromised by electromagnetic interference from sources such as CRT monitors, AC power cabling, and some electrical signals from ultrasound transducers. Also, ferrous and highly conductive metals can distort the magnetic field, causing geometric errors in the tracking information. However, by ensuring that the immediate scanning environment is free of electrical interference and metals, high quality 3D images can be obtained [24, 25, 26, 27, 28, 29, 30].
- Speckle Decorrelation: Free-hand scanning techniques described above require an external sensor to measure the relative positions and orientations of the acquired 2D images. However, the relative positions of adjacent 2D images can also be measured using the well-known phenomenon of speckle decorrelation. When a source of coherent energy interacts with scatterers, the reflected spatial energy pattern will appear as a speckle pattern. This speckle is characteristic of US images and can be used in 3D US imaging.

If two 2D US images are acquired from the same location and orientation, then their speckle patterns will be the same. However, if one image is acquired a short distance away from the other, then the degree of decorrelation in the speckle will be proportional to the distance between the two images [31]. Since the relationship between the degree of decorrelation and distance will depend on several transducer parameters, accurate determination of the separation of the images requires calibration of the relationship between distance and various transducer parameters.

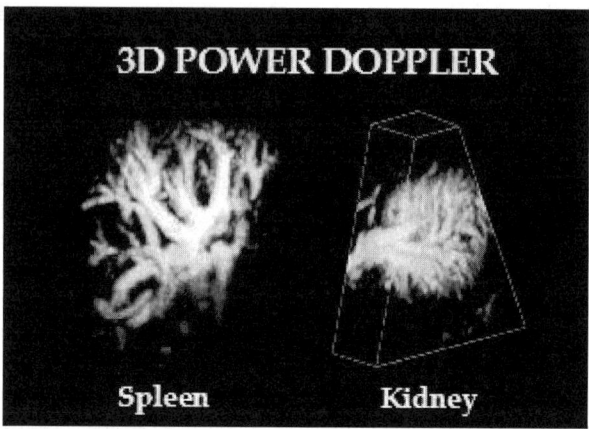

**Figure 8.3.** Two 3D power Doppler images obtained with free-hand scanning without position sensing and displayed using volume rendering. On the left is a spleen and on the right, a kidney.

**Untracked Free-Hand Scanning**

In this approach, the operator moves the transducer in a steady motion, while a series of 2D images are acquired. Because the position and orientation of the transducer are not recorded, the linear or angular spacing between the acquired images is assumed in the reconstructing 3D image. Thus, geometric measurements such as distance or volume may still be inaccurate. If the transducer's motion is uniform and steady, then very good 3D images may be reconstructed, as illustrated by Figure 8.3 [7].

## 8.3 3D US Image Reconstruction

Reconstruction of 3D US images refers to the process of placing the acquired 2D images into the 3D image in their correct relative positions. Two methods have been used: feature-based and voxel-based reconstruction.

### 8.3.1 Reconstruction of Features

In this approach, each acquired 2D image is segmented and classified into the desired features that are to be reconstructed. Typical examples have been implemented by manually or automatically outlining of the boundaries of the ventricles in echocardiographic images or of the fetus in obstetrical images. The boundary of each structure is then represented by a mesh and displayed.

The advantage of this reconstruction approach is that the 3D image is reduced to a description of surfaces in the form of a mesh, enabling the use of common hardware and software tools to manipulate and display these surfaces in real time. The main disadvantage of this approach relates to the segmentation and classification process, which identifies and stores information only about the anatomical boundaries. Thus, important anatomical information is lost, such as subtle pathological features related to tissue image texture.

### 8.3.2 Reconstruction of a Cartesian Volume

The most common 3D US reconstruction approach uses the set of acquired 2D images to build a 3D grid of voxels. This is accomplished by placing the pixels in the acquired 2D images into their correct locations within the 3D image. The values (color or grayscale) of the voxels not sampled are calculated by interpolation from their nearest neighbors. In this approach, all the original information is preserved in the 3D image, and the original 2D images can be recovered. However, if the angular or spatial spacing between the acquired 2D images is not chosen properly, the scanning process will not sample the volume adequately, resulting in degradation of resolution in the reconstructed 3D image. Sampling the volume properly and avoiding gaps results in large image files that can range from 16 MB to 96 MB [32, 6].

### 8.3.3 Viewing 3D US Images

#### Multiplanar Reformatting: Cube View

The most common approach for viewing 3D US images is based on multiplanar reformatting (MPR) of the 3D image. In the cube view approach, the 3D image is presented as a polyhedron, and the appropriate 2D US images are "painted" on each face by texture mapping (Figures 8.2 and 8.4). The polyhedron can be rotated and any face may be moved, either parallel or obliquely to its original location, while the appropriate US data is continuously texture-mapped in real time onto the new face. Thus, the operator always has 3D image-based cues relating the plane being examined to the rest of the anatomy [11, 19, 2].

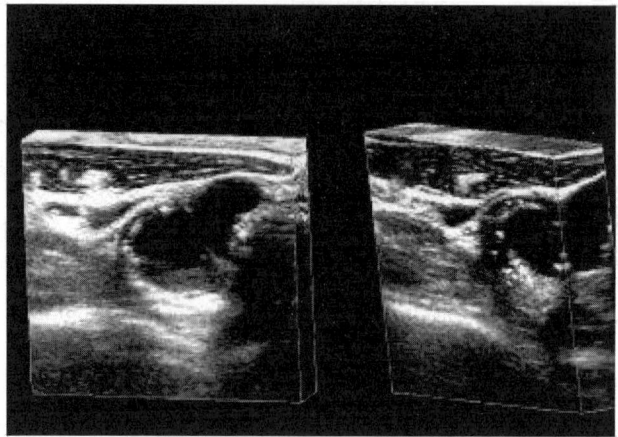

**Figure 8.4.** Two multi-planar rendered views of a 3D US image of the carotid arteries with a plaque at the entrance to the internal carotid artery.

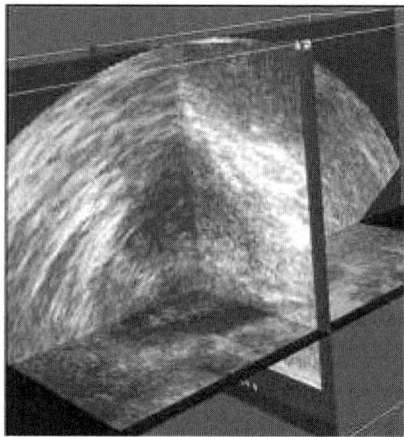

**Figure 8.5.** A 3D US image of a prostate displayed using the MPR technique as intersecting orthogonal planes.

## MPR: Orthogonal Planes

In this MPR approach, three perpendicular planes are displayed simultaneously, with graphical cues indicating their relative orientations (Figure 8.5). For easier appreciation, these planes are typically made perpendicular to each other. User interface tools are provided to allow the operator to select any plane, and move it parallel or obliquely to the original, to provide the desired view of the anatomy [33, 34, 2].

Chapter 8 Visualization and Segmentation in 3D Ultrasound 251

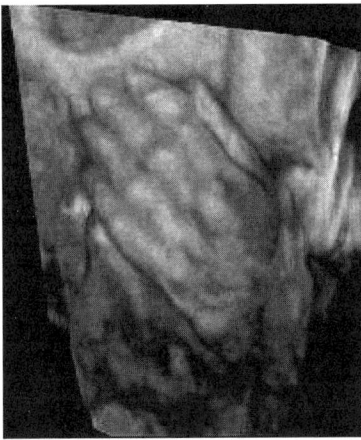

**Figure 8.6.** 3D US image displayed using the VR technique showing a fetal hand.

**Volume Rendering (VR)**

The volume rendering technique presents a display of the entire 3D image after projection onto a 2D plane. Image projection is typically accomplished via ray-casting techniques [35]. Although many VR algorithms have been developed, 3D US imaging currently makes use of primarily two approaches: maximum (minimum) intensity projection, and translucency rendering. This approach has been primarily used to display fetal (Figure 8.6) and vascular anatomy (Figure 8.3)[7].

## 8.4 3D US-Guided Prostate Brachytherapy

Prostate cancer is the most commonly diagnosed malignancy in men over 50, and was found at autopsy in 30% of men at age 50, 40% at age 60, and almost 90% at age 90 [36]. However, when it is correctly diagnosed at an early stage, prostate cancer is curable, and even at later stages, surgical treatment can still be effective.

One of the most promising options for treatment of early prostate cancer is prostate brachytherapy, which is used to implant 80–100 radioactive seeds (e.g., 125I or 103Pd) in or near the prostate and avoid radiation-sensitive normal-tissue structures [37]. Because the radiation dose produced by each seed falls off rapidly with distance, the correct positioning of the seeds is crucial to the success of the procedure. For each patient, an implantation dose preplan is generated based on a CT or 3D US image of the prostate. During a later outpatient procedure, the patient is positioned in approximately the same position as for the preimplantation image, and brachytherapy seeds are then implanted in the prescribed positions using a needle. Real-time transrectal US (TRUS) is used to ensure accurate placement of the seeds.

Because of possible changes in prostate size and shape between the preimplantation and implantation procedures, as well as possible changes in the patient position between the preimplantation and implantation images, errors in seed placement can occur. Performing preimplant dose planning and seed implantation during the same session could avoid these problems. This could be done by using a 3D US imaging system with rapid scanning and immediate viewing of the 3D prostate anatomy. In addition, this system could be used to provide improved image guidance during the procedure. Thus, an efficient brachytherapy procedure could be achieved by integrating 3D US imaging with the accurately determined positions of the prostate, needles and seeds.

In the following discussion, we review prostate, needle, and seed segmentation techniques developed by the investigators and present the most novel technique in details.

## 8.5 Prostate Segmentation in 3D US Images

In prostate brachytherapy, assigning the appropriate therapy dose requires accurate knowledge of the prostate shape and its volume. Traditionally, this is done via manual planimetry, which not only is time-consuming and tedious to perform, but also highly operator dependent, causing variability in the results. In order to determine the shape of the prostate and measure its volume more accurately and consistently, an automated or semi-automated technique is required.

### 8.5.1 Previous Work

Most currently available prostate segmentation algorithms focus on segmentation of 2D US images. [38] proposed the Laplacian-of-Gaussian edge operator followed by an edge-selection algorithm, which requires the user to select several initial points to form a closed curve. Their method correctly identified most of the boundary in a 2D US prostate image. This technique was extended using four texture energy measures associated with each pixel in the image [39]. An automated clustering procedure was used to label each pixel in the image with the label of its most probable class. Although good results were reported for 2D US prostate images, the algorithm was computationally intensive, requiring about 16 minutes to segment the prostate boundary in 2D with a 90-MHz SUN SPARCstation.

Aarink et al. [40, 41, 42, 43] published a series of papers on prostate segmentation. First, they used cross-sections of the prostate obtained with a well-defined distance. In each image, the contour of the prostate was determined using edge detection techniques. After locating the prostate, the area of the prostate was calculated for each image and the volume of the prostate was determined by multiplying the summation of these areas by the distance between the cross-sections. In 1994, they proposed a practical clinical method to determine the 2D prostate contour, which comprised three steps: edge detection, edge enhancement and selection, and edge linking by interpolation. Although they reported good segmentation results in 1996, their method could not ensure robust and ac-curate segmentation due to the speckled noise and image shadows.

In 1997, Liu et al. [44] presented an algorithm based on their radial bas-relief (RBR) edge detector. First the RBR detector highlighted the edges in the image, and then binary processing and area labeling were used to segment the boundary. Their results showed that RBR performed well with a good-quality image, and marginally for poor-quality images. The RBR approach was able to extract a skeletonized image from an US image automatically. However, many spurious branches were created introducing ambiguity in defining the actual prostate boundary. In 1998, Kwoh et al. (1998) [45] extended the RBR technique by fitting a Fourier boundary representation to the detected edges, resulting in a smooth boundary, but this technique required careful tuning of algorithm parameters.

Pathak et al. [46, 47, 48] also developed an edge guidance delineation method for deformable contour fitting in a 2D ultrasound image and statistically demonstrated a reduction in the variability in prostate segmentation.

In 1999, Knoll et al. [49] proposed a technique for elastic deformation of closed planar curves restricted to particular object shapes using localized multiscale contour parameterization based on the wavelet transform. The algorithm extracted only important edges at multiple resolutions and ignored other information caused by noise or insignificant structures. This step was followed by a template-matching procedure to obtain an initial guess of the contour. This wavelet-based method constrained the shape of the contour to predefined models during deformation. They reported that this method provided a stable and accurate fully automatic segmentation of 2D objects in ultrasound and CT images.

In 2001, Garfinkel et al. [50] used a deformable model to segment the prostate from 3D US images. Their approach required the user to initialize the model by outlining the prostate in 40–70% of the 2D slices of each prostate, using six to eight vertices for each 2D contour, and then an initial 3D surface was generated. The running time of the algorithm was about 30 sec on a SUN Ultra 20. They compared algorithm and manual segmentation results by computing the ratio of the common pixels that were marked as prostate by both methods. The results showed an accuracy of nearly 89% and a three- to six-fold reduction in time compared to a totally manual outlining. No editing of the boundary was possible to improve the results.

Although a variety of segmentation methods have been proposed, none segments the prostate in the 3D US image directly and allows editing of the results. Without the use of a complete 3D prostate image, prostate shape information is not used efficiently, leading to an inaccurate and time-consuming segmentation. To overcome this shortcoming, an alternative direct 3D segmentation method is more promising.

In this section, we describe a semi-automatic 3D prostate segmentation approach, which uses an ellipsoid 3D mesh model of the prostate to initialize the procedure, and used the discrete dynamic contour (DDC) approach to refine the contour in 3D. Experiments demonstrated that the average difference between our algorithm and manually segmented 3D prostate boundaries varied from 0.08 mm to 0.5 mm while, the volume difference varied from 6% to 10%. The computational time for the whole 3D prostate is about 60 sec on a Pentium III 400-MHz PC computer.

### 8.5.2 3D Prostate Segmentation Based on the Deformable Ellipsoid Model

The deformable model, also called the snake or active contour, was first proposed by Kass et al. in 1987 [51, 52], and has become a widely used technique in medical image analysis [53, 54, 55, 56, 57, 58, 59, 60].

Miller et al. [61] extended this technique and developed a geometrically deformable model, which was further extended by Lobregt and Viergever to the discrete dynamic contour [62]. They used force analysis at contour vertices to replace the energy minimization of the whole contour so that a boundary was found by driving an initial contour to the true object boundary. Because the internal and external forces are evaluated only at the vertices, instead of the trajectory of the connected edge segments, the DDC method is much faster than the traditional deformable model approach.

In our previous paper [63], we reported on the development of an algorithm to fit the prostate boundary in a 2D image using the DDC with model-based initialization. A cubic interpolation function was used to estimate the initial 2D contour from four user-selected points, which was then deformed automatically to fit the prostate boundary. However, diagnosis and therapy planning of prostate cancer typically require the prostate volume and its 3D shape. Constructing a 3D prostate boundary from a sequence of 2D contours can be time-consuming or subject to errors.

Based on the 3D triangle mesh deformable model [64], we developed a deformable ellipsoid model for 3D prostate segmentation. It comprises three steps: (1) 3D mesh initialization of the prostate using an ellipsoid model; (2) automatic deformation to refine the prostate boundary; and (3) interactive editing of the deformed mesh, after which step 2 is repeated.

**Initialization**

In order to initialize the prostate mesh, the user manually selects six control points $(x_n, y_n, z_n), n = 1, 2, ..., 6$, on the "extremities" of the prostate in the 3D US image. Typically, the user selects an approximate central transverse

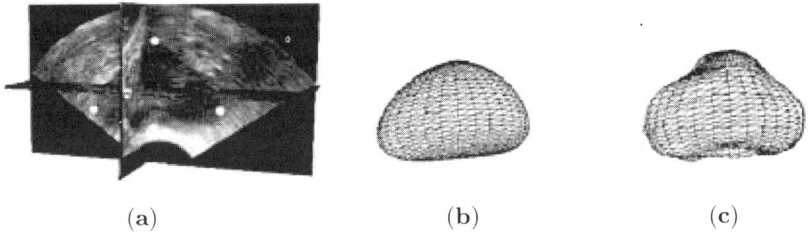

**Figure 8.7.** 3D prostate segmentation using the deformable ellipsoid model: (a) 3D US image with five of the six user-selected control points shown in white, (b) initial mesh, (c) final deformed mesh.

prostate cross-section and then places two points near the prostate's lateral extremes and two near its top and bottom on the central axis. The last two points are placed near the prostate's apex and base. A 3D US prostate image with five of the six initialization points is shown in Figure 8.7(a). These control points are used to estimate an initialization ellipsoid, which is parameterized as follows [65]:

$$\mathbf{r}(\eta, \omega) = \begin{bmatrix} x \\ y \\ z \end{bmatrix} = \begin{bmatrix} x_0 + a\, \cos(\eta)\cos(\omega) \\ y_0 + b\, \cos(\eta)\sin(\omega) \\ z_0 + c\, \sin(\eta) \end{bmatrix}, \quad -\frac{\pi}{2} \leq \eta \leq \frac{\pi}{2}, -\pi \leq \omega \leq \pi, \quad (8.1)$$

where $(x_0, y_0, z_0)$ is the center of the ellipsoid and $a, b,$ and $c$ are the lengths of the semi-major axes in the $x$, $y$, and $z$ directions, respectively. The length, width, and height of the prostate are assumed to be approximately oriented along the $x-$, $y-$, and $z-$axes of the 3D US image, as shown in Figure 8.7(a). First, from the pair of control points with extreme $x$ values, we estimate $x_0$ as half the sum the $x$ coordinates and $a$ as half the absolute difference. Similarly, we estimate $y_0$ and b from the pair with extreme $y$ values, and $z_0$ and $c$ from the pair with extreme $z$ values. The vector $\mathbf{r}(\eta, \omega)$ then describes the surface coordinates of the ellipsoid as a function of the azimuthal angle $\omega$ and elevational angle $\eta$, defined in the usual manner with respect to the $x$-axis and the $x$-$y$ plane. Thus, by stepping the angles $\omega$ and $\eta$ through an appropriate grid of angular values, a set of regularly spaced points is obtained on the surface of the ellipsoid. The ellipsoid's surface is then represented by a mesh of triangles connecting these points [66].

Usually, the ellipsoid generated as described above does not pass through the six control points, nor does it fit the prostate boundary very well. To obtain a better fit, the ellipsoid is warped using the thin-plate spline transformation [67]. Using this transformation, the six ends of the ellipsoid's major axes are mapped onto the corresponding control points. The resulting mesh, as shown in Figure 8.7(b), then becomes the initial mesh for the deformation step.

**Contour Deformation**

The process for finding the 3D prostate boundary is almost identical to the 2D segmentation technique described by [63], but extended to 3D by calculating the internal force $\mathbf{f}_i^{int}$ acting on each vertex $i$ from Equation (8.2)

$$\mathbf{f}_i^{int} = \left( \left( \frac{1}{6} \sum_i \hat{e}_{ij} \right) \cdot \hat{\mathbf{r}}_j \right) \cdot \hat{\mathbf{r}}_j \tag{8.2}$$

$$e_{ij} = p_i - p_j \tag{8.3}$$

where $e_{ij}$ is the edge vector pointing towards vertex $i$ from one of its six neighboring vertices $j$, and $\hat{\mathbf{r}}$ denotes the unit radial vector at vertex $i$. These equations are iterated until either all the vertices have reached equilibrium, or the number of iterations has reached a preset limit. In our experiment, we found that sufficient accuracy was generally achieved within 40 iterations (see Figure 8.7(c)).

**Contour Editing**

The initialization procedure may place some mesh vertices far from the actual prostate boundary. The deformation process may not drive these vertices towards the actual boundary because the image force exerts a strong effect only near the edge of the prostate, where the image intensity changes rapidly. In these cases, the user can edit the mesh by dragging a vertex to its desired location. To avoid rapid transition in the boundary, adjoining vertices within a user-defined radius are also automatically deformed using the thin plate spline transformation. After editing, the automatic deformation process resumes, with the iteration count reset to zero.

## 8.6 Needle Segmentation in 3D US Images

### 8.6.1 Introduction

As discussed in Section 8.4, brachytherapy radioactive seeds are delivered by an implantation needle as it is withdrawn. Although the needle insertion can be guided with a real-time 2D US imaging system, it is our experience from prostate cryosurgery [20] that the 3D trajectory of the needle cannot be fully ascertained, because lateral deflection is poorly detected. However, in order to use 3D US guidance to overcome this problem, it is necessary to be able to segment the needle in the 3D US image, in near real time. This task is made difficult by artifacts in the US image caused by speckle, shadowing, refraction, and reverberation, and by the low contrast of the needle when it is not parallel to the US transducer. Because of these difficulties, traditional local operators such as edge detectors are inadequate for finding the needle boundary. The challenge is to find a needle segmentation approach that is insensitive to these image artifacts, yet is fast, accurate, and robust.

In this section, we describe just such an approach, which also requires minimal manual initialization. This approach is motivated by four observations: (1) the needle image lies along a straight line; (2) the needle is more conspicuous in a projection image than in the original 3D US image; (3) after segmenting the needle in a 2D projection image, we know that the needle lies in the plane defined by two vectors: the projection direction and the needle direction in the projection image; (4) therefore, if we use two orthogonal projections, we know that the needle must lie along the intersection of the two corresponding planes.

Our approach is composed of four steps: (1) volume cropping, using a priori information to restrict the volume of interest; (2) volume rendering, to form the 2D projection images; (3) 2D needle segmentation, in a projection image; and, (4) 3D needle segmentation, or calculation of the 3D needle endpoint coordinates.

Experiments with US images of both agar and turkey breast test phantom (see Figure 8.8) demonstrated that our 3D needle segmentation method could be performed in near real time (about 10 frames-per-second with a 500 MHz personal computer equipped with a commercial volume-rendering card to calculate the 2D projection images). The root-mean-square accuracy in determining needle lengths and endpoint positions was better than 0.8 mm, and about 0.5 mm on average, for needle insertion lengths ranging from 4.0 mm to 36.7 mm.

### 8.6.2 Volume cropping

The 3D US image is usually large with displays complex echogenicity (see Figures 8.8 and 8.9 (a). These attributes result in increased computational time required to perform the image segmentation. In addition, they also increase the likelihood that some background structures have a similar range of voxel intensities as the needle voxels, increasing the difficulty of segmenting the needle. Even in a 2D projection (volume-rendered) image, in which a background structure overlaps that of the needle, automatic segmentation of the needle is much more difficult. However, if background structures are well separated from the needle in the 3D US image, volume cropping can effectively alleviate this problem. Furthermore, cropping will also reduce the size of the volume to be rendered, significantly reducing computation time.

Based on a priori information about the needle location and orientation, volume cropping restricts the volume of interest to the 3D region that must contain the needle. This information is composed of the approximate insertion point, insertion distance, insertion direction, and their uncertainties. This information can be made available when the needle is inserted manually or by a motorized device under computer control, or whenever it is localized using

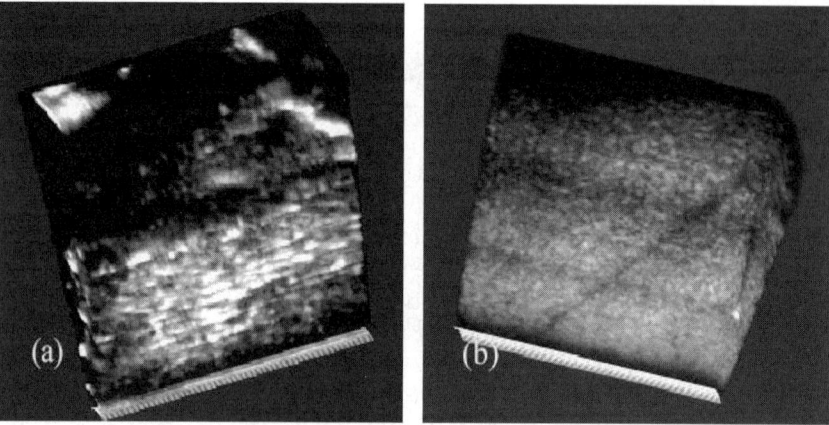

**Figure 8.8.** Views of 3D US images of the two types of phantoms used to experimentally evaluate our automatic needle segmentation algorithm. The 3D images are displayed using the multi-planar technique: (a) agar, (b) turkey breast.

an acoustic, articulated arm, electromagnetic, or optical tracking device [36]. Our approach proceeds as follows.

Let $\mathbf{L}'_1 = (x'_1, y'_1, z'_1)$ and $\mathbf{L}'_2 = (x'_2, y'_2, z'_2)$ be the 3D image coordinates of the needle's insertion point and end point, respectively, and let $L'_0 = |\mathbf{L}'_2 - \mathbf{L}'_1|$ be its inserted length. Then the 3D needle vector from insertion point to tip can be written as $\mathbf{L}'_0 = \mathbf{L}'_2 - \mathbf{L}'_1 = L'_0 \mathbf{P}'$, where the unit vector $\mathbf{P}'$ is the needle direction. We then assume that the uncertainty in the needle insertion point can be estimated and that $x'_1 = x_1 \pm \Delta x$, $y'_1 = y_1 \pm \Delta y$, $z'_1 = z_1 \pm \Delta z$, $L'_0 \leq L_0$, and that the angle between $\mathbf{P}'$ and a given approximate needle direction $\mathbf{P} = (a, b, c)$ is at most $\phi < \pi/2$. Then every needle voxel $(x, y, z)$ must lie within the cropped volume defined by the 3D range:

$$[x_{\min}, x_{\max}] \times [y_{\min}, y_{\max}] \times [z_{\min}, z_{\max}] \tag{8.4}$$

where

$$x_{\min} = \min(x_1 - \Delta x, x_1 - \Delta x + (a - 2\sin(\phi/2)) \cdot L_0), \tag{8.5}$$

$$x_{\max} = \max(x_1 + \Delta x, x_1 + \Delta x + (a + 2\sin(\phi/2)) \cdot L_0), \tag{8.6}$$

$$y_{\min} = \min(y_1 - \Delta y, y_1 - \Delta y + (b - 2\sin(\phi/2)) \cdot L_0), \tag{8.7}$$

$$y_{\max} = \max(y_1 + \Delta y, y_1 + \Delta y + (b + 2\sin(\phi/2)) \cdot L_0), \tag{8.8}$$

$$z_{\min} = \min(z_1 - \Delta z, z_1 - \Delta z + (c - 2\sin(\phi/2)) \cdot L_0), \tag{8.9}$$

$$z_{\max} = \max(z_1 + \Delta z, z_1 + \Delta z + (c + 2\sin(\phi/2)) \cdot L_0). \tag{8.10}$$

An example of volume cropping is given in Figure 8.9(b).

### 8.6.3 Volume rendering

Volume rendering is used extensively in exploring 3D images. In this process, projection images are generated by casting parallel (or divergent) rays through a 3D image volume, where they accumulate both luminance $c(I)$ and opacity $\lambda(I) \leq 1$ as a function of the local voxel intensity $I$ [68]. The grayscale intensity of the projected 2D image plane is then the final accumulated luminance C after the ray has passed through the volume. In general, the projected 2D image of the needle will have the greatest contrast when both of the functions $c(I)$ and $\lambda(I)$, called the transfer functions, match the distribution of $I$ in the needle voxels. Using phantom test objects, we found that this distribution could be accurately modelled as a Gaussian distribution with mean $\bar{I}$ and standard deviation $\sigma_I$, where the parameters $\bar{I}$ and $\sigma_I$ depend on the image appearance of the object. In our experiments, we defined the transfer functions as

$$C(I) = \lambda(I) = \exp\left(-\frac{1}{2}\left(\frac{I-\bar{I}}{\sigma_I}\right)^2\right). \tag{8.11}$$

Using these Gaussian transfer functions, the 2D projection image is then specified by the projection direction, which in our case is always perpendicular to the a priori approximate needle direction **P**.

### 8.6.4 2D Needle Segmentation

In the 2D projection image, we initially search for objects (connected groups of pixels) with pixel intensities exceeding a fixed threshold of 25 (on an 8-bit scale of 0 to 255). This threshold was chosen experimentally to include virtually all needle pixels, while excluding most non-needle pixels. Because of the volume cropping operation, we may safely assume that the object with the largest extent in the **P** direction is the needle. (We note that, because the projection direction is always chosen perpendicular to **P**, the vector **P** always lies in the projected image plane.) Then, using a flood-fill algorithm, the pixels in this object are assigned a value of 1, while all other pixels are assigned a value of 0, resulting in a binary 2D projection image of the needle as shown in Figure 8.9(d).

We assume that the projection image has 2D Cartesian coordinates $(u, v)$, chosen so that $\hat{u} = \mathbf{P}$. Then, the projected needle image lies within an angle $\phi < \pi/2$ of the $u$-axis, so we can perform a least-squares fit of the binary projected needle image, to find the estimated projected needle vector **L**. With knowledge of the needle direction, we determine its endpoints $(u_1, v_1)$ and $(u_2, v_2)$, with $u_1 < u_2$, so that $(u_1, v_1)$ corresponds the needle's insertion point and $(u_2, v_2)$ corresponds to its end point.

### 8.6.5 3D Needle Segmentation

Given the a priori approximate needle direction **P**, we define two other unit vectors **Q** and **R** such that $\{\mathbf{P}, \mathbf{Q}, \mathbf{R}\}$ are mutually orthogonal, and form the

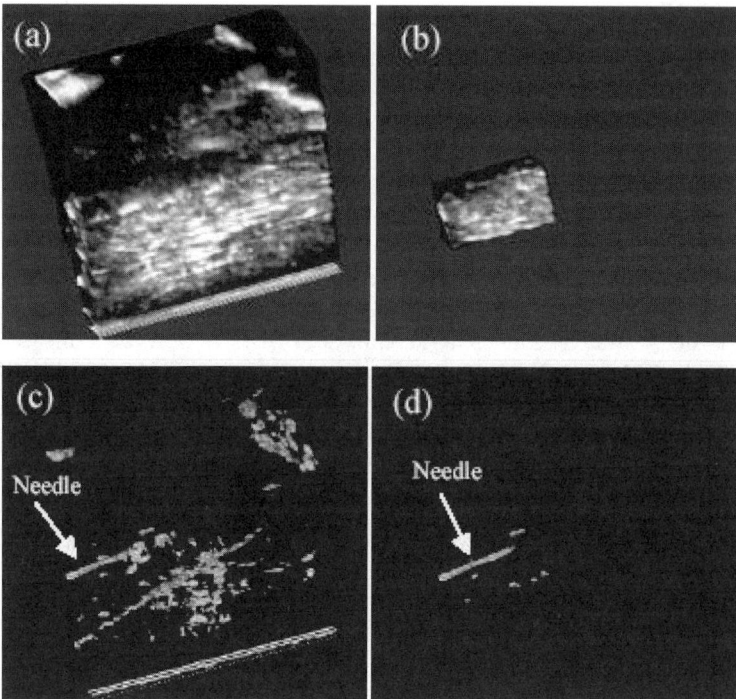

**Figure 8.9.** Comparison of volume cropped and rendered images of a turkey breast phantom. (a) The MPR display of 3D US turkey phantom image; (b) Cropped volume of (a); (c) Rendered image of (a); (d) rendered image of (b).

basis vectors of a right-handed 3D Cartesian coordinate system (Figure 8.10). We then form two projection images in the orthogonal directions $\mathbf{Q}$ and $\mathbf{R}' = -\mathbf{R}$.

For the 2D projection image projected in the direction $\mathbf{Q}$, i.e., with $\hat{u} \times \hat{v} = -\mathbf{Q}$, we choose $\hat{u} = \mathbf{P}$. Then $\hat{v} = \hat{u} \times \mathbf{Q} = \mathbf{R}$. From the 2D needle segmentation, we then find that the endpoints of the estimated projected needle vector $\mathbf{L}$ are

$$(p_n, r_n)_Q = (u_n, v_n), \quad n = 1, 2. \tag{8.12}$$

For the 2D projection image projected in the direction $\mathbf{R}' = -\mathbf{R}$, i.e., with $\hat{u} \times \hat{v} = -\mathbf{R}' = \mathbf{R}$, we again choose $\hat{u} = \mathbf{P}$. Then $\hat{v} = \hat{u} \times \mathbf{R}' = -\hat{u} \times \mathbf{R} = \mathbf{Q}$. From the 2D needle segmentation, we then find that the endpoints of the estimated projected needle vector $\mathbf{L}$ are:

$$(p_n, q_n)_{R'} = (u_n, v_n), \quad n = 1, 2 \tag{8.13}$$

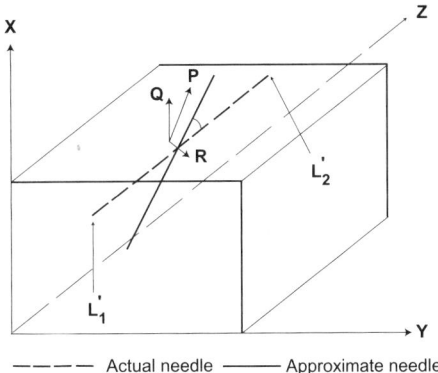

**Figure 8.10.** 3D coordinate systems used in 3D needle segmentation.

We now suppose that a known point $\mathbf{O} = (x_0, y_0, z_0)$ in the cropped volume (e.g., its center) is projected onto the origin of the $u$-$v$ plane in both projection images, and define

$$p_n = (p_{nQ} + p_{nK})/2, \quad n = 1, 2, \tag{8.14}$$

$$q_n = q_{nR'}, \quad n = 1, 2, \tag{8.15}$$

$$r_n = r_{nQ}, \quad n = 1, 2. \tag{8.16}$$

Then, the estimated 3D coordinates of the needle's endpoints are given by:

$$\mathbf{L}''_n = \mathbf{O} + p_n \mathbf{P} + q_n \mathbf{Q} + r_n \mathbf{R}, \quad n = 1, 2, \tag{8.17}$$

and the estimated length of the needle is given by:

$$\mathbf{L}''_0 = |\mathbf{L}''_2 - \mathbf{L}''_1|. \tag{8.18}$$

Finally, the errors $\delta_n$ in $\mathbf{L}_n$, $n = 0, 1, 2$, are given by

$$\delta_0 = |\mathbf{L}''_0 - \mathbf{L}'_0|, \tag{8.19}$$

$$\delta_n = |\mathbf{L}''_n - \mathbf{L}'_n|, \quad n = 1, 2, \tag{8.20}$$

where the actual needle endpoints $\mathbf{L}'_1$ and $\mathbf{L}'_2$ and needle length $\mathbf{L}'_0 = |\mathbf{L}'_2 - \mathbf{L}'_1|$ are determined by manually segmenting the needle in the 3D US image.

## 8.7 Brachytherapy Seed Segmentation in 3D US Images

After completing the brachytherapy seed implantation procedure, the seeds' actual locations must be evaluated by CT or MRI imaging. Using this information, a dosimetric analysis (post-plan) is then performed to determine whether the dose coverage of the implant is satisfactory. If it is not, then additional seeds can be implanted in the underdosed area. However, if the seed placement could be evaluated with a 3D US imaging system (the same system used to provide image guidance during the procedure), then a post-plan could be performed immediately, and the additional seeds could then be implanted intraoperatively as an extension of the regular procedure, instead of in a separate procedure after the post-plan CT or MRI examination is completed. Moreover, if 3D US image guidance with simultaneous seed placement verification could be carried out, the treatment plan could be continuously adjusted during the procedure, as the seeds are implanted, thereby minimizing the required number of needle insertions, and maximizing the effectiveness of the radiation therapy.

However, performing seed segmentation in 3D US images is extremely difficult, for four reasons: First, calcifications and other small structures with high echogenicity can mimic the bright appearance of a brachytherapy seed in an US image, making positive seed identification difficult, unlike the situation in a CT or MRI image (see Figure 8.11). Second, many seeds are implanted into the prostate, typically 80-100, unlike the situation described in the preceding section, where a single needle is being segmented. Third, because a seed is cylindrically shaped (about 0.8 mm in diameter and 4 mm in length, similar to a grain of rice), the brightness in the 3D US image can vary greatly, depending on its orientation relative to the ultrasound transducer (being much brighter when the seed is oriented parallel to the transducer, due to specular reflection).Fourth, the images of the seeds are superimposed on a highly cluttered US image background.

### 8.7.1 Volume Cropping

By using the 3D prostate segmentation described in Section 8.5, we can crop the 3D US image to a volume that contains only the prostate and its immediate surroundings, where all the radioactive seeds are located. This not only greatly eases the task of seed segmentation, but also saves a great deal of computational time.

### 8.7.2 Adaptive Thresholding

The voxels in the cropped volume are segmented using an adaptive thresholding technique. Suppose that the 3D US image has $N$ gray levels, and let $\{h(n), n = 0, \cdots, N-1\}$ be the gray-level histogram of the cropped volume. Then, because the images of the seeds are bright, the adaptive threshold $T$ is defined as

$$T = \max\left\{ t \,\Bigg|\, \sum_{n=t}^{255} h(n) \geq n_0, \ t = 0, 1, \cdots, 255 \right\}, \tag{8.21}$$

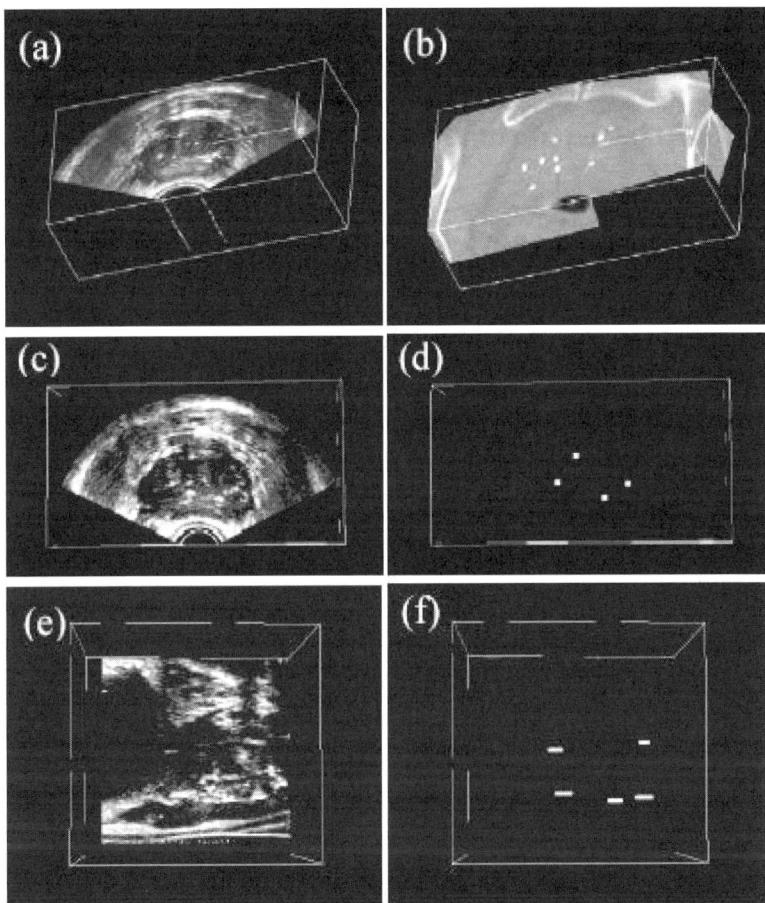

**Figure 8.11.** Seed segmentation results: (a) and (b) are the 3D US and the corresponding CT image, displayed using the MPR technique. The size of the 3D US image is 383,383,246 voxels; (c) is a lateral slice, and (e) is a sagittal slice of the 3D US image; (d) and (f) are the seed segmentation results of (c) and (e), with black background and white seeds.

where $n_0$ is the number of seed candidates. In implementing this technique, the value of $n_0$ must be chosen carefully. It must be large enough to ensure that the seed candidates include as many of the seeds as possible (for a high detection rate), but small enough to exclude as many false candidates as possible (for a low false positive rate). A suitable value for $n_0$ is 1% of the number of voxels in the cropped volume. The cropped volume is converted to a binary image, with the voxels whose gray levels equal or exceed $T$ being assigned a value of 1, and the remaining voxels a value of 0.

### 8.7.3 Seed Candidate Extraction

In the binary image obtained by adaptive thresholding, each connected group of voxels is generally considered to be a seed candidate. Morphological operations are used to separate or join these groups and may be employed when deemed appropriate.

### 8.7.4 Seed Identification

Each seed candidate is labeled and analyzed to determine its size $s$ in voxels, its mean gray-level $\bar{I}$, the direction of its principal axis, and the angle $\theta$ between the principal axis and the planned seed line. Then, the candidates are classified according to three criteria. Candidates which meet all these criteria are identified as seeds and the rest are classified as nonseed bright objects, such as calcifications or other small echogenic structures. The criteria are:

1. $s_{\min} \leq s \leq s_{\max}$,
2. $\bar{I} \geq I_{\min}$,
3. $\theta \leq \theta_{\max}$,

where $s_{\min}$ is determined empirically and $s_{\max}$ is calculated from the physical dimensions of the seed and the voxel dimensions of the 3D US image; $I_{\min}$ is found from a set of manually segmented seed images; and $\theta_{\max}$ depends on the statistical distribution of $\theta$ in a typical line cluster of seed candidates. In our case, we used $\theta_{\max} = 12$.

Some of our seed segmentation results are shown in Figure 8.11. Figures 8.11(a) and (b) are views of the 3D US image and the corresponding CT image, both displayed using the multiplanar reformatting technique. These images demonstrate the 3D US image contains much lower seed contrast and much higher background complexity, compared to the CT image. In its current form, the seed segmentation algorithm has a seed detection rate of about 80%, and a false positive rate of about 11 %.

## 8.8 Conclusion

Three-dimensional ultrasound imaging is becoming a mature technology that has been used in many different diagnostic and therapeutic/surgical applications. In this chapter we described the principles of 3D US imaging and its applications in image-guided prostate brachytherapy. Our focus has been on the discussion of 3D segmentation of the prostate, needle and seeds to be used in 3D US-guided prostate brachytherapy. Additional developments are still required to exploit the full potential of 3D US imaging capabilities. For example, the development of a real-time 3D US imaging system would increase the efficiency and accuracy of needle biopsies. Also, improving 3D visualization techniques, so that they are more intuitive, would allow physicians to become more comfortable with advanced computational tools, resulting in the wider availability of this technology. Finally, faster, more accurate, and less variable segmentation and classification techniques using 3D US images would allow better estimation of organ and tumor volumes, resulting in improved treatment planning and therapy monitoring of patients.

## References

[1] Fenster, A., Downey, D.B., Cardinal, H.N.: Topical review: Three-dimensional ultrasound imaging. Phys. Med. Biol. **46** (2001) R67–R99
[2] Nelson, T.R., Downey, D.B., Pretorius, D.H., Fenster, A.: Three-dimensional ultrasound. Lippincott-Raven, Philadelphia (1999)
[3] Fenster, A., Downey, D.B.: Three-dimensional ultrasound imaging: A review. EEE Engineering in Medicine and Biology **15** (1996) 41–51
[4] Smith, S.W., Trahey, G.E., von Ramm, O.T.: Two-dimensional arrays for medical ultrasound. Ultrason Imaging **14** (1992) 213–33
[5] Tvon Ramm, O.T., Smith, S.W., Pavy, H.G.J.: High-speed ultrasound volumetric imaging system. part ii. parallel processing and image display. IEEE Trans. Ultrason Ferroelec Freq. Contr. **38** (1991) 109–115
[6] Downey, D.B., Fenster, A.: Three-dimensional power doppler detection of prostatic cancer. Am. J. Roentgenol **165** (1995a) 741
[7] Downey, D.B., Fenster, A.: Vascular imaging with a three-dimensional power doppler system. Am. J. Roentgenol **165** (1995b) 665–8
[8] Silverman, R.H., Rondeau, M.J., Lizzi, F.L., Coleman, D.J.: Three-dimensional high-frequency ultrasonic parameter imaging of anterior segment pathology. Ophthalmology **102** (1995) 837–43
[9] Guo, Z., Fenster, A.: Three-dimensional power doppler imaging: a phantom study to quantify vessel stenosis. Ultrasound Med. Biol. **22** (1996) 1059–69
[10] Picot, P.A., Rickey, D.W., Mitchell, R., Rankin, R.N., Fenster, A.: Three-dimensional color doppler imaging of the carotid artery. In: SPIE Proceedings: Image Capture, Formatting and Display. Volume 1444., Atlanta, GA (1991) 206–213
[11] Picot, P.A., Rickey, D.W., Mitchell, R., Rankin, R.N., Fenster, A.: Three-dimensional color doppler imaging. Ultrasound Med. Biol. **19** (1993) 95–104
[12] Pretorius, D.H., Nelson, T.R., Jaffe, J.S.: Three-dimensional sonographic analysis based on color flow doppler and grayscale image data: a preliminary report. J. Ultrasound Med. **11** (1992) 225–32
[13] Bamber, J.C., Eckersley, R.J., Hubregtse, P., Bush, N.L., Bell, D.S., Crawford, D.C.: Data processing for 3D ultrasound visualization of tumour anatomy and blood flow. SPIE **1808** (1992) 651–663

[14] Guo, Z., Moreau, M., Rickey, D.W., Picot, P.A., Fenster, A.: Quantitative investigation of in vitro flow using three-dimensional color doppler ultrasound. Ultrasound Med. Biol. **21** (1995) 807–16
[15] Cardinal, H.N., Gill, J.D., Fenster, A.: Analysis of geometrical distortion and statistical variance in length, area, and volume in a linearly scanned 3d ultrasound image. IEEE Transactions on Medical Imaging **19** (2000) 632–651
[16] Delabays, A., Pandian, N.G., Cao, Q.L., Sugeng, L., Marx, G., Ludomirski, A., Schwartz, S.L.: Transthoracic real-time three-dimensional echocardiography using a fanlike scanning approach for data acquisition: methods, strengths, problems, and initial clinical experience. Echocardiography **12** (1995) 49–59
[17] Downey, D.B., Fenster, A.: Three-dimensional orbital ultrasonography. Can. J Ophthalmol **30** (1995c) 395–8
[18] Elliot, T.L., Downey, D.B., Tong, S., McLean, C.A., Fenster, A.: Accuracy of prostate volume measurements in vitro using three-dimensional ultrasound. Acad. Radiol. **3** (1996) 401–6
[19] Tong, S., Downey, D.B., Cardinal, H.N., Fenster, A.: A three-dimensional ultrasound prostate imaging system. Ultrasound Med. Biol. **22** (1996) 735–46
[20] Chin, J.L., Downey, D.B., Mulligan, M., Fenster, A.: Three-dimensional transrectal ultrasound guided cryoablation for localized prostate cancer in nonsurgical candidates: a feasibility study and report of early results. J. Urol. **159** (1998) 910–4
[21] Tong, S., Cardinal, H.N., Downey, D.B., Fenster, A.: Analysis of linear, area and volume distortion in 3d ultrasound imaging. Ultrasound Med. Biol. **24** (1998) 355–73
[22] Brinkley, J.F., McCallum, W.D., Muramatsu, S.K., Liu, D.Y.: Fetal weight estimation from lengths and volumes found by three-dimensional ultrasonic measurements. J. Ultrasound Med. **3** (1984) 163–8
[23] King, D.L., King, D.J., Shao, M.Y.: Three-dimensional spatial registration and interactive display of position and orientation of real-time ultrasound images. J. Ultrasound Med. **9** (1990) 525–32
[24] Bonilla-Musoles, F., Raga, F., Osborne, N.G., Blanes, J.: Use of three-dimensional ultrasonography for the study of normal and pathologic morphology of the human embryo and fetus: preliminary report. J. Ultrasound Med. **14** (1995) 757–65
[25] Detmer, P.R., Bashein, G., Hodges, T., Beach, K.W., Filer, E.P., Burns, D.H., Strandness, D.E.J.: Three-dimensional ultrasonic image feature localization based on magnetic scanhead tracking: in vitro calibration and validation. Ultrasound Med. Biol. **20** (1994) 923–36
[26] Ganapathy, U., Kaufman, A.: 3d acquisition and visualization of ultrasound data. In: SPIE Proc. Of Visualization in biomedical computing SPIE. Volume 1808. (1992) 535–545

[27] Gilja, O.H., Detmer, P.R., Jong, J.M., Leotta, D.F., Li, X.N., Beach, K.W., Martin, R., Strandness, D.E.J.: Intragastric distribution and gastric emptying assessed by three-dimensional ultrasonography. Gastroenterology **113** (1997) 38–49

[28] Hughes, S.W., D'Arcy, T.J., Maxwell, D.J., Chiu, W., Milner, A., Saunders, J.E., Sheppard, R.J.: Volume estimation from multiplanar 2D ultrasound images using a remote electromagnetic position and orientation sensor. Ultrasound Med. Biol. **22** (1996) 561–72

[29] Leotta, D.F., Detmer, P.R., Martin, R.W.: Performance of a miniature magnetic position sensor for three-dimensional ultrasound imaging. Ultrasound Med. Biol. **23** (1997) 597–609

[30] Nelson, T.R., Pretorius, D.H.: Visualization of the fetal thoracic skeleton with three-dimensional sonography: a preliminary report. Am. J. Roentgenol **164** (1995) 1485–8

[31] Tuthill, T.A., Krucker, J.F., Fowlkes, J.B., Carson, P.L.: Automated three-dimensional us frame positioning computed from elevational speckle decorrelation. Radiology **209** (1998) 575–82

[32] Chin, J.L., Downey, D.B., Onik, G., Fenster, A.: Three-dimensional prostate ultrasound and its application to cryosurgery. Tech. Urol. **2** (1996) 187–93

[33] Kirbach, D., Whittingham, T.A.: Three-dimensional ultrasound - the Kretztechnik-Voluson approach. European J. Ultrasound **1** (1994) 85–89

[34] Zosmer, N., Jurkovic, D., Jauniaux, E., Gruboeck, K., Lees, C., Campbell, S.: Selection and identification of standard cardiac views from three-dimensional volume scans of the fetal thorax. J Ultrasound. Med. **15** (1996) 25–32

[35] Levoy, M.: Volume rendering, a hybrid ray tracer for rendering polygon and volume data. IEEE Computer Graphics and Applications **10** (1990) 33–40

[36] Garfinkel, L., Mushinski, M.: Cancer incidence, mortality, and survival trend in four leading sites. Stat. Bull. **75** (1994) 19–27

[37] Nag, S.: Principles and Practice of Brachytherapy. Futura Publishing Company, Inc, Amonk, NY (1997)

[38] Richard, W.D., Grimmell, C.K., Bedigian, K., Frank, K.J.: A method for 3D prostate imaging using transrectal ultrasound. Comput. Med. Imaging Graphics **17** (1993) 73–79

[39] Richard, W.D., Keen, C.G.: A method for 3D prostate imaging using transrectal ultrasound. Comput. Med. Imaging Graphics **20** (1996) 131–140

[40] Aarnink, R.G., Giesen, R.J.B., Huynen, A.L., Debruyne, F.M.J., Wijkstra, H.: Automated prostate volume determination. In: Proc. of IEEE Eng. Med. Biol. Soc. Volume 14. (1992) 2146–2147

[41] Aarnink, R.G., Giesen, R.J.B., Huynen, A.L., De la Rosette, J.J., Debruyne, F.M.J., Wijkstra, H.: A practical clinical method for contour de-

termination in ultrasonographic prostate images. Ultrasound Med. Biol. **20** (1994) 705–717
[42] Aarnink, R., Giesen, R.J.B., Huynen, A.L., De la Rosette, J.J., Debruyne, F.M.J., Wijkstra, H.: Automated prostate volume determination with ultrasonographic imaging. J. Urol. **153** (1995) 1549–1554
[43] Aarnink, R.G., Giesen, R.J.B., Huynen, A.L., De la Rosette, J.J., Debruyne, F.M.J., Wijkstra, H.: Edge detection in prostatic ultrasound images using integrated edges maps. Ultrasounics **36** (1998) 635–642
[44] Liu, Y.J., Ng, W.S., Teo, M.Y., Lim, H.C.: Computerised prostate boundary estimation of ultrasound images using radial bas-relief method. Med. Biol. Eng. Comput. **35** (1997) 445–454
[45] Kwoh, C.K., Teo, M.Y., Ng, W.S., Tan, S.N., Jones, L.M.: Outlining the prostate boundary using the harmonics method. Med. Biol. Eng. Comput. **36** (1998) 768–771
[46] Pathak, S.D., Aanink, R.G., De la Rosette, J.J., Chalara, V., Wijkstra, H., Haryror, D.R., Debruyne, F.M.J., Kim, Y.: Quantitative three-dimensional transrectal ultrasound (TRUS) for prostate imaging. In: Proc. of SPIE. Volume 3335. (1998) 83–92
[47] Pathak, S.D., Chalara, V., Haryror, D.R., Kim, Y.: Edge-guided delineration of the prostate in transrectal ultrasound images. In: Proc. of the 1st Joint Meeting of the Biology Engineering Society and IEEE Engineering in Medicine and Biology Society, Atlanta, GA (1999) 1056
[48] Pathak, S.D., Chalara, V., Haryror, D.R., Kim, Y.: Edge-guided boundary delineation in prostate in ultrasound images. IEEE TMI **19** (2000) 1211–1219
[49] Knoll, C., Alcaniz, M., Grau, V., Monserrat, C., Juan, M.C.: Outlining of the prostate using snakes with shape restrictions based on the wavelet transform. Pattern Recognition **32** (1999) 1767–1781
[50] Garfinkel, L., Mushinski, M.: A three-dimensional deformable model for segmentation of human prostate from ultrasound images. Med. Phys. **28** (2001) 2147–2153
[51] Kass, M., Witkin, A., Terzopoulos, D.: Snakes: Active contour models. International Journal of Computer Vision **1** (1987) 321–331
[52] Terzopoulos, D., Fleischer, K.: Deformable models. The Visual Computer **4** (1988) 306–331
[53] Singh, A., Goldgof, D., Terzopoulos, D.: Deformable Models in Medical Image Analysis. IEEE Press, Los Alamitos, CA (1998)
[54] McInerney, T., Terzopoulos, D.A.: A dynamic finite-element surface model for segmentation and tracking in multidimensional medical images with applications to cardiac 4D image analysis. Computer Medical Imaging Graphics **19** (1995) 69–83
[55] McInerney, T., Terzopoulos, D.: Deformable models in medical image analysis: A survey. Medical image Analysis **1** (1996) 91–108

[56] McInerney, T., Terzopoulos, D.: Topology adaptive deformable surfaces for medical image volume segmentation. IEEE Transactions on Medical Imaging **18** (1999) 840–850
[57] McInerney, T., Terzopoulos, D.: T-snakes: Topology adaptive snakes. Medical Image Analysis **4** (2000) 73–91
[58] Keeve, E., Kikinis, R.: Deformable modeling of facial tissue. In: Proceedings of the First Joint BMES/EMBS Conference. Volume 1.1. (1999) 502
[59] Sitek, A., Klein, G.J., Gullberg, G.T., Huesman, R.H.: Deformable model of the heart with fiber structure. IEEE Transactions on Nuclear Science **49** (2002) 789–793
[60] Metaxas, D.N., Kakadiaris, I.A.: Elastically adaptive deformable models. IEEE Transactions on Medical Imaging **24** (2002) 1310–1321
[61] Miller, J.V., Breen, D.E., Lorensen, W.E., O'Bara, R.M., Wozny, M.J.: Geometrically deformable models: A method to extract closed geometric model from volume data. Computer Graphics **25** (91) 217–226
[62] Lobret, S., Viergever, M.A.: A discrete dynamic contour model. IEEE Trans. Med. Image **14** (1995) 12–24
[63] Ladak, H.M., Mao, F., Wang, Y., Downey, D.B., Steinman, D.A., Fenster, A.: Prostate boundary segmentation from 2D ultrasound images. Med. Phys. **27** (2000) 1777–1788
[64] Gill, J.D., Ladak, H.M., Steinman, D.A., Fenster, A.: Accuracy and variability assessment of a semiautomatic technique for segmentation of the carotid arteries from three-dimensional ultrasound images. Med Phys **27** (2000) 1333–1342
[65] Solina, F., Bajcsy, R.: Recovery of parametric models from range images: The case for superquadrics with global deformation. IEEE Transactions on Pattern Analysis and Machine Intelligence **12** (1990) 131–147
[66] Schroeder, W.J., Martin, K.M., Avila, L.S., Law, C.C.: The VTK Users Guide. Kitware, Inc (1998)
[67] Bonilla-Musoles, F., Raga, F., Osborne, N.G., Blanes, J.: Principal warps: Thin-plate splines and the decomposition of deformations. , IEEE Trans. Pattern Analysis and Machine Intelligence **11** (1989) 567–585
[68] Lichtenbelt, B., Crane, R., Naqvi, S.: Introduction to Volume Rendering. Prentice-Hall, Upper Saddle River, NJ (1998)

# Chapter 9

# Time-Frequency Analysis in Terahertz-Pulsed Imaging

Elizabeth Berry[1], Roger D Boyle[2], Anthony J Fitzgerald[3], and James W Handley[4]

[1] Academic Unit of Medical Physics, University of Leeds, e.berry@leeds.ac.uk
[2] School of Computing, Centre of Medical Imaging Research
  roger@comp.leeds.ac.uk
[3] Teraview Ltd., Cambridge, UK tony.fitzgerald@teraview.co.uk
[4] School of Computing, Centre of Medical Imaging Research
  jwh@comp.leeds.ac.uk

**Summary.** Recent advances in laser and electro-optical technologies have made the previously underutilized terahertz frequency band of the electromagnetic spectrum accessible for practical imaging. Applications are emerging, notably in the biomedical domain. In this chapter the technique of terahertz-pulsed imaging is introduced in some detail. The need for special computer vision methods, which arises from the use of pulses of radiation and the acquisition of a time series at each pixel, is described. The nature of the data is a challenge since we are interested not only in the frequency composition of the pulses, but also how these differ for different parts of the pulse. Conventional and short-time Fourier transforms and wavelets were used in preliminary experiments on the analysis of terahertz-pulsed imaging data. Measurements of refractive index and absorption coefficient were compared, wavelet compression assessed, and image classification by multi dimensional clustering techniques demonstrated. It is shown that the time-frequency methods perform as well as conventional analysis for determining material properties. Wavelet compression gave results that were robust through compressions that used only 20% of the wavelet coefficients. It is concluded that the time-frequency methods hold great promise for optimizing the extraction of the spectroscopic information contained in each terahertz pulse, for the analysis of more complex signals comprising multiple pulses or from recently introduced acquisition techniques.

## 9.1 Introduction

The terahertz (110 GHz to 10 THz) band of the electromagnetic spectrum, between microwaves and the infrared, has until recently been unexplored as a significant imaging tool. Recent advances in laser and electro-optical technologies now make the band accessible for practical use, and applications, notably in the medical domain, are emerging. Previously, terahertz radiation was generated either by using thermal sources that produced weak and incoherent radiation, as conventionally used in far infrared Fourier transform spectroscopy, or by highly complex and bulky equipment such as free electron lasers or optically pumped gas lasers [1, 2]. Similarly, incoherent detection methods were used, which were able to record only the intensity of the terahertz electric field. The most sensitive detectors of this type were liquid helium cooled bolometers, which give a relatively noisy signal and have low sensitivity. The key advances that have made terahertz imaging a practical proposition have been in the fields of ultrashort pulsed lasers, nonlinear optics and crystal growth techniques [3]. These have resulted in sources of bright, coherent, broadband terahertz pulses and enabled coherent room temperature detection [4]. The advantage of coherent detection methods is that it is possible to record not only the intensity, but also a time-resolved amplitude of the electric field: a time series. In turn this leads to the possibility of obtaining a spectrum by Fourier transformation of the time domain signal, and opens up a wealth of spectroscopic analytic techniques, including those that rely on measuring changes in the phase of the measured signal.

In parallel with the development of pulsed techniques, work has been undertaken in the development of continuous wave terahertz imaging [5, 6], which allows precise tuning to a particular frequency. As monochromatic radiation is used the data acquired are simpler than in the pulsed case, and we shall not be considering these systems and the corresponding data further in this chapter. Advances have also led to the design of compact free-electron laser systems [7].

Terahertz-pulsed imaging is a development of terahertz time domain spectroscopy [8, 9, 10]. These workers have had success in measuring, in the terahertz band, the dielectric and optical properties of a range of materials including water, polar, and nonpolar liquids, gases, semiconductors and dielectrics. Terahertz-pulsed imaging involves projecting broadband pulses at a sample and either detecting them after transmission through the sample (transmission-based imaging) or detecting their reflections (reflection-based imaging). In the extension to imaging, the spectroscopic response of a sample is mapped by recording the transmitted or reflected broadband terahertz pulse at a series of contiguous pixel locations [11, 12]. The simplest images are generated by acquiring data at only one time point during the pulse, and plotting the amplitude of the signal at that time. Potentially more useful images can be generated by calculating parameters associated with the full time series at each pixel, and displaying those values using color look up tables [13]. Selected parametric terahertz-pulsed images of a wax-embedded melanoma section of thickness 1 mm, which has been prepared with the standard techniques used in histopathology, are shown in Plate V. The melanoma is outlined in the photograph in Plate V(a). The terahertz-imaged section was 7 mm * 7 mm corresponding with the lower right quadrant of the photograph. The numerical values associated with the color scale are different for each of (b), (c), and (d).

## Chapter 9 Time-Frequency Analysis in Terahertz-Pulsed Imaging

The images shown in Plate V were acquired using the technique as it was introduced in 1995. Since that time, workers have introduced alternative acquisitions designed to improve one or more aspects of the measurement. Many of these are based on adaptation of mature algorithms from other fields. Dark-field imaging [14] was introduced to generate images where the image contrast arises from the differential scattering or diffraction of radiation. An alternative approach to diffraction imaging has been developed [15], with the emphasis on solution of the inverse problem to predict the aperture shape responsible for a measured diffraction field. Extraction of information about the location of buried structures was first obtained in 1997 [16], using time-of-flight reflection measurements analogous to B-mode ultrasound. In addition to generation of slice images in a plane perpendicular to the object surface, knowledge of depth of objects of interest allows spectroscopic measurement to be made only from the relevant location or the reconstruction of a slice parallel to the surface. The technique is most appropriate for objects where both negligible dispersion and absorption can be assumed. This assumption is not applicable to biomedical subjects, and led to the use of techniques from two related fields. Retaining a reflection geometry, techniques from geophysics have been applied leading to estimates of thickness and refractive index [17]. In contrast, by using a transmission geometry, the filtered back projection methods that underpin medical imaging techniques such as x-ray computed tomography have been applied to parametric projection images and the reconstruction of strong interfaces successfully demonstrated [18].

In spite of the increasing complexity of data acquisition, all the techniques retain a feature in common that sets them apart from other imaging methods. The data acquired consist of a time series rather than a single value, and new image analysis techniques are needed to ensure that all the information present is used. In our own area of interest, because the depth of penetration in human tissue is of the order of millimeters [19] it is likely that the first practical human in vivo imaging will involve data acquired in reflection, with transmission techniques being reserved for imaging of samples in vitro. While in the latter case it may be possible to simplify the geometry by sample preparation, in the former, unknown and complex tissue arrangements are expected.

There has been only limited work on applying computer vision techniques to terahertz images. Some [13] have suggested using specific "display modes" for certain applications, for example, ensuring that parameters are calculated from the part of the spectrum corresponding with absorption lines of particular molecules, and the range of parameters available for display was illustrated by others [20]. The first application of multidimensional classification techniques to terahertz data has also been described [21]. Mittleman et al. [22] introduced the idea of using wavelet-based techniques, and this idea was taken up by others [23, 24] for pulse denoising. We return to the topic of computer vision in terahertz imaging in Section 9.1.3, and introduce our own work in this field.

### 9.1.1 Equipment for Terahertz Pulsed Imaging

A schematic layout of a transmission terahertz-pulsed imaging system is shown in Figure. 9.1. The technique is based on the pump- and probe-technique of optical spectroscopy. An ultrafast infrared laser beam, giving femtosecond pulses, is split in two. One part is used as the pump beam to generate picosecond terahertz pulses, whilst the other forms part of the coherent detection system and is used as a probe beam to detect the amplitude of the terahertz electric field after it has interacted with the sample or subject.

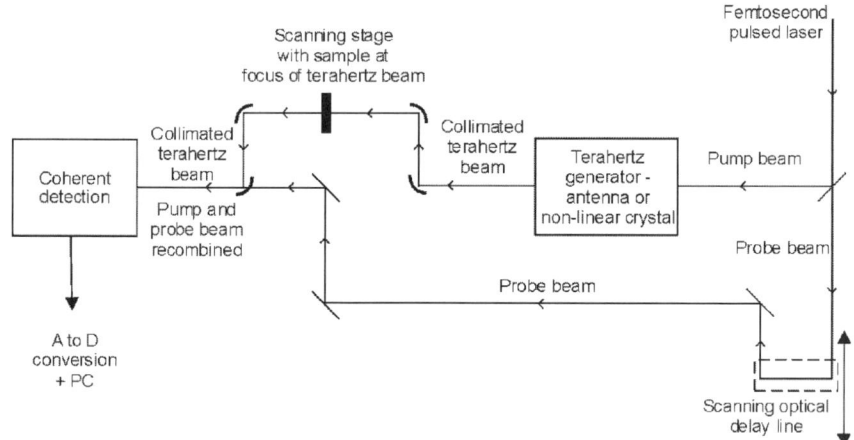

**Figure 9.1.** Schematic layout of a transmission terahertz-pulsed imaging system.

There are two commonly used techniques for generating the pulses of terahertz radiation using the pump beam. In one, a voltage-biased photoconductive antenna [25] is illuminated with pulses from the ultrafast infrared laser. Alternatively, the technique of optical rectification or optical mixing, may be used to yield pulses containing frequencies up to 70 THz, which cross the border between the terahertz band and the far infrared. The infrared pulses are used to illuminate a crystal with high nonlinear susceptibility [4, 26]. The resulting terahertz beam is directed onto the sample or subject using parabolic mirrors. The transmitted terahertz pulse profile is measured at a discrete number of time points by scanning using an optical delay stage. The spatial scanning of the object for image formation may most simply be performed using raster scanning of either the terahertz beam or of the sample itself, but this is time consuming. Alternative, faster, schemes are under development. A promising method involves the illumination by the pump beam of a larger area representing many pixels; a multielement array detector such as a charge coupled device is used for detection [12, 27, 28]. More complete descriptions of terahertz imaging systems are available [29].

Many of the systems in use for research are laboratory-based and occupy an area of up to 3 m * 2 m, but more compact and portable systems are under development. An example is shown in Figure 9.2 which is a commercially available reflection system for use in dermatology. The box shown is 100 cm × 60 cm × 100 cm in size. The subject of interest is placed on a window on the top of the instrument, or examined using the probe attachment.

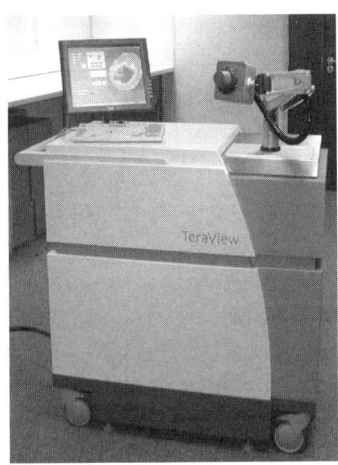

**Figure 9.2.** The TPI Scan$^{(TM)}$: A complete terahertz imaging system including near-infrared laser and terahertz imaging optics for imaging biological tissue. Courtesy of TeraView Ltd, Cambridge, UK.

In Figure 9.3, a commercially available system is shown, which can be used in both reflection and transmission modes [30, 17].

A further system is under development by the Zomega Technology Corporation.

### 9.1.2 Potential Applications

In addition to the possibility of characterizing materials by spectroscopy, the penetration characteristics of terahertz radiation have also guided researchers towards potential applications. For example, polar liquids absorb strongly in the terahertz band; an example of such a liquid is water. Metals are opaque to terahertz radiation, whilst non-metals such as plastics and paper products are transparent, as are non-polar substances. Dielectrics have characteristic absorption features peculiar to each material. The exploitation of these penetration characteristics is discussed in more detail in the following sections.

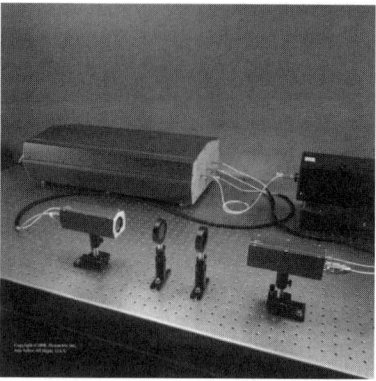

**Figure 9.3.** The Picometrix T-Ray$^{(TM)}$ 2000, the first commercial time domain terahertz spectroscopy and imaging system. Courtesy of Picometrix, Inc., Ann Arbor, MI.

## Biomedical Imaging Applications

Our particular interest is in biomedical applications for terahertz imaging. The perceived advantages of using the terahertz band for biomedical applications include its sensitivity to the presence of water which may be of use for detecting or characterizing disease state, the lack of a hazard from ionization, relatively less Rayleigh scattering than for infrared and visible radiation, and the possibility of characteristic "signatures" from different tissues in health and in disease [31, 32, 33]. These characteristic signals may result from water content or other chemical features related to the composition or functional properties of the tissues.

- *Sensitivity to water*: An excellent example of utilization of the sensitivity of terahertz frequency radiation to the presence of water molecules is a study concerning the noninvasive continuous measurement of leaf water content [22, 34]. The technique may find further applications in agriculture and manufacturing [35], for assessing the moisturizing effects of cosmetics, and characterizing the flow of water through engineered materials or tissues. There is potential for distinguishing healthy and diseased tissue by its water content, where it may also be possible to differentiate bound and free water molecules [36]. We return to this last point later. The drawback of the high attenuation of terahertz frequency radiation by water is the concomitant limited depth of penetration (up to a few millimeters). However, even without special devices to deliver and detect the radiation inside the body, there is still a number of promising applications that do not require the radiation to travel far compared with the aforementioned depth of penetration. There are hopes that terahertz frequency imaging will be of value for in vivo characterization of dermatological conditions, or the early diagnosis of tooth decay.

- *Safety issues*: In common with many of the more recent introductions for medical imaging (ultrasound, magnetic resonance imaging, infrared techniques) terahertz frequency radiation is free from the ionization hazard associated with x-rays and nuclear medicine techniques. There are published guidelines regarding safe exposures, for example [37]. These were based on measurements made using wavelengths under 10.6 µm and pulse durations over 1.4 ns [38] and it is believed that under those conditions the damage mechanism is thermal. However there is also the possibility of resonant absorption mechanisms and thermomechanical and thermochemical effects for pulses of the type used in terahertz-pulsed imaging, and work is underway to investigate this [39, 40]. We have estimated that skin exposure for current pulsed systems using electronic and optical generation methods, where the average power of each pulse is under a milliwatt, will give exposures that are well within the limits set in the guidelines [41]. It is likely that the more powerful systems, such as those based on free electron lasers, may carry with them a hazard associated with heating.
- *Rayleigh scattering*: As the amount of Rayleigh scattering decreases with the fourth power of the wavelength, it is expected that terhertz frequency radiation should be scattered less than visible and near-infrared frequencies, which would be advantageous for imaging. This has been borne out by experiment. In a direct comparison of imaging using terahertz and near-infrared pulses, higher image contrast was obtained using the terahertz pulse although the near-infrared pulse was of higher power [42]. The difference could be explained by wavelength-dependent scattering.
- *Characteristic tissue "signatures"*: In Fourier transform infrared (FTIR) spectroscopy the word "signature" describes the presence of a characteristic absorption peak at a particular wavelength, which indicates the presence of a specific molecular bond. In some vitro applications for terahertz pulsed imaging, particularly of pure samples, signatures of this kind may be present. For the reasons outlined below we do not expect to see a single characteristic spectral absorption feature associated with each tissue in terahertz-pulsed imaging in vivo, or of tissue samples in vitro. Instead, we hypothesize that spectra from different tissues will have different shapes (or signatures), perhaps best described by a combination of absorption characteristics. The reason we do not expect to see sharp absorption peaks in spectra from tissues is because the samples will contain a complex mixture of several molecules. For each molecule, there may be several chemical environments, which will lead to smoothing of the spectral features. Finally, the presence of water, with its strong absorption, will potentially mask the other molecular absorptions.
- *Proposed biomedical applications*: Many applications have been proposed, ranging from studies at the genetic level, for example, investigating the hybridization state of DNA [43] to in vivo measurements of the thickness of skin [36]. Two Europe-wide projects are leading work in this field: first, Terahertz Bridge (http://www.frascati.enea.it/THz-BRIDGE/) which is

following a streamline of increasing complexity from biomolecules to cell membranes, cell nuclei, and tissues, and, secondly, Teravision (http://www.teravision.org) concentrating on imaging of intact tissue in vitro and in vivo.

Imaging, and computer vision, is most likely to be used for in vivo applications where knowledge of the spatial distribution of the chemically specific spectroscopic measurements is of value. A number of groups have demonstrated image contrast between tissues, though using, in general, very small numbers of samples. These demonstrations include data from pork and chicken [32, 31, 44, 45, 14], human tooth enamel and dentine [46], human skin in vivo [36], Spanish Serrano ham [24], histopathologically prepared human liver [47], and canine tumor [14]. A study of healthy tissue that includes repeated measurements from dehydrated tooth samples from seven individuals and freshly excised tissue samples from two donors has been reported [19]. The first study to include more samples in both health and disease, and thus having a higher power for hypothesis testing, is that of [48]. They imaged 15 samples of human healthy tissue and basal cell carcinoma.

These early results have led to optimism that early dental caries and skin cancers may be detectable using terahertz-pulsed imaging. Although there are alternative modalities that can be applied to each application, none has the necessary high sensitivity and specificity [49]. Other suggestions for terahertz-pulsed imaging include in vivo imaging of breast tumors, based on promising results obtained using microwaves [50] and in vitro results from terahertz-pulsed imaging [23]. Wound healing is an area attracting much interest, as terahertz imaging offers the potential of a noncontact measurement technique that could be used through a dry dressing.

## Nonbiomedical Imaging Applications

- *Security and military applications*: These applications take advantages of the chemical specificity of the technique, and terahertz imaging has been advocated for remote scanning for biological agents such as anthrax or explosives [51]. Ranging studies to simulate radar of larger objects have been performed using terahertz radiation and scale models [52], and nonimaging versions of the technology are expected to have applications in communications.
- *Quality control*: Most packaging materials are transparent to terahertz radiation, so terahertz radiation has been proposed as a nonionizing substitute for quality control using x-rays in several industries including the food and textile industries. The connections in packaged integrated circuits may be assessed [22].
- *Characterization of semiconductors, gas identification*: Semiconductor characterization has been a particularly fertile area because dielectrics have a characteristic absorption dependent on the polarity and optical phonon

Chapter 9 Time-Frequency Analysis in Terahertz-Pulsed Imaging     279

resonances peculiar to that material. Gas identification at terahertz frequencies [53] is also a good application because the emission and absorption lines of rotational and vibrational excitations of lighter molecules are strong in that part of the spectrum. Spectral lines are generally sharper in gaseous than in solid or liquid states, and this makes possible the precise localization of gas emission or the characterization of combustion flames.
- *Research applications*: The nondestructive and noncontact nature of the radiation has attracted the interest of those wishing to investigate valuable artifacts. There are potential uses for studying fossils in paleontology, visualizing through the surface layers of art works and other antiquities, and determining the content of books without the need to touch or disturb delicate pages.

Reviews that describe potential applications, with an emphasis on the biomedical, include those by Mittleman et al. [22] and Koch [54]. Siegel [55] gives a complete historical overview, including the passive imaging methods that are used for applications in astronomy and Anderton et al. [56] consider military applications.

### 9.1.3 Terahertz-Pulsed Imaging and Computer Vision

Terahertz-pulsed imaging presents challenges to analysis because in the acquired data set, each pixel contains a time series representing the measured terahertz frequency pulse. We have investigated some novel data processing approaches.

**Time–Frequency Analysis**

The aim in terahertz-pulsed imaging is to extract useful comparisons between a well-understood reference pulse, and those detected after transmission or reflection. Figure 9.4 shows three pulses — a reference pulse and two different transmission responses, one from nylon and one from cortical bone. Figure 9.5 shows these pulses after Fourier transformation. Notice that the spectra are very different, as are the pulses; they capture some of the information about the pulses that is obvious. For example, the high-frequency (noise) activity in the reference, and the reduced activity, or "power" in the transmitted pulses. However, features which, to the human eye, are dominant, such as the pulse delay, cannot be seen in this representation.

In order to describe the different overall shapes and characteristic absorption features that characterize the different materials, analysis of waveforms is usually based on Fourier decomposition [57]. Most of the applications listed in Section 9.1.2 have used analyses of this type. For example, Kindt and Schmuttenmaer [10] present a series of plots showing the refractive index and linear absorption coefficient of polar liquids plotted against frequency. On its own, however, Fourier transformation is somewhat crude since the measure is global

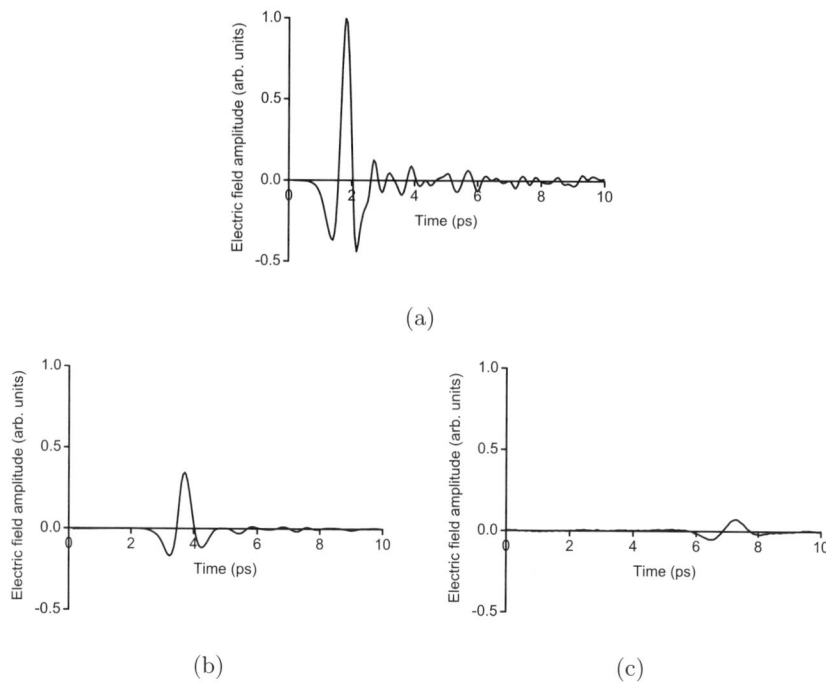

**Figure 9.4.** Three terahertz pulses. (a) A reference pulse. (b) The pulse after transmission through 1 mm of nylon. (c) The pulse after transmission though 1 mm of cortical bone.

to the signal. We seek mechanisms for deriving more local, time-specific, features. This is of particular importance for data comprising a series of pulses, at different times, representing interactions with different boundaries. The work in this chapter uses transmission data, from a single layer of material. Figure 9.4 illustrates that the reference pulse peak suffers a delay (along the time axis) and a flattening, or spreading, that are characteristic of the particular part of the sample under inspection. Ordinary Fourier transforms do not capture the time-dependent qualities of the information. There are also noticeable effects in the remainder of the signal that can be perceived as changes to the frequency make-up, which are dependent on the temporal instant. For example, the broadening of the main pulse suggests that the higher frequencies are no longer present, due to absorption, reflection or scattering, as they are responsible for the sharpness of the peak. It is hard to say, however, if the higher frequencies have been removed from the pulse uniformly over time.

An analytic technique was required that would perform the frequency space decomposition that Fourier transforms provide, but in a time-dependent manner. We have applied two approaches, the short-time Fourier transform and wavelets. The theoretical bases for these approaches are described in Sec-

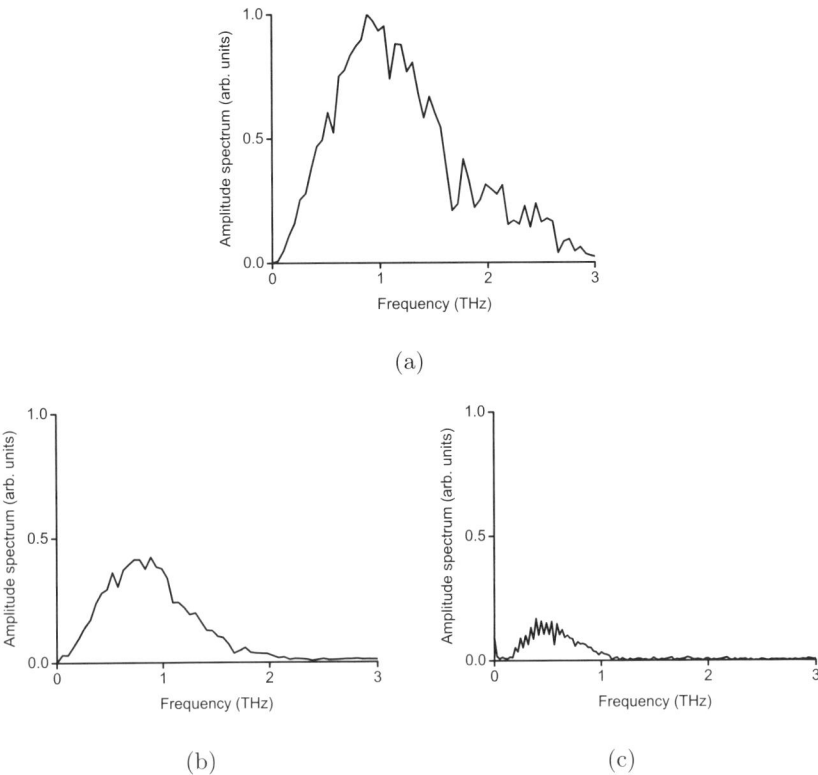

**Figure 9.5.** Frequency domain representation of three terahertz pulses. **(a)** A reference pulse. **(b)** The pulse after transmission through 1 mm of nylon. **(c)** The pulse after transmission though 1 mm of cortical bone.

tions 9.2.1 and 9.2.2. Wavelets were first used for terahertz imaging by Mittleman et al. [22], and have been used in other areas of biomedical signal processing [58, 59]. The techniques can generically be described as time-frequency techniques.

We have investigated the use of time-frequency techniques in two ways. The first of these would, in a generalized framework for computer vision [60], be described as a segmentation task. Previous time-frequency segmentation work in biomedicine has been performed on breast RF data and on neurophysiological signals [61, 62]. For the terahertz data, both conventional and time-frequency methods were used to determine the refractive index and absorption coefficient of samples of nylon and resin, and the results compared. The second application is an example of a preprocessing or signal processing step. Time-frequency techniques have been applied for noise reduction of the acquired time series, following work on related modalities [63, 64, 65] and for compression. Compression may be required because very large data sets can result from the need for spectroscopic detail at high spatial resolution. The effect of compression was determined by calculating the refractive index and absorption coefficient of nylon with and without compression. The refractive index and absorption coefficient of a material are key factors for terahertz-pulsed imaging. They determine the amplitude and timing of a transmitted, reflected or propagated pulse. Most parameters used for parametric images are strongly related to one or other property.

## Clustering

In general, terahertz-pulsed imaging data are not amenable or accessible to the customary image processing and computer vision approaches unless these 3D data are first reduced, by the production of parametric images, to the two or three spatial dimensions used by such algorithms. However, workers in FTIR spectroscopy have reported successes with the analysis of complex samples by applying classification techniques to the acquired spectra [66]. This appears to be a promising approach for the terahertz frequency spectra we expect from tissue. They used prior knowledge of cellular structure in normal tissue and in the presence of carcinoma. The latter state is characterized by absence of particular cells, e.g., absence of goblet and mucin from colorectal adenocarcinomas, which has an effect on the shape of the spectrum. Classification success using training data, but without prior knowledge of biology, has recently been reported for terahertz imaging data [21].

Clustering methods fall into the category of segmentation in computer vision [60]. Here, image classification using clustering techniques (both for the full time series and for parameters derived from it) was applied to several terahertz images. These included a synthetic image with a known true classification, an acquired image where a classification was available from another imaging modality, and acquired images without knowledge of the true classification.

## High-Level Processes

Higher-level processes used in computer vision, such as shape representation and shape extraction via motion, texture, etc., will be relevant for the analysis of parametric images or for classified images. Such analysis is not presented here. The ultimate aim of any computer vision analysis route is to extract "understanding" of some description from the data being processed. The understanding in biomedical terahertz imaging will be highly specific. Examples include the identification of regions of low mineralization representing early tooth decay, and associating different regions of an image with states of health and disease.

Thus, the purpose of this work was to apply data processing methods designed to suit the nature of terahertz-pulsed imaging data:

- To compare time-frequency techniques with conventional Fourier methods for extracting optical properties of materials.
- To determine the maximum degree of wavelet compression that would lead to no significant alteration in measured optical properties.
- To demonstrate clustering by multi-dimensional techniques.

In the following section we present a brief theoretical overview to support the methods that are described in Section 9.3.

## 9.2 Theory

### 9.2.1 Short-Time Fourier Transform (STFT)

The Fourier series representation of a real valued periodic function $f(t)$, with period $T$ (so $f(t+T) = f(t)$) has Fourier series representation

$$f(t) = \sum_{-\infty}^{\infty} a_k e^{ik\omega t}, \qquad (9.1)$$

where $\omega = 2\pi/T$ is the *fundamental frequency* and the Fourier coefficients are given by

$$a_k = \frac{1}{T} \int_{t_0}^{t_0+T} f(t) e^{-ik\omega t} dt. \qquad (9.2)$$

This representation provides a decomposition of the function into frequency harmonics, whose contribution is given by the coefficients $a_k$. This decomposition is of great use in the analysis of functions since it betrays many useful properties; for example, very sharp changes contribute very high harmonic information, while slow variation is associated with low harmonics. Similarly, noise effects are often characterized by high frequency components.

More generally, for a nonperiodic function, the Fourier transform of $f(t)$ is given by

$$\hat{f}(\omega) = \int_{-\infty}^{\infty} f(t) e^{-i\omega t} dt. \qquad (9.3)$$

This transform may be inverted, where $\hat{f}$ and $f$ are a Fourier transform pair, by

$$f(t) = \frac{1}{2\pi} \int_{-\infty}^{\infty} \hat{f}(\omega) e^{i\omega t} d\omega. \qquad (9.4)$$

However, the Fourier transform, and Fourier methods generally, are global, in the sense that they operate on the whole period of a function. In (9.2) we need full knowledge of $f(t)$ in order to extract the coefficients, and all values of $f(t)$ contribute to them. Any perturbation of $f(t)$ at any point will affect all the $a_k$. This leaves it of limited value when the effects under scrutiny are local, in the sense that there are time dependencies in the frequency content of the signal. This is very much the case in the study of terahertz data — for example, in Figures 9.4(b) and 9.4(c), it is clear that the major peak of the pulse has been delayed, but by different lengths of time. The lack of information on time delays is evident in Figure 9.5. While it is possible to use the Fourier transform to estimate time delays by determining the phase difference between two pulses, the estimates tend to be inaccurate. The transform is a periodic function with $2\pi$ phase increments concealed by the periodicity, and inaccuracies arise from the limitations of the phase unwrapping algorithm used to estimate time delay. We seek, therefore, a combination of time and frequency analyses to permit the extraction of local effects using the power of the Fourier approach. This can be done by extracting windows of the original function before performing the Fourier analysis — this is the approach of the short-time Fourier transform. The simplest way to extract windows from a function is to multiply it by another function, such as a rectangular window of width $2\tau$, described by (9.5).

$$\begin{aligned} f_b(t) &= f(t)\Phi(t-b) \\ &= \begin{cases} f(t) & t\epsilon[b-\tau, b+\tau] \\ 0 & \text{otherwise.} \end{cases} \end{aligned} \quad (9.5)$$

The value of $f_b(t)$ can then be subjected to Fourier analysis in the normal manner.

The short-time Fourier transform (STFT) of the function $f(t)$ with respect to the window function $\Phi(t)$ calculated at the frequency $\xi$ and the time $\beta$ is then

$$G_\Phi f(\beta, \xi) = \int_{-\infty}^{\infty} f(t)\Phi(t-\beta)e^{-i\xi t}dt. \quad (9.6)$$

Contrast this with (9.4); we have added a second variable $\beta$ to locate the transform in time. This is sometimes written, using the bar notation for the complex conjugate, as

$$G_\Phi f(\beta, \xi) = \int_{-\infty}^{\infty} f(t)\overline{\Phi_{\beta,\xi}(t)}dt, \quad (9.7)$$

where

$$\Phi_{\beta,\xi}(t) = \Phi(t-\beta)e^{i\xi t}. \quad (9.8)$$

It should be clear that the rectangular window function (9.5) is not best chosen for our purpose; the hard limiting step edges will cause any subsequent Fourier transform to include high-frequency components that are properties of the step rather than of the function. For this reason it is more common to use smooth window functions such as Gaussians. In particular, the Gabor transform uses the window function

$$g_\alpha(t) = \frac{1}{2\pi\alpha}e^{t^2/4\alpha} \quad (9.9)$$

for some $\alpha > 0$.

In this work, we have used a simple Gaussian window (parameterized by its standard deviation) in all applications of the STFT. Applications of the STFT to two of the pulses of Figure 9.4 are in Figure 9.6. It can be seen that, in addition to the reduction in power, the transform has successfully captured the delay in the information in the transmitted pulse, at each frequency.

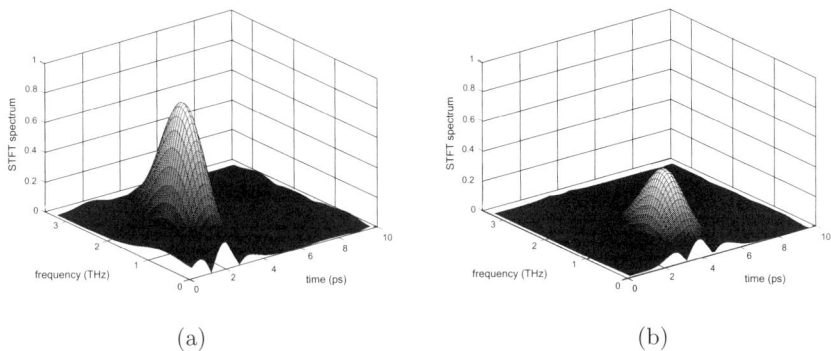

**Figure 9.6.** The STFT applied to: **(a)** the pulse of Figure 9.4(a) and **(b)** the pulse of Figure 9.4(b).

It is important to realize that it is impossible simultaneously to obtain good time resolution and good frequency resolution. Imagine a box drawn on a time-frequency graph, a narrow tall box has good time but poor frequency resolution, while a short wide one has bad time but good frequency resolution. The uncertainty principle determines the minimum area of this box, and thereby the usefulness of the technique. Conversely, the fact that the function is windowed permits real-time application of the STFT (since only limited information is needed), although this advantage is not relevant in this application. Various other transforms, which we will not consider here, exist with the aim of localizing the study of frequency, in particular the Wigner Ville transform [67]. The theory of Fourier transforms and series in discrete and continuous forms is described exhaustively elsewhere [68], and their application to signal and image processing is similarly fully described in other texts [60].

### 9.2.2 Wavelets

Wavelets overcome the shortcomings of the STFT by providing a basis for function representation that varies in frequency and time (translation). Where a number of STFT calculations would be required to include a range of window sizes in both frequency and time, this is achieved in a single wavelet operation.

This basis is derived from a mother function $\Psi(t)$ which is dilated (scaled) and translated to construct the family of basis functions. We write

$$\Psi_{b,a}(t) = \frac{1}{\sqrt{a}}\Psi\left(\frac{t-b}{a}\right)a \qquad (9.10)$$

so $a$ has the effect of dilating, or scaling, $\Psi$, and $b$ translates.

These effects are illustrated in Figure 9.7 for a very simple mother function, the Haar. The scaling parameter $a$ is clearly influencing the frequency of the function – $\frac{1}{a}$ is a measure of frequency.

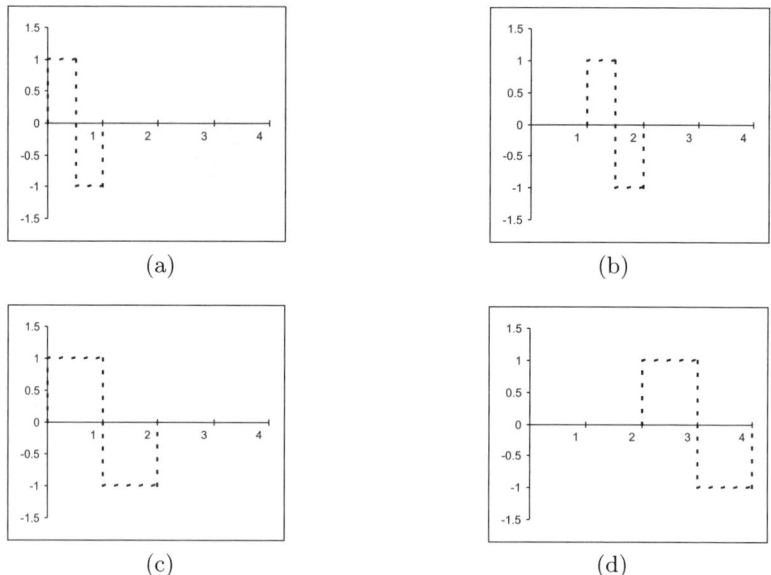

**Figure 9.7.** The Haar mother function with **(a)** a = 1, b = 0; **(b)** a = 1, b = 1; **(c)** a = 2, b = 0; **(d)** a = 2, b = 1.

Given this family of basis functions, we can now represent an arbitrary function $f(x)$ by correlating it with the scaled and translated versions of the mother. The continuous wavelet transform of $f(x)$ with respect to the mother $\Psi(t)$ is given by

$$W_\Psi f(b,a) = \int_{-\infty}^{\infty} f(t)\overline{\Psi_{b,a}(t)}dt. \qquad (9.11)$$

A wavelet transform of the terahertz pulse of Figure 9.4(b) is shown in Figure 9.8. Notice the degree of detail here; wavelet representations can be difficult to interpret without practice and concentration, however, as for Figure 9.6, it is straightforward to observe the time delay of the high-frequency component.

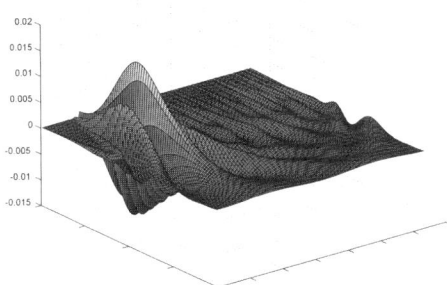

**Figure 9.8.** The Morlet wavelet transform of the terahertz pulse in Figure 9.4(b). The scale axis runs from front to back, and the time (or translation) axis from left to right.

The wavelet transform is invertible, provided

$$\hat{\Psi}(0) = 0 \tag{9.12}$$

where $\hat{\Psi}$ is the Fourier transform of $\Psi$ – (9.4). To recapture $f$, a two-dimensional integration over both parameters a and b is required;

$$f(t) = \frac{1}{C_\Psi} \int_{-\infty}^{\infty} \int_{-\infty}^{\infty} \frac{1}{a^2} [W_\Psi f(b,a)] \Psi_{b,a}(t) da db. \tag{9.13}$$

Here $C_\Psi$ is a constant dependent on the mother, given by

$$C_\Psi = \int_{-\infty}^{\infty} \frac{|\hat{\Psi}(\omega)|^2}{\omega} d\omega. \tag{9.14}$$

We require this constant to be finite. This is known as the admissibility condition that restricts the class of functions that may be chosen as wavelets. Note in particular that of necessity $\hat{\Psi}(0) = 0$.

The functions $f$ we observe are of course discrete, and so require a discretized continuous wavelet transform in the same manner as a discrete Fourier transform is defined. In fact, the continuous transform contains a lot of redundancy; this, together with the computational load of a simple discretization, has led to the development of very efficient subsamplings to provide the discrete wavelet transform (DWT). This is normally done by taking the scale parameter a to be of the form $2^s$ for an integer $s$, and $b = k2^{-s}$. Then, (9.11) becomes

$$W_\Psi f(k2^{-s}, 2^{-s}) = 2^{s/2} \int_{-\infty}^{\infty} f(t) \Psi(2^s t - k)(t) dt. \tag{9.15}$$

If $F$ is discretized, with a sampling rate chosen as 1 for convenience, this becomes

288 Elizabeth Berry et al.

$$W_\Psi f(k2^{-s}, 2^{-s}) = 2^{s/2} \sum_n f(n)\Psi(2^s n - k). \tag{9.16}$$

Note that in computing this, we only need to know the function values where the corresponding wavelet value is nonzero.

### 9.2.3 Computer Vision Background

**Wide-Band Cross-Ambiguity Function, WBCAF**

Significant relevant theory on signal pulses was developed in the study of radar, in which we are often interested in correlating a signal with time-shifted copies of itself. The aim is to extract time delay and Doppler shift, thereby making good estimates of signal delay, and implicitly distance of travel [67]. The relevant ambiguity function of a signal is given by

$$A_f(\xi, x) = \int_{-\infty}^{\infty} f\left(b + \frac{x}{2}\right) \overline{f\left(b - \frac{x}{2}\right)} e^{-i\xi b} db. \tag{9.17}$$

The transforms in which we are interested perform a correlation of the observed data with individual basis functions, not with itself. The wide-band cross-ambiguity function, WBCAF [69], of a function $f_2(t)$ against a reference $f_1(t)$ is defined as

$$WBCAF_{f_1} f_2(\tau, \sigma) = \frac{1}{\sqrt{\sigma}} \int_{-\infty}^{\infty} f_2(t) \overline{f_1 \frac{t - \tau}{\sigma}} dt, \tag{9.18}$$

where $f_1$ has been delayed by $\tau$ and scaled by $\sigma$. This permits the extraction of the appropriate $\tau$ (time delay) for each scale $\sigma$, by locating the value of $\tau$ that provides the maximum value of correlation at that $\sigma$.

Note, however, the similarity between (9.11) and (9.18). The WBCAF resembles the continuous wavelet transform, using the reference function as the wavelet mother. This is an expensive computation, since the reference $f_1$ needs repeated resampling (interpolation, filtering, subsampling), which can generate a prohibitive overhead. To overcome this, Young [70] has developed the wavelet-based wide-band cross-ambiguity function

$$WB - WBCAF_{f_1} f_2(\tau, \sigma) = \int_{-\infty}^{\infty} \int_{-\infty}^{\infty} W_\Psi f_2(a, b) \overline{W_\Psi f_1\left(\frac{a}{\sigma}, \frac{b - \tau}{\sigma}\right)} \frac{dadb}{a^2}, \tag{9.19}$$

where $W_\Psi$ is the continuous wavelet transform with respect to the mother $\Psi$. Using an arbitrary mother permits efficient precomputation of the wavelets for all relevant $\tau$ and $\sigma$

The WB-WBCAF permits useful measurements to be made on an output pulse. Fixing the scale at 1, determining the translation which provides the maximal response gives a good estimate of the time delay associated with the pulse. This measurement, which may be considered as a cross-correlation or the application of a matched filter, provides the parameter that gives the best match with the reference. Note that we cannot determine the time delay as easily at different scales, since the scaling operation also causes a shift in the reference. We can normalize the WB-WBCAF and the WBCAF using (9.18) at various values of $\sigma$

$$\frac{MAX_\tau(WBCAF_{f_1}f_2[\tau,\sigma])}{MAX_{\tau'}(WBCAF_{f_1}f_1[\tau',\sigma])}. \quad (9.20)$$

This parameter, which we shall call the WBCAF absorption, can then be taken to be proportional to the relative transmission of the band of frequencies corresponding to scale $\sigma$. The reasoning here is that $WBCAF_{f_1}f_2(\tau,\sigma)$ is the spectral content of $f_2$ with respect to $f_1$ at $\sigma$ and $\tau$, while similarly $WBCAF_{f_1}f_1(\tau',\sigma)$ is the spectral content of $f_1$. It is necessary to use different values of $\tau$ because the pulses experience different time delays, and must first be aligned in time to ensure that the WBCAF absorption is calculated using the same part of each pulse. Finding the maximum value of the WBCAF in $\tau$ achieves this. The ratio of these two should then give the power ratio of $f_1$ to $f_2$ at that $\sigma$.

**Compression**

The task of data compression has been approached via function transforms in many ways [60]. At their simplest, the coefficients of a function's Fourier transform are often negligibly small (corresponding to harmonics that scarcely appear in the function). One approach to compression is to derive a far more compact representation of the transform by neglecting these coefficients. The inverse Fourier transform of this compressed transform will then provide a good approximation to the original function, since the information lost is insignificant.

The same approach may be taken with the wavelet transform, where a similar observation may be made about coefficients of small magnitude. Figure 9.9 shows a histogram of the frequency of occurrence of values of the wavelet coefficients for the terahertz pulse in Figure 9.4(b). It is clear that the majority of these are relatively insignificant. Wavelet compression could be achieved by setting the smallest value coefficients to zero.

**Clustering**

Multidimensional data are often easily represented by clusters. The centers (usually centroids) of these clusters may then be used as exemplars. If the exemplars are indeed good examples of the data clustered around them, a highly compact codebook representation of the data becomes available. Terahertz imaging datasets can be interpreted as large, high-dimensional vectors; at their simplest, each "pixel" (time series) can be taken as a vector of length

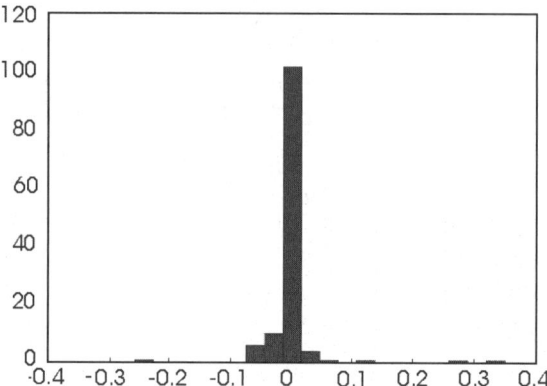

**Figure 9.9.** The frequency of occurrence of values of the wavelet coefficients for the terahertz pulse in Figure 9.4(b).

equal to the number of time samples it provides. Each vector is then termed a "feature." Alternatively, we might look for clustering of other features derived from the pulses, such as Fourier or wavelet coefficients. A clustering of these vectors might then seek out physical similarities in the sample under inspection. Clustering has received much attention in the literature. The simplest approach (and widely used) is the k-means algorithm [71], which may be summarized as the following

1. Select $N$, the number of clusters to be formed.
2. Choose initial cluster centers $v_1, v_2, \ldots, v_N$.
3. Determine for each data point its closest "center."
4. Recalculate centers as centroids of allocated data points.
5. Iterate from step 3 until no change.

It should be clear that this algorithm may be applied generally to any multidimensional data. It has been the subject of significant work, in particular to determine favorable values for $N$, and the initialization $v_1, v_2, \ldots, v_N$ [72].

### 9.2.4 Transmission of Terahertz Radiation

In this work we have assumed the following simple model for the transmission of the terahertz pulses.

**Refractive Index**

When considering the time delay caused by transmission through a thickness $x$ of material, compared with transmission through the same distance in vacuo. Let the velocity in vacuo be $c$, and in the medium $v$. Then the difference in time taken by radiation following the two paths, $\Delta t = (x/v - x/c)$. But the refractive index $n$, is defined as $n = c/v$, so

$$\Delta t = x(n-1)/c. \tag{9.21}$$

Thus, if the time delay is measured for a range of sample thickness, $n$ can be found from the slope of a plot of $\Delta t$ against $x$. The true refractive index of a material varies with frequency, but the single value found by this technique is a single broadband measurement and will be peculiar to the system on which measurements were made.

**Absorption Coefficient**

The absorption of a beam of radiation of incident intensity $I_0$, transmitted through a thickness $x$, is described by the Beer–Lambert law,

$$\frac{I}{I_0} = e^{-\mu x} \text{ or } \ln\left(\frac{I}{I_0}\right) = -\mu x \tag{9.22}$$

where $\mu$ is the linear absorption coefficient. If scattering is considered to be negligible, a plot of $\ln(I/I_0)$ against $x$ is linear with slope $-\mu$, as for a given material and with the simple geometry of the samples used here, reflection losses will be constant and do not affect the slope of the graph. The Beer–Lambert expression was used in the time domain to give an estimate of the broadband absorption coefficient. In this work we also apply it in the frequency domain to give the absorption coefficient as a function of frequency, and in the STFT and WBCAF analyses. In the latter cases the relevant linear absorption coefficient is derived by replacing $I/I_0$ with an estimate of maximum transmittance (over $t$ or $\tau$) derived from the STFT or WBCAF; in the case of the WBCAF this is defined by (9.20).

## 9.3 Methods

### 9.3.1 Optical Properties of Materials

We wanted to use samples whose optical properties were known not to vary with time, and in spite of our interest in biomedical applications, this ruled out the use of tissue samples. The acquisition time was long enough for tissue samples to dehydrate and change the values we were trying to measure. So specially manufactured test objects were used instead. Two step-wedges were manufactured by rapid prototyping [73]. One, from nylon (Duraform polyamide, nylon 12) by the selective laser sintering process. The other made of resin by stereo-lithography. The test objects had steps of known thickness ranging from 0.1 mm to 7 mm and are shown in Figure 9.10.

Transmission data were acquired from both step wedges, using the pulsed terahertz imaging system at JW Goethe-Universität, Frankfurt. The time series recorded at each pixel comprised 128 points, separated by 0.15 ps for the nylon step wedge and at 0.2 ps for the resin step wedge. Parametric images representing the step wedge data are shown in Figure 9.11.

**Figure 9.10.** Nylon (left) and resin (right) step wedges. Each block measures approximately 4 cm × 3 cm × 1 cm.

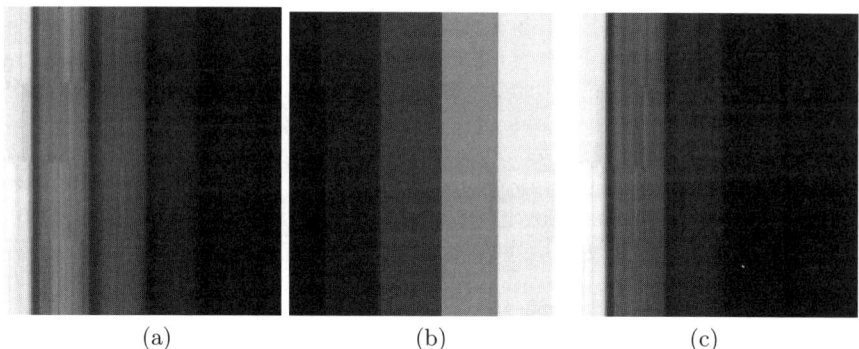

**Figure 9.11.** Parametric terahertz-pulsed images taken from the nylon step wedge data set. The white end of the gray scale represents large values, and as in Figure 9.12 the thinnest step is at the left. **(a)** Pulse amplitude relative to reference pulse amplitude, in time domain. **(b)** Time delay between transmitted pulse peak and peak of reference. **(c)** Transmittance (ratio of transmitted and incident intensities after Fourier transformation of pulses) at 1.2 THz.

### Refractive Index, Broadband, and as a Function of Frequency

Two methods for finding the broadband refractive index were used — the conventional time domain analysis and the WBCAF analysis. STFT analysis was used to find the variation of refractive index with frequency. For the conventional analysis, the time delay was estimated for each pixel in the time domain. For the wavelet-based analysis, the WBCAF was calculated at scale 1, and the time delay was the value of $\tau$ that maximized the correlation (9.18). For the STFT analysis, the STFT was performed using a Gaussian window of width 1.5 ps. For each frequency, the time delay was the time at which the value of the STFT was a maximum. In each case, the mean value for the time delay was found within a region of interest, approximately $40 * 40$ pixels, one in each step. Time delay was plotted against step depth, and the refractive index was found using (9.21). The standard error of the slope was calculated.

**Attenuation: Broadband and as a Function of Frequency or Scale**

For each material, a region of interest approximately $40 * 40$ pixels was defined in each step of the wedge.

1. A broadband estimate of transmittance was made, using the square of the measured peak amplitude in the time domain $I/I_0$. These results were plotted against step thickness and (9.22) used to calculate the broadband linear absorption coefficient
2. Following Fourier transformation, the transmittance $I(\omega_i)/I_0(\omega i)$ was calculated at several values of frequency. At each frequency, (9.22) was used to calculate the linear absorption coefficient, and these results were plotted against frequency.
3. Following short-time Fourier transformation, the maximum transmittance, $\max_t(I(\omega_i)/I_0(\omega_i))$ was calculated at several values of frequency. At each frequency, this value was substituted for $I/I_0$ in (9.22), and was used to calculate the STFT linear absorption coefficient. These results were plotted against frequency.
4. Equation (9.20) was used to calculate the proposed WBCAF-absorption at 13 values of scale. At each scale, this value was substituted for $I/I_0$ in (9.22), and was used to calculate the WBCAF linear absorption coefficient. These results were plotted against scale.

In each case the standard deviation of the linear absorption coefficient was determined from the plot against step thickness. The results of the analyses were not directly comparable because the wavelet-based technique gives results in terms of scale rather than frequency, but it was possible to inspect the plots of absorption parameters to gain a qualitative impression of their ability to discriminate between materials.

### 9.3.2 Signal Compression

The conventional calculations to determine the refractive index and absorption coefficient associated with the nylon step wedge, described in Section 9.3.1, were repeated using a range of reduced versions of the wavelet transform.

These reductions were compressions of the transform data obtained by setting to zero the smallest (in magnitude) $p\%$ of coefficients, before performing the calculations. This common form of compression does not, of course, result in the data requiring $(100 - p)\%$ space, since the resulting sparse arrays require indexing information to locate the surviving coefficients. Nevertheless, given that we experimented with values of $p$ in excess of 50, the potential for compression of the (real valued) data is clear.

The quality of the results generated was easy to assess in the case of the refractive index; taking the uncompressed transform data as a gold standard, a simple numerical comparison permitted an evaluation of the result extracted from the compressed form. The absorption coefficient measurement was generated at a range of terahertz frequencies. These results were compared with those from the uncompressed data by calculating the Pearson correlation coefficient (measuring the degree of linear relationship between data sets), the root mean square difference, and Student's paired $t$-test probability, which provides a confidence estimate for a set of pairs of observations being matching pairs [74].

### 9.3.3 Clustering Demonstrations

#### Synthetic Image of Tooth Slice

Our first experiment on clustering was designed to determine the importance of initialization on the success of the classification, and to compare various choices of feature vector. We used a synthetic image data set because in that case the class of each pixel is known, and this can be used to determine the success of the classification. A $50 * 50$ pixel image of a slice of a tooth was generated. Each pixel was set to belong to one of three classes comprising tooth enamel, tooth dentine, and air, and these were distributed in a realistic configuration by tracing the outlines from an image of a tooth slice. This is illustrated in Figure 9.12(a). A typical time series for each material was taken from a real data set, and noise was modeled by adding normally distributed noise, with values selected at random from the distribution, to each time point of the time series at each pixel. The noise was taken from a single Gaussian distribution; the mean of the distribution was taken from inspection of time series from the background of the real image, and the standard deviation chosen empirically. The validity of this noise model is discussed in Section 9.5. In this preliminary experiment partial volume effects were not incorporated into the model. Three parametric images representing the synthetic data are shown in Figures 9.12(b)–(d).

The number of clusters was set to three, representing air, enamel, and dentine, and standard $k$-means clustering with random initialization was used [75]. Four different feature vectors were used: for 1–3 the vector dimension was 64 or 128 depending on the number of time samples in the data:

1. The time series.

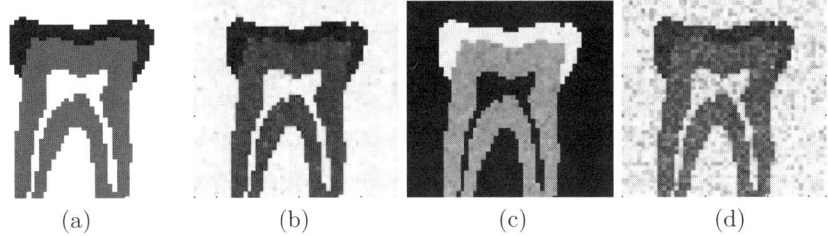

(a)  (b)  (c)  (d)

**Figure 9.12.** (a) Allocation of classes in the synthetic image of a tooth slice. Black represents enamel, gray dentine, and white air. (b)–(d) Parametric terahertz-pulsed images taken from the synthetic data set. (b) Pulse amplitude relative to reference pulse amplitude, in time domain. (c) Time delay between transmitted pulse peak and peak of reference. (d) Transmittance (ratio of transmitted and incident intensities after Fourier transformation of pulses) at 0.85 THz.

2. Fast Fourier transform (FFT) of time series.
3. Discrete wavelet transform (DWT) of time series.
4. A three dimensional vector using three parameters calculated from the time series. These were the integral phase shift between 0.5 and 1 THz, the integral phase shift between 1 and 1.5 THz and the absorbance $A = -\log_{10}(I/I_0)$ at 1 THz. The three parameters were normalized to be univariate within a unit hypercube to ensure that differences in units between them did not bias the outcome.

The result of using random initialization was, as expected, that many classifications were "unsuccessful," for example, many of the air pixels were wrongly classed as enamel or dentine, or where only two classes resulted. Future work will concentrate on refining the initialization, but for these initial experiments we simply repeated each classification several times, and termed it "successful" if the result was three contiguous regions broadly occupying the relevant locations. For each "successful" classification the number of misclassified pixels in the image was determined as a percentage of the total number of pixels.

**Terahertz-Pulsed Image of Tooth Slice**

The same methods of classification were used on a nonsynthetic image of a dehydrated tooth slice of thickness approximately 200 μm. Transmission data were acquired from an area 22.2 mm ∗ 9 mm, using the pulsed terahertz imaging system at the University of Leeds. The image array was 56 ∗ 56 pixels, and the time series recorded at each pixel comprised 64 points separated by 0.15 ps. Three parametric images representing the tooth data are shown in Figure 9.13.

Unlike the synthetic tooth data, there are no known classes for this data. To allow the results to be assessed against an independent modality we acquired a radiograph of the tooth slice using a dental x-ray system operating

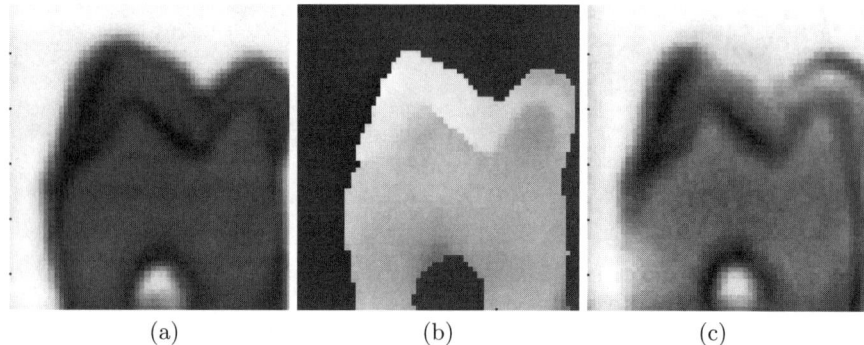

**Figure 9.13.** Parametric terahertz-pulsed images taken from the tooth slice data set. **(a)** Pulse amplitude relative to reference pulse amplitude, in time domain. **(b)** Time delay between transmitted pulse peak and peak of reference. **(c)** Transmittance (ratio of transmitted and incident intensities after Fourier transformation of pulses) at 0.85 THz.

at 60 kV. Using a commercial biomedical image processing package (Analyze, AnalyzeDirect.com, Lenexa, Kansas, USA), this image was registered to a parametric image derived from the terahertz-pulsed image data. Interactive region growing was used to define the enamel, dentine, and air regions on the registered x-ray image. These definitions were used to produce an outline that was overlaid on the clustering results, to give a visual indication of the accuracy of the classification. The percentage of misclassified pixels was determined. Further quantification was not attempted because the tooth slice was not of perfectly uniform thickness, being thinner at the top right, and this would lead to errors in the classification based on automatic clustering.

**Terahertz-Pulsed Images of Histopathological Samples**

Classification techniques were run on two further data sets. In these cases unregistered optical images were available for comparison, and it is important to note that clinically important differences will not necessarily be seen in the photograph. The image data sets were acquired from wax embedded histopathologically prepared sections of tissue, of thickness 1 mm. The preparation involved dehydration and fixing with formalin, and this means that terahertz radiation was able to penetrate the tissue more readily than would be the case if the tissue were fresh. Transmission data were acquired from both, using the pulsed terahertz imaging system at the University of Leeds. The first example was of basal cell carcinoma and the second melanoma. In each case the time series recorded at each pixel comprised 64 points separated by 0.15 ps. For the basal cell carcinoma the image array was 56 * 56 pixels over an area 10 mm * 10 mm, for the melanoma the image array was 20 by 18 pixels over an area 5 mm * 4.5 mm.

$k$-means clustering on the fast Fourier transform coefficients of the time series was applied to the basal cell carcinoma data, using five and eight classes. For the melanoma, the fast Fourier transform coefficients were again used, this time with both $k$-means clustering and vector quantization, each for five classes. Three parametric images representing the basal cell carcinoma and melanoma histopathological data are shown in Figure 9.14.

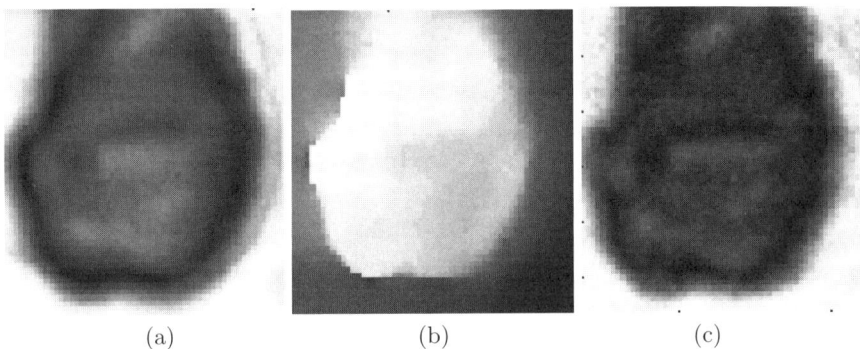

(a) (b) (c)

**Figure 9.14.** Parametric terahertz-pulsed images taken from the histopathological basal cell carcinoma (top) and melanoma (below) data sets. **(a)** Pulse amplitude relative to reference pulse amplitude, in time domain. **(b)** Time delay between transmitted pulse peak and peak of reference. **(c)** Transmittance (ratio of transmitted and incident intensities after Fourier transformation of pulses) at 2 THz.

Matlab (Version 6, The MathWorks, Inc., Cambridge, UK) was used for the post-acquisition image and signal analysis work described in this section.

## 9.4 Results

### 9.4.1 Optical Properties of Materials

**Refractive Index: Broadband and as a Function of Frequency**

Results for the broadband refractive index are shown in Table 9.1, and the variation of refractive index with frequency from the STFT analysis is in Figure 9.15. It can be seen that the refractive index decreases with increasing frequency, and is higher for resin than for nylon, which is consistent with the broadband results.

These results illustrate that any image reconstruction algorithm requiring an assumption of negligible dispersion would be valid for nylon but not for resin.

**Table 9.1.** Results of refractive index measurements, mean ± standard deviation

|  | Nylon | Resin |
| --- | --- | --- |
| Conventional time domain | 1.603 ± 0.004 | 1.66 ± 0.01 |
| WBCAF at scale 1 | 1.597 ± 0.005 | 1.64 ± 0.02 |

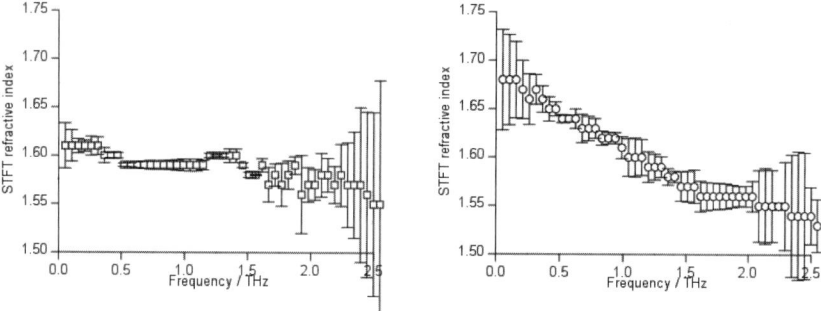

**Figure 9.15.** STFT refractive index against frequency for nylon (left) and resin (right). Error bars show ± one standard deviation.

## Attenuation: Broadband and as a Function of Frequency

From the conventional broadband analysis, for nylon the linear absorption coefficient was 6.8± 0.5 cm$^{-1}$ and for resin 11.8± 1.6 cm$^{-1}$. Figure 9.16(a) shows the variation of linear absorption coefficient with frequency, using conventional Fourier analysis. The two materials may be differentiated throughout the frequency range by their linear absorption coefficient. Figure 9.16(b) shows the variation of the STFT linear absorption coefficient with frequency. As was the case for the conventional Fourier analysis the two materials may be differentiated throughout the frequency range by their STFT absorption coefficient. The values diverge with increasing frequency as the curves diverge for the conventional analysis, but the analysis is robust to a higher frequency. This is likely to be a consequence of the windowed nature of the transform, and suggests that the STFT may be valuable for analysis of noisy data such as that acquired from tissue in our system between 1.5 and 2.5 THz.

Figure 9.17 shows the variation of the WBCAF linear absorption coefficient with frequency. As was the case for the conventional Fourier analysis the two materials may be differentiated throughout the scale range by their WBCAF absorption coefficient, which is higher for resin than for nylon. The values diverge as the scale decreases, which corresponds with the way in which the curves diverge with increasing frequency for the conventional and STFT analyses. For nylon, the WBCAF linear absorption coefficient is almost constant with scale, in contrast to the behavior with frequency of the linear absorption coefficient calculated both by conventional methods and using the STFT.

# Chapter 9 Time-Frequency Analysis in Terahertz-Pulsed Imaging

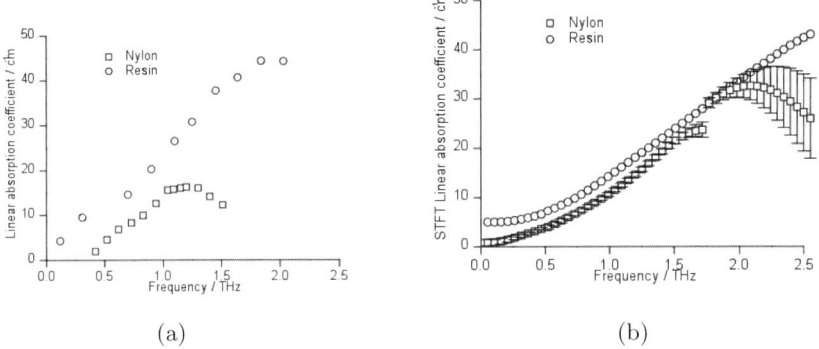

**Figure 9.16.** (a) Conventional Fourier analysis to show the variation of the linear absorption coefficient with frequency (b) STFT analysis to show the variation of the STFT linear absorption coefficient with frequency. The error bars have been omitted from the resin STFT data points — they were of approximately the size of the symbol, where shown they represent ± one standard deviation.

Thus, for some materials, reconstruction methods that require an assumption of negligible changes in absorption with frequency could appropriately be tackled using a wavelet approach.

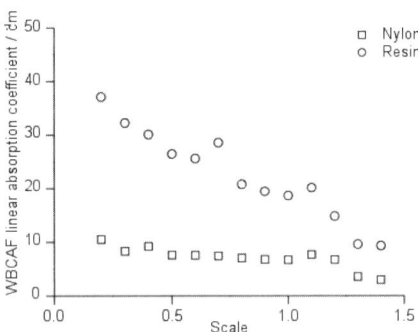

**Figure 9.17.** Results of WBCAF analysis to show the variation of the WBCAF linear absorption coefficient with scale.

## 9.4.2 Signal Compression

### Refractive Index

Compression to very high levels has a negligible impact on the broadband refractive index calculations (Table 9.2). This measurement is robust until 90%

**Table 9.2.** Broadband refractive index, $n$, for different compression levels.

| Coefficients removed / % | 0 | 50 | 80 | 90 | 95 |
|---|---|---|---|---|---|
| $n$ | 1.603 | 1.603 | 1.604 | 1.603 | 1.61 |
| Standard deviation of $n$ | 0.002 | 0.002 | 0.003 | 0.005 | 0.01 |

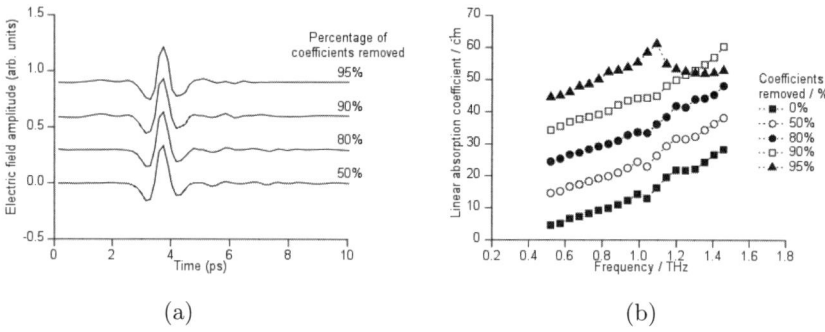

**Figure 9.18.** (a) The terahertz pulse of Figure 9.4(b) transmitted through 1 mm of the nylon step-wedge compressed by removal of 50%, 80%, 90%, and 95% of the smaller coefficients. The pulses are shown offset in the amplitude direction for clarity. (b) Linear absorption coefficient against frequency for nylon using conventional Fourier analysis, at various degrees of wavelet compression. To allow comparison, the results for the various degrees of compression have been offset from the uncompressed result at intervals of $10\,\text{cm}^{-1}$.

of the coefficients are removed, at which point a small error is introduced. The terahertz pulse of Figure 9.4(b) is shown in Figure 9.18 reconstructed from the discrete wavelet transform after the truncation of 50%, 80%, 90%, and 95% of the smaller coefficients. As expected, the major feature of the pulse (the main peak) is unaffected, while the truncation removes more and more of the smaller scale features, which we might expect to be largely noise.

**Absorption Coefficient**

The results of the statistical tests of similarity between the uncompressed and compressed observations across a range of frequencies from 0.52 to 1.46 THz at intervals of 0.05 THz, for various truncations, are shown in Table 9.3. Figure 9.18(b) shows the absorption coefficient at selected frequencies, for various truncations. More detailed results are presented elsewhere [76]. Up to about 1 THz, the curves coincide. However, note that the spectral feature at 1.1 THz is lost when 90% of the coefficients are removed, and significant errors appear at 95%. The statistical tests all suggested that there is no significance in the differences as far as 10% compression (90% of the coefficients are removed), the global nature of the statistic masked the disappearance of a local feature, which could be of importance.

**Table 9.3.** Results of statistical tests comparing absorption coefficient cm$^{-1}$ at various degrees of compression.

| Coefficients removed / % | 0% | 50% | 80% | 90% | 95% |
|---|---|---|---|---|---|
| Pearson correlation | 1.0 | 0.999 | 0.997 | 0.992 | 0.495 |
| RMSD | 0 | 0.09 | 0.61 | 0.97 | 7.05 |
| P (paired $t$) | — | 0.43 | 0.85 | 0.40 | 0.06 |

### 9.4.3 Clustering

**Synthetic Image of Tooth Slice**

Clustering performed on the synthetic tooth image produced promising results. Table 9.4 records the percentage of misclassified pixels as a result of clustering the entire pulse data, the FFT of the pulse data, the DWT, and a 3D feature vector. The high quality of these results is partly because the issue of initialization was not addressed. The basic $k$-means algorithm is very susceptible to initialization, and some runs produced very poor results that were simply excluded in these experiments. We have not yet experimented with more intelligent initialization procedures, but once this is done we would expect results approaching this quality.

**Table 9.4.** Percentage of misclassified pixels: Tooth slice.

| Synthetic image of tooth slice | % misclassified |
|---|---|
| Time series | 0 |
| FFT coefficients | 0.04 |
| DWT coefficients | 0 |
| Feature vector | 1 |

**Terahertz-Pulsed Image of Tooth Slice**

Results for the image of the 200 µm tooth slice are presented in Table 9.5.

Figure 9.19 illustrates classifications resulting from time series, FFT, and feature-vector clustering. The pulse relative amplitude image is shown as an indicator of where physical boundaries in the sample lie. The boundaries segmented from the radiograph are shown in white as an overlay.

**Table 9.5.** Percentage of misclassified pixels: Tooth slice.

| Terahertz pulsed image of tooth slice | % misclassified |
|---|---|
| Time series | 19 |
| FFT coefficients | 23 |
| DWT coefficients | 19 |
| Feature vector | 13 |

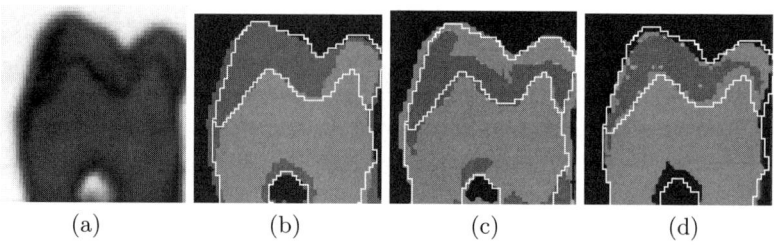

(a)      (b)      (c)      (d)

**Figure 9.19.** The tooth slice data set. (**a**) Parametric terahertz-pulsed image showing pulse amplitude relative to reference pulse amplitude, in time domain. (**b**) Result of clustering using the whole time series in the time domain. (**c**) Result of clustering using the FFT coefficients. (**d**) Result of clustering using the 3D feature vector. In (b)–(d) air is shown in black, enamel in dark gray, and dentine in light gray. The white outline indicates the boundaries of dentine and enamel that were identified interactively on the radiograph of the tooth slice.

**Terahertz-Pulsed Images of Histopathological Samples**

Results from performing clustering on the hisopathological data sets are shown in Figures 9.20 and 9.21. For the basal cell carcinoma data a feature is emphasized in white at the center of Figure 9.20(b) which is not apparent in the photograph. As stated previously, one does not expect clinically important features necessarily to be obvious in the photograph. The feature is also evident in the transmittance terahertz image shown earlier in Figure 9.14(c), but use of the FFT coefficients in the clustering has defined it more clearly.

For the melanoma data set the classes returned by the two types of clustering differ. Again they do not correspond with the regions of the photograph, nor were they expected to. The mid-gray feature shown by clustering using the FFT coefficients in Figure 9.21(b) is also apparent to a lesser extent in each of the parametric images in Figure 9.14.

## 9.5 Discussion

Terahertz-pulsed imaging is a relatively new modality that records large datasets that are difficult to visualize. Conventional Fourier and time domain techniques are widely used for analysis, but there is no standardization on the

Chapter 9 Time-Frequency Analysis in Terahertz-Pulsed Imaging 303

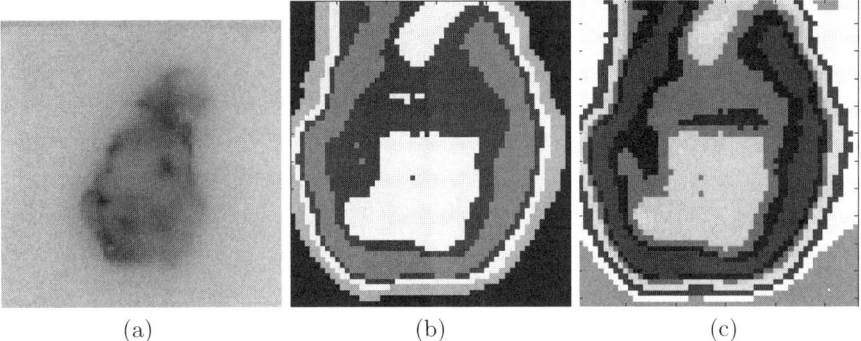

(a) (b) (c)

**Figure 9.20.** The basal cell carcinoma data set. (a) Photograph showing the histopathologically prepared sample. (b) Result of clustering into five classes using the using the FFT coefficients. (c) Result of clustering into eight classes using the using the FFT coefficients.

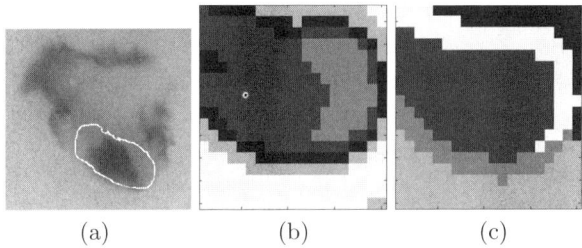

(a) (b) (c)

**Figure 9.21.** The melanoma data set. (a) Photograph showing the histopathologically prepared sample, with the melanoma outlined. The terahertz imaged area corresponds to the lower right quadrant of this photograph. (b) Result of $k$-means clustering into five classes using the using the FFT coefficients. (c) Result of vector quantization clustering into five classes using the FFT coefficients.

methods or presentation of data. A powerful addition to the analysis would be the ability to consider time and frequency independently, but this is not possible using the Fourier transform, so time-frequency techniques such as the STFT and wavelets have been considered. The limitations of the uncertainty principle on the STFT means that it would be necessary to run the analysis many times to get a complete picture of the variations with time and frequency. Wavelet analysis, although still constrained by the uncertainty principle, can give a more complete picture from a single analysis. Wavelet analysis has proved successful in related areas where analysis of time series is required, including ultrasound and radar. The analysis is particularly attractive for use with terahertz pulses because they can be described very efficiently by wavelets, where the Fourier transform is a less efficient basis. This is encouraging for applications such as compression and noise reduction, and may be an advantage in other analytical methods. Wavelet techniques are also attractive experimentally, because there is an established theoretical background and robust software is available.

In this work we investigated a range of potential alternatives to conventional Fourier analysis. We have shown that, for our test materials, the STFT and wavelet analyses give, within experimental error, the same result for refractive index as the conventional method. The simple WBCAF analysis involves estimation of time delays only at scale 1 because changing the scale results in an associated time shift. Extension to higher scales is an area for future work. The absorption coefficients produced by conventional and wavelet analysis are not directly comparable, because the conventional analysis produces values at given frequencies, while the wavelet results are plotted against scale, each of which represents a range of frequencies. However, we see a similar pattern of differences between two materials in this simple case of transmission imaging, and believe that the opportunity to focus on particular time periods with wavelets will be helpful for more complex images. In addition, a material that showed increasing absorption coefficient with frequency by Fourier techniques demonstrated an almost constant WBCAF absorption coefficient with scale. This may be a property that could be exploited in reconstruction techniques that require an assumption of negligible variation in absorption with frequency . It is a limitation of the work presented here that only two materials were studied, and neither of these was biomedical. In such material, biological variation is expected, and the value of the optical parameters will vary from sample to sample [19]. Nor will tissue be neatly arranged as a step wedge, so a smaller number and range of thicknesses would be available. Adaptation of existing techniques for refractive index estimation, which have been successful for characterizing semiconductors [77, 78], will be needed.

An understanding of the imaging process is often fundamental to successful analysis of medical imaging data [79, 80]. There are still many factors that need to be included in the analysis, perhaps by including a model of the imaging process such as diffraction, scattering, pulse shape, partial volume effects, frequency dependent amplitude and frequency-dependent spatial resolution. It should also be possible to exploit models of how the pulses are expected to appear [81], especially with advances in pulse shaping [82].

In this work a very simple model of the noise in terahertz pulsed images was used when making the synthetic tooth data set. Recent analysis has shown that a single Gaussian does not correctly model the noise, and improving the noise model is an area of current research. These are very early results, but clearly do not rule out the use of time-frequency techniques in this field. The robust results from these initial experiments in the simplest case of using transmission data through a single layer give us confidence for applying adaptations of the methods to data where time localization will be important for identifying which pulse corresponds with which boundary.

The use of wavelets for signal compression is a well known application, and it would be of real practical benefit in terahertz pulsed imaging. Our initial work presented here has highlighted the difficulties associated with quantifying the effect of such compression. Statistical tests showed no significant difference between pulses where, by eye, it could be seen that a potentially relevant spectral feature had been lost. Testing of the effect of compression is therefore application specific, and one must address the question of whether the final conclusion has been affected by the process. In our particular example, we sought changes to the measured values of refractive index, and in this case the loss of the spectral feature was not important. We suspect that this will not be the case when terahertz-pulsed imaging moves into clinical and laboratory use.

The idea of using clustering methods that use more of the information in the time series at each pixel than just two or three parameters is very attractive, and has been used successfully in other fields [66]. Our work here is very preliminary, and the next stage is to perform a study on large numbers of samples for which we have a gold standard classification by another technique, for evaluation. However, from these early results it appears that robust segmentation, using features derived from the terahertz pulses is a real possibility. The incorporation of knowledge about likely signal changes from particular disease processes could be built in to the classification scheme, and classified data may form the input for filtered back projection.

Terahertz-pulsed imaging is in its infancy, and in order to speed up the imaging process many of the early systems do not collect the whole time series at each pixel. Even when data from the whole pulse are available, users display or use parametric images for further calculation without an understanding of the most appropriate choice of parameter to emphasize a feature of interest. In this work, we have presented preliminary simple experiments to investigate techniques that made use of all the information in each pulse. The results from these justify further investigation in this area and our aim is to provide tools for the user that will optimize their use of the spectroscopic information hidden in each terahertz signal.

## 9.6 Acknowledgments

We are grateful for financial support for this work under the Teravision project by the European Union (IST-1999-10154), and by the Engineering and Physical Sciences Research Council via a project grant (GR/N39678), a P.h.D studentship and conference support (GR/R85280). Biomedical terahertz imaging research in our institution is under the direction of Professors J.M. Chamberlain and M.A. Smith. We thank them, N.N. Zinov'ev and R.E. Miles at the University of Leeds, and K. Siebert and T. Loeffler of the Physikalisches Institut der JW Goethe-Universität, Frankfurt, for their expertise in terahertz imaging that has allowed us to pursue computer vision work in the field. A.J. Marsden manufactured the step wedges; D.J. Wood, F. Carmichael and S. Strafford provided dental specimens and expertise; J. Bull and M. Fletcher carried out x-ray imaging; J. Biglands performed additional software work; and G.C. Walker commented on the manuscript.

## References

[1] Allen, S.J., Craig, K., Galdrikian, B., Heyman, J.N., Kaminski, J.P., Scott, J.S.: Materials science in the far-IR with electrostatic based FELs. Nuclear Instruments & Methods in Physics Research Section A; Accelerators Spectrometers Detectors and Associated Equipment **35** (1995) 536–539

[2] Jaroszynski, D.A., Ersfeld, B., Giraud, G., Jamison, S., Jones, D.R., Issac, R.C.: The Strathclyde terahertz to optical pulse source (TOPS). Nuclear Instruments & Methods in Physics Research, Section A: Accelerators Spectrometers Detectors and Associated Equipment **445** (2000) 317–31

[3] Grischkowsky, D.R., Mittleman, D.M.: Introduction. In Mittleman, D., ed.: Sensing with Terahertz Radiation. Springer-Verlag, Berlin (2003) 1–38

[4] Auston, D.H., Nuss, M.C.: Electrooptic generation and detection of femtosecond electrical transients. IEEE Journal of Quantum Electronics **24** (1988) 184–197

[5] Kleine-Ostmann, T., Knobloch, P., Koch, M., Hoffmann, S., Breede, M., Hofmann, M.: Continuous-wave THz imaging. Electronics Letters **37** (2001) 1461–146

[6] Siebert, K.J., Quast, H., Leonhardt, R., Loeffler, T., Thomson, M., Bauer, T.: Continuous-wave all-optoelectronic terahertz imaging. Applied Physics Letters **80** (2002) 3003–3005

[7] Gallerano, G.P., Doria, A., Giovenale, E., Renieri, A.: Compact free electron lasers: From Cerenkov to waveguide free electron lasers. Infrared Physics & Technology **40** (1999) 161–174

[8] Grischkowsky, D., Keiding, S., van Exter, M., Fattinger, C.: Far-infrared time-domain spectroscopy with terahertz beams of dielectrics and semiconductors. Journal of the Optical Society of America, B: Optical Physics **7** (1990) 2006–201

[9] van Exter, M., Grischkowsky, D.R.: Carrier dynamics of electrons and holes in moderately doped silicon. Phys. Rev. B **41** (1990) 12140–12149

[10] Kindt, J.T., Schmuttenmaer, C.A.: Far-infrared dielectric properties of polar liquids probed by femtosecond terahertz-pulse spectroscopy. Journal of Physical Chemistry **100** (1996) 10373–10379

[11] Hu, B.B., Nuss, M.C.: Imaging with terahertz waves. Optics Letters **20** (1995) 1716–171

[12] Wu, Q., Hewitt, T.D., Zhang, X.C.: Two-dimensional electro-optic imaging of THz beams. App. Phy. Lett. **69** (1996) 1026–1028

[13] Herrmann, M., Tani, M., Sakai, K.: Display modes in time-resolved terahertz imaging. Japanese Journal of Applied Physics, Part 1: Regular Papers, Short Notes & Review Papers **39** (2000) 6254–625

[14] Loeffler, T., Bauer, T., Siebert, K.J., Roskos, H.G., Fitzgerald, A., Czasch, S.: Terahertz darkfield imaging of biomedical tissue. Optics Express **9** (2001) 616–62

[15] Ruffin, A.B., Van Rudd, J., Decker, J., Sanchez-Palencia, L., Le Hors, L., Whitaker, J.: Time reversal terahertz imaging. IEEE Journal of Quantum Electronics **38** (2002) 1110–111

[16] Mittleman, D.M., Hunsche, S., Boivin, L., Nuss, M.C.: T-ray tomography. Optics Letters **22** (1997) 904–90

[17] Dorney, T.D., Symes, W.W., Baraniuk, R.G., Mittleman, D.M.: Terahertz multistatic reflection imaging. Journal of the Optical Society of America, A: Optics Image Science and Vision **19** (2002) 1432–144

[18] Ferguson, B., Wang, S.H., Gray, D., Abbot, D., Zhang, X.C.: T-ray computed tomography. Optics Letters **27** (2002) 1312–131

[19] Berry, E., Fitzgerald, A.J., Zinovev, N.N., Walker, G.C., Homer-Vanniasinkam, S., Sudworth, C.D.: Optical properties of tissue measured using terahertz-pulsed imaging. Proceedings of SPIE: Medical Imaging **5030** (2003)

[20] Loeffler, T., Siebert, K.J., Czasch, S., Bauer, T., Roskos, H.G.: Visualization and classification in biomedical terahertz-pulsed imaging. Physics in Medicine and Biology **47** (2002) 3847–3852

[21] Ferguson, B., Wang, S., Gray, D., Abbott, D., Zhang, X.C.: Identification of biological tissue using chirped probe THz imaging. Microelectronics Journal **33** (2002) 1043–105

[22] Mittleman, D.M., Jacobsen, R.H., Nuss, M.C.: T-ray imaging. IEEE Journal of Selected Topics in Quantum Electronics **2** (1996) 679–69

[23] Mickan, S., Abbott, D., Munch, J., Zhang, X., van Doorn, T.: Analysis of system trade-offs for terahertz imaging. Microelectronics Journal **31** (2000) 503–51

[24] Ferguson, B., Abbott, D.: De-noising techniques for terahertz responses of biological samples. Microelectronics Journal **32** (2001) 943–95

[25] Auston, D.H., Cheung, K.P., Valdmanis, J.A., Kleinman, D.A.: Coherent time-domain far-infrared spectroscopy with femtosecond pulses. Journal of the Optical Society of America A: Optics Image Science and Vision **1** (1984) 1278

[26] Zhang, X.C., Jin, Y., Hu, B.B., Li, X., Auston, D.H.: Optoelectronic study of piezoelectric field in strained-layer superlattices. Superlattices and Microstructures **12** (1992) 487–490

[27] Shan, J., Weling, A.S., Knoesel, E., Bartels, L., Bonn, M., Nahata, A.: Single-shot measurement of terahertz electromagnetic pulses by use of electro-optic sampling. Optics Letters **25** (2000) 426–42

[28] Ruffin, A.B., Decker, J., Sanchez-Palencia, L., Le Hors, L., Whitaker, J.F., Norris, T.B.: Time reversal and object reconstruction with single-cycle pulses. Optics Letters **26** (2001) 681–68
[29] Mittleman, D.M., ed.: Sensing with Terahertz Radiation. Springer-Verlag, Berlin (2003)
[30] Zimdars, D.: Commercial T-ray systems accelerate imaging research. Laser Focus World **37** (2001) 91
[31] Arnone, D.D., Ciesla, C.M., Corchia, A., Egusa, S., Pepper, M.: Applications of terahertz (THz) technology to medical imaging. Proceedings of SPIE **3828** (1999) 209–219 Terahertz Spectroscopy and Applications 11; JM Chamberlain (ed.).
[32] Mittleman, D.M., Gupta, M., Neelamani, R., Baraniuk, R.G., Rudd, J.V., Koch, M.: Recent advances in terahertz imaging. Applied Physics B-Lasers and Optics **68** (1999) 1085–109
[33] Smye, S.W., Chamberlain, J.M., Fitzgerald, A.J., Berry, E.: The interaction between terahertz radiation and biological tissue. Physics in Medicine and Biology **46** (2001) R101–R112
[34] Hadjiloucas, S., Karatzas, L.S., Bowen, J.W.: Measurements of leaf water content using terahertz radiation. IEEE Transactions on Microwave Theory and Techniques **47** (1999) 142–149
[35] Boulay, R., Gagnon, R., Rochette, D., Izatt, J.R.: Paper sheet moisture measurements in the far-infrared. International Journal of Infrared and Millimeter Waves **5** (1984) 1221–1234
[36] Cole, B.E., Woodward, R., Crawley, D., Wallace, V.P., Arnone, D.D., Pepper, M.: Terahertz imaging and spectroscopy of human skin, invivo. Commercial and Biomedical Applications of Ultrashort Pulse Lasers; Laser Plasma Generation and Diagnostics **4276** (2001) 1–10
[37] Institute, A.N.S.: American National Standard for Safe Use of Lasers (ANSI Z136.1 - 2000). Laser Institute of America, Orlando, FL (2000)
[38] Sliney, D.H., Wolbarsht, M.L.: Laser Safety Standards: Evolution and Rationale. Safety with Lasers and Other Optical Sources. Plenum Press, New York (1980)
[39] Clothier, R.H., Bourne, N.: Effect of THz exposure on human primary keratinocyte differentiation and viability. Journal of Biological Physics **29** (2003) 179–185
[40] Scarfi, M.R., Romano, M., Di Pietro, R., Zeni, O., Doria, A., Gallerano, G.P.: THz exposure of whole blood for the study of biological effects on human lymphocytes. Journal of Biological Physics **29** (2003) 171–176
[41] Berry, E., Walker, G.C., Fitzgerald, A.J., Chamberlain, J.M., Smye, S.W., Miles, R.E.: Do *in vivo* terahertz imaging systems comply with safety guidelines? Journal of Laser Applications **15** (2003) 192–198
[42] Han, P.Y., Tani, M., Usami, M., Kono, S., Kersting, R., Zhang, X.C.: A direct comparison between terahertz time-domain spectroscopy and far-infrared Fourier transform spectroscopy. Journal of Applied Physics **89** (2001) 2357–235

[43] Bolivar, P.H., Brucherseifer, M., Nagel, M., Kurz, H., Bosserhoff, A., Buttner, R.: Label-free probing of genes by time-domain terahertz sensing. Physics in Medicine and Biology **47** (2002) 3815–3821
[44] Bezant, C.D.: Application of THz Pulses in Semiconductor Relaxation and Biomedical Imaging Studies. PhD thesis, Department of Physics (2000)
[45] Han, P.Y., Cho, G.C., Zhang, X.C.: Time-domain transillumination of biological tissues with terahertz pulses. Optics Letters **25** (2000) 242–24
[46] Ciesla, C.M., Arnone, D.D., Corchia, A., Crawley, D., Longbottom, C., Linfield, E.H.: Biomedical applications of terahertz-pulse imaging. Commercial and Biomedical Applications of Ultrafast Lasers II **3934** (2000) 73–8
[47] Knobloch, P., Schildknecht, C., Kleine-Ostmann, T., Koch, M., Hoffmann, S., Hofmann, M.: Medical THz imaging: an investigation of histopathological samples. Physics in Medicine and Biology **47** (2002) 3875–388
[48] Woodward, R.M., Cole, B.E., Wallace, V.P., Pye, R.J., Arnone, D.D., Linfield, E.H.: Terahertz-pulse imaging in reflection geometry of human skin cancer and skin tissue. Physics in Medicine and Biology **47** (2002) 3853–3863
[49] Fitzgerald, A.J., Berry, E., Zinovev, N.N., Walker, G.C., Smith, M.A., Chamberlain, J.M.: An introduction to medical imaging with coherent terahertz frequency radiation. Physics in Medicine and Biology **47** (2002) R67–R8
[50] Hagness, S.C., Taflove, A., Bridges, J.E.: Two-dimensional FDTD analysis of a pulsed microwave confocal system for breast cancer detection: Fixed-focus and antenna-array sensors. IEEE Transactions on Biomedical Engineering **45** (1998) 1470–147
[51] Wang, S., Ferguson, B., Mannella, C., Abbott, D., Zhang, X.C.: Powder detection using THz imaging. In: Proceedings of Conference on Lasers and Electro-Optics, Long Beach, CA (2002) 132
[52] McClatchey, K., Reiten, M.T., Cheville, R.A.: Time-resolved synthetic aperture terahertz impulse imaging. Applied Physics Letters **79** (2001) 4485–448
[53] Jacobsen, R.H., Mittleman, D.M., Nuss, M.C.: Chemical recognition of gases and gas mixtures with terahertz waves. Optics Letters **21** (1996) 2011–201
[54] Koch, M.: Biomedical applications of THz imaging. In Mittleman, D., ed.: Sensing with Terahertz Radiation. Springer-Verlag, Berlin (2003) 295–316
[55] Siegel, P.H.: Terahertz technology. IEEE Transactions on Microwave Theory and Techniques **50** (2002) 910–928
[56] Anderton, R.N., Appleby, R., Borrill, J.R., Gleed, D.G., Price, S., Salmon, N.A. In: Prospects of Imaging Applications [Military]. IEE (1997) 4/1–4/10

[57] Papoulis, A.: The Fourier Integral and Its Application. McGraw-Hill, New York (1962)
[58] Akay, M., ed.: Time Frequency and Wavelets in Biomedical Signal Processing. IEEE Press and John Wiley & Sons (1998)
[59] Xu, X.L., Tewfik, A.H., Greenleaf, J.F.: Time-delay estimation using wavelet transform for pulsed-wave ultrasound. Annals of Biomedical Engineering **23** (1995) 612–621
[60] Sonka, M., Hlavac, V., Boyle, R.: Image Processing, Analysis and Machine Vision. second edn. Brooks/Cole Publishing Company, Pacific Grove, CA (1999)
[61] Georgiou, G., Cohen, F.S., Piccoli, C.W., Forsberg, F., Goldberg, B.B.: Tissue characterization using the continuous wavelet transform, part II: Application on breast RF data. IEEE Transactions on Ultrasonics Ferroelectrics and Frequency Control **48** (2001) 364–373
[62] Sun, M., Sclabassi, R.J.: Wavelet feature extraction from neurophysiological signals. In Akay, M., ed.: Time Frequency and Wavelets in Biomedical Signal Processing. IEEE Press and John Wiley & Sons (1998) 305–321
[63] Ching, P.C., So, H.C., Wu, S.Q.: On wavelet denoising and its applications to time delay estimation. IEEE Transactions on Signal Processing **47** (1999) 2879–2882
[64] Coifman, R.R., Wickerhauser, M.V.: Experiments with adapted wavelet de-noising for medical signals and images. In Akay, M., ed.: Time Frequency and Wavelets in Biomedical Signal Processing. IEEE Press and John Wiley & Sons, Piscataway, NJ (1998) 323–346
[65] Sardy, S., Tseng, P., Bruce, A.: Robust wavelet denoising. IEEE Transactions on Signal Processing **49** (2001) 1146–115
[66] Lasch, P., Naumann, D.: FT-IR microspectroscopic imaging of human carcinoma thin sections based on pattern recognition techniques. Cellular and Molecular Biology **44** (1998) 189–20
[67] Carmona, R., Hwang, W.L., Torresani, B.: Practical Time-Frequency Analysis. Academic Press, San Diego (1998)
[68] Gioswami J. C., Chan, A.K.: Fundamentals of Wavelets: Theory, Algorithms, and Applications. John Wiley and Sons, New York (1999)
[69] Weiss, L.G.: Wavelets and wideband correlation processing. IEEE Signal Processing Magazine **11** (1994) 13–32
[70] Young, R.K.: Wavelet Theory and Its Applications. Kluwer, Boston (1993)
[71] MacQueen, J.: Some methods for classification and analysis of multivariate observations. In: Proceedings of 5th Berkeley Symposium 1. (1967) 281–297
[72] Kaufmann, L., Rousseeuw, P.J.: Finding groups in data: An introduction to cluster analysis. John Wiley & Sons, New York (1990)
[73] Webb, P.A.: A review of rapid prototyping (RP) techniques in the medical and biomedical sector. Journal of Medical Engineering & Technology **24** (2000) 149–15

[74] Goulden, C.H.: Methods of Statistical Analysis. second edn. John Wiley and Sons, New York (1956)
[75] Hartigan, J.: Clustering Algorithms. John Wiley and Sons, New York (1975)
[76] Handley, J.W., Fitzgerald, A.J., Berry, E., Boyle, R.D.: Wavelet compression in medical terahertz-pulsed imaging. Physics in Medicine and Biology **47** (2002) 3885–389
[77] Duvillaret, L., Garet, F., Coutaz, J.L.: A reliable method for extraction of material parameters in terahertz time-domain spectroscopy. IEEE Journal of Selected Topics in Quantum Electronics **2** (1996) 739–74
[78] Dorney, T.D., Baraniuk, R.G., Mittleman, D.M.: Material parameter estimation with terahertz time-domain spectroscopy. Journal of the Optical Society of America, A: Optics Image Science and Vision **18** (2001) 1562–157
[79] Highnam, R., Brady, M.: Mammographic Image Analysis. Kluwer Academic Publishers, Dordrecht (1999)
[80] Cotton, S., Claridge, E., Hall, P.: Noninvasive skin imaging. Information Processing in Medical Imaging **1230** (1997) 501–50
[81] Duvillaret, L., Garet, F., Roux, J.F., Coutaz, J.L.: Analytical modeling and optimization of terahertz time-domain spectroscopy experiments using photoswitches as antennas. IEEE Journal of Selected Topics in Quantum Electronics **7** (2001) 615–62
[82] Lee, Y., Meade, T., Norris, T.B., Galvanauskas, A.: Tunable narrow-band terahertz generation from periodically poled lithium niobate. Applied Physics Letters **78** (2001) 3583–358

# Index

$L^2$ distance, 177
$l_2$-norm, 95, 96
3D AAM segmentation, 215
3D scanning systems, 244
3D segmentation, 214
4D analysis, 215

AAM two-dimensional, 205
absorbance, 295
absorption, 173
absorption coefficient, 291
  linear, 279
ACE, 187
acoustic tracking, 247
acquisition, 218
active contour, 254
adaptive target classification, 125
adaptive thresholding, 262
admissibility condition, 287
affinity, 225
algorithm
  $k$-means, 290
  6D recognition, 50
  crimmins, 41
  flood-fill, 259
  model construction, 47, 53
  MTI object detection, 142
  SAR recognition, 48
  similarity-computation, 16
  two-class classification, 116
ambiguity function, 288
angiography, 201
  magnetic resonance, 215–227
  x-ray, 216, 218

apex slice, 206
arteries and veins
  separation, 217
arteriovenous separation, 221
artery
  femoral, 221
  iliac, 216, 221
artery–vein separation, 218, 226
articulated arms, 247
articulation, 59, 82
atherosclerotic plaque
  assessment, 229
  lesions, 228
atmospheric transmission, 78
attenuation, 293
automatic target recognition, 115
azimuth invariance, 42
azimuthal variance, 61

background projection matrix, 126
background subtraction, 144
barycentric coordinates, 208
basal cell carcinoma, 296
basis vectors, 122
Bayesian statistical inference, 90
beer-lambert law, 291
best view, 147
bifurcation detection, 216
binary mask generation, 222
binning, 89
black-blood imaging, 197
blackbody, 172
Boltzmann's constant, 172
border detection, 230

313

# Index

brachytherapy radioactive seeds, 256
brachytherapy seed segmentation, 261
bright-blood imaging, 198

calibration, 171, 172
cardiac
   function, 198
   imaging, 193
   gating, 228
   magnetic resonance, 203
   segmentation, 204
   ventricles, 206
cardiovascular
   image analysis, 193
   MRI, 196
centering matrix, 95
centroid, 104
CFAR, 116
cholesky decomposition, 95
cine MRI, 199
clustering, 156, 252, 282, 289
clutter, 1, 2, 7, 9
clutter region, 9
CND, 149
coherent detection system, 274
combined distortion, 9
complex conjugate, 284
compression, 141, 154, 289
computer–aided design, 80
concave weight function, 58
conditional PDF, 20
conflict resolution, 221, 224
confusion matrix, 57
connection cost, 220
consistency region, 8
contour
   deformation, 256
   editing, 256
convex weight function, 59
coordinate system
   cylindrical, 209
   polar, 72
   rectilinear, 72
   reference, 47
core, 216, 217, 219
coronary
   angiograms, 216
   arteries, 216
coronary artery, 203

covariance matrix, 126
covariants, 102
cross-correlation, 288
CTA, 216
cube view approach, 249

data distortion models, 6, 8
DDC, 254
deformable ellipsoid model, 254
deformable model, 205
dentine, 295
detection sensors
   coherent, 74
   direct, 74
diffraction imaging, 273
dilation
   bounded space, 222
distance metric, 149
distortion group
   real, 25
   synthetic, 24
distribution
   binomial, 13
   hypergeometric, 20
   plank, 168
   probability, 4, 19
   spectral, 168
dosimetric analysis, 261
double inversion recovery, 228
DPCM, 156
dynamic programming 3D, 221

echo time, 197
echocardiographic, 249
echocardiography, 243
edge
   detection, 253
   enhancement and selection, 253
   linking, 253
edge-selection algorithm, 252
effective clutter region, 9
effective size, 14
eigenfaces, 184
eigenspace classification, 148
eigenvalues, 124, 149
eigenvectors, 117, 124, 149
elastic deformation, 253
electromagnetic reflectance, 116
electromagnetic spectrum, 167

electronic stabilization, 143
emissivity, 168, 173, 181
empirical performance estimation, 2
enamel, 295
entropy rate, 159
equal-error-rate, 186
equinox database, 170
error function, 210
Euclidean distance, 149
Euclidean norm, 127
evidence accrual, 90

face recognition, 167
Fast 3D GRE, 202
fast spin echo imaging, 228
feature vectors, 294
features, 83
fidelity, 82
first bounce, 76
Fisherfaces, 184
FLIR, 116
flow-based imaging, 201
focus of expansion, 148
fourier
  coefficients, 283
  decomposition, 279
  methods
    global, 283
    local, 283
Fourier transform
  spectroscopy, 272
frequency space decomposition, 280
FSE, *see* spin-echo imaging
FTIR spectroscopy, 282
full generalized form, 98
fundamental frequency, 283
fuzzy connectivity, 225

gadolinium contrast, 200
gas identification, 278
gaussian random noise, 119
generalized cylinders, 84
genetic algorithms, 90
geometric hashing, 48, 88
gradient descent optimization, 210
gradient matrix, 211
GRE, 198

Haar transform, 159

hanging togetherness, 225
HARP, 200
Helmert submatrix, 95
high-level processes, 282
high-pass filter, 159
Hilbert–Schmidt estimator, 5
histogram
  all-similarity, 8, 16
  model collision, 54
  peak-similarity, 8, 16
histopathological samples, 296
Hough transform, 89
HYDICE, 115
hyperspace, 108
hyperspectral cube, 130
hyperspectral imaging, 115, 118
hyperspectral sensor, 127

ICA-based feature extraction, 122
IFOV, 80
illumination invariance, 175
incoherent detection, 272
independent component analysis, 115, 117, 120
indexing, 108
infrared imaging, 167
initialization ellipsoid, 255
insignificant coefficients, 157
intensity projection, 251
intensity ratio, 220
interclass variance, 149
interventional procedures, 242
intraclass variance, 149
invariance, 92
invariant function, 102
inversion recovery, 197
iterative refinement, 211

Kalman filter, 215
Kirchoff's law, 173
knowledge-based segmentation, 221
Kullback–Leibler divergence, 177

ladar
  3D, 72
  flash, *see* 3D ladar
  FPA, 75
  projecton
    orthographic, 75
    perspective, 75

sensors, 74
laser, 71
laser radars, 71
last bounce, 76
learning process, 90
likelihood-ratio test, 4
linear discriminant analysis, 184
linear distance, 220
linear interpolation, 182
linear motion, 244
lipschitz signatures, 147
local feature analysis, 184
look-up table, 47, 272
low-pass filter, 159
LWIR, 168

magnetic field tracking, 247
magnetic resonance, 228
  angiography, 215
magnitude invariance, 45
manual planimetry, 252
Markov model, 159
Markov random fields, 7
matched filter, 86
matching
  image-based, 85
  model-based
    alignment-based, 87
    voting-based, 87
  pixel–level, 86
maximum intensity projection, 217
mean curvature, 217
mean spectral curve, 130
mean squared prediction error, 156
mechanical assemblies, 244
medial axis transform, 217
medialness, 217, 219
melanoma, 296
memory, 163
mixed-pixel, see subpixel
model
  objects, 1
  set, 24
  similarity, 52
monostatic system, 75
morphometrics, see statistical shape analysis
mother function, 286
MPR, 250

MR wall imaging, 229
MRA, 215
MRI, 194, 204
MSTAR, 3, 24, 41
MTI, 142
multiple-look-angle SAR, 62
multiresolution, 155
multivariate linear regression, 211
MWIR, 168
myocardial perfusion, 200
myocardial tagging, 200

navigator gating, 196
needle insertion point, 258
needle segmentation, 256
  2D, 259
  3D, 259
needle vector, 260
negative evidence, 90
negative gain, 84
NIST, 171
nongaussian, 121
nuclear magnetic resonance, 194
NURBS, 84

object detection, 142, 147
object recognition
  feature-based, 4
  model-based, 1
object similarity, 7, 9
object–image relations, 100
obscuration, 77, 81
observations, 50
occlusion, 1, 2, 7, 9, 60
operating conditions, 72
optical properties, 291
optical rectification, 274
optical spectroscopy, 274
optimal vessel path, 223
orthogonal projection, 257
  matrix, 126
orthogonal subspace projection, 126

partial acquisition, 202
PCR
  lower bound, 21
  upper bound, 22
PDF, 17
PDM, 204

Index    317

peak signal-to-noise ratio, 155
penetration, 275
performance bounds, 17
performance–prediction method, 6
persistence, 156
PET, 242
phantom test objects, 259
phase-contrast imaging, 201
Planck's constant, 172
plaque morphology, 229
PMF, 44
point cloud, 83
point correspondence, 81, 99
point distribution model
   2D, 205
   3D, 206
positive evidence, 47
postprocessing, 219
prediction method, 3
predictor coefficients, 156
preimplant dose planning, 252
principal component analysis, 117, 184
principal components, 91
priori, 257, 259
probability of correct recognition(PCR), 3
probe, 185
Procrustean metrics, 94
Procrustes analysis, 206
prostate brachytherapy, 251
prostate mesh, 254

quality control, 278
quaternion, 207
   transformation, 210

radiofrequency, 194
range image, 74
range profile, 76
Rayleigh scattering, 277
RBR edge detector, 253
real model, 49
real-time imaging, 203
reference pulse, 279
reflectance, 77
reflectance spectra, 117
reflective sensor, 80
refractive index, 290
   broadband, 292

resonance, 194
respiratory triggering, 196
responsivity, 174
ROC curve, 4, 39, 50, 59, 64, 135, 145
ROI, 40
rotation, 80
run-length coding, 157
run-time, 163

sampling lattice, 72
sampling rate, 287
scalar quantization, 155, 158
scaling parameter, 286
scanning
   tilt approach, 245
   tracked free-hand, 246
   untracked free-hand, 248
scatterer amplitude, 44
scatterer location invariance, 44
scattering centers, 40
scintillation, 43, 77
seed
   candidate extraction, 264
   identification, 264
   implantation, 252
   point, 219
segmentation, 128
semiconductor characterization, 278
sensor model, 99
sensor panning motion, 144
shape
   factor, 51
   intensity vector, 210
   metrics, 93
   parameters, 210
   recognition, 85
   representations, 83
signal compression, 293
signal-to-noise ratio (SNR), 129
signatures, 277
similarity
   function, 8
   group, 94
   histograms, 15
singular value decomposition, 97
size-based detection, 148
smaller-aspect separation, 227
snake contour, 254
spatial distribution, 278

spatial sampling degradation, 246
spatial variance, 142
speckle decorrelation, 247
spectral
   dissimilarity, 127
   feature vector, 130
   reflectance curves, 130
   template, 136
   variability, 123, 125, 130
spectrometer, 118
spin angular momentum, 194
spin-echo imaging, 197
statistical shape analysis, 93
step wedge, 292
STFT, 283
subpixel, 118
subspace
   estimation, 133
   modeling, 126
superquadrics, 84
surface
   positioning errors, 214
   resampling, 80
synthetic aperture radar, 1, 24, 39

target bias, 63
target classification, 137
template matching, 85, 126
temporal variance, 142
terahertz
   frequency band, 271
   pulsed imaging, 271, 274
   radiation, 271
      sensitivity, 276
      transmission, 290
test set, 24
tetrahedron, 208
threshold, 128
tilt motion, 245
time delay, 288
time–frequency analysis, 279
TOF, 202
transducer, 243
transesophageal, 245
transfer functions, 259
translation, 79
translation space, 48
translucency rendering, 251
transmission responses, 279

transrectal, 245
tree
   growing, 219
   root, 220
   vasculature, 220
   vessel, 219
tree construction, 220
tree search, 91
tree segmentation
   venous, 226
   vessel, 227
tree-structure generation, 222
triangular mesh, 83
TRUS, 252
two-point radiometric calibration, 175

ultrafast infrared laser, 274
ultrasonography, 241
ultrasound imaging, 241
   2D, 242
   3D, 241, 243
   3D reconstruction, 248
      feature-based, 249
      voxel-based, 249
      conventional, 242
uncertain instance, 8
uncertainty, 2, 6, 8, 257
uncertainty principle, 285
uniform similarity model, 13
unknown objects, 82
unoccluded confuser, 61, 65
unsupervised segmentation, 133

variance analysis, 142, 143
vasculature assessment, 219
vector quantization, 297
ventricular segmentation, 206
vessel
   central axes, 217
   segmentation, 216, 219
      segment labeling, 223
vitro applications, 277
vivo applications, 278
VLWIR, 168
volume
   cropping, 257, 262
   rendering, 259
volume rendering, 251
vote process, 17

votes, 3, 47

warping, 208
wavelet, 155, 285
wavelet transform, 155
    continuous, 286, 288
    discrete, 287

WBCAF, 288
weight functions, 55
weighted voting, 55
window function, 284

zerotree, 156
    coding and DPCM coding, 155